SOIL AND ENVIRONMENT: NORTHERN IRELAND

SOIL AND ENVIRONMENT: NORTHERN IRELAND

Editor and Main Author
J.G. Cruickshank
Principal Soil Surveyor, DANI

Published by:
Agricultural and Environmental Science Division, DANI and The Agricultural and Environmental Science Department, The Queen's University of Belfast.

Published in 1997 by Agricultural and Environmental
Science Division, DANI and The Agricultural and
Environmental Science Department, The Queen's
University of Belfast.
Newforge Lane, Belfast BT9 5PX

© J.G. Cruickshank 1997

This book is copyrighted under the Berne Convention. All rights
reserved. Apart from any fair dealing for the purpose of private study,
research, criticism or review, as permitted under the Copyright Act,
1956, no part of this publication may be reproduced, stored in a
retrieval system, or transmitted in any form or by any means,
electronic, electrical, chemical, mechanical, optical, photocopying,
recording, or otherwise, without prior written permission. All enquiries
should be addressed to the publisher.

ISBN 0 85389 699 2

Designed by Nicholson and Bass

Printed in Great Britain by
Nicholson and Bass,
Belfast, Northern Ireland

SOIL AND ENVIRONMENT: NORTHERN IRELAND

Contents

The Contributors		vi
Acknowledgements		vii
Foreword		ix
1	Introduction – The Soil-forming Environment	1
2	The Soils: Nature and Variability. Analytical data for the main soil series	11
3	Geological Background for Soil Development R.A.B. Bazley	57
4	Climate N.L. Betts	63
5	Rivers, Drainage Basins and Soils D.N. Wilcock	85
6	Land Cover – based on 'CORINE' Land Cover Programme R.W. Tomlinson	99
7	Economic and Policy Environment for Farming Joan E. Moss	119
8	The Soil Maps – description and interpretation of soil distributions	133
9	Applications of Soil Survey	165
	A Soil carbon pools – Margaret M. Cruickshank	165
	B Acidification of soils – C. Jordan and Jane R. Hall	171
	C. HOST – hydrology of soil types – A.J. Higgins	177
	D. Applications of Soil Survey in Northern Ireland: Inventory and Prospect	187
10	Appendices	
	(a) Field mapping methods	193
	(b) Soil analytical methods	194
	(c) Soil attribute database	195
	(d) Soil fertility tables	197
	(e) Soil texture triangle	199
	(f) References	200
	(g) Selective Index	211

SOIL AND ENVIRONMENT: NORTHERN IRELAND

The Contributors

Dr. R.A.B. Bazley, Director, Geological Survey of Northern Ireland.

Dr N.L. Betts, School of Geosciences, The Queen's University of Belfast.

J.G. Cruickshank, Principal Soil Surveyor, Agricultural and Environmental Science Division, Department of Agriculture (NI).

Margaret M. Cruickshank, Honorary Research Fellow, School of Geosciences, The Queen's University of Belfast.

Jane R. Hall, Institute of Terrestrial Ecology, Huntingdon, Cambridgeshire.

A.J. Higgins, Agricultural and Environmental Science Division, Department of Agriculture (NI).

Dr C. Jordan, Agricultural and Environmental Science Division, Department of Agriculture (NI).

Dr Joan E. Moss, Centre for Rural Studies, The Queen's University of Belfast.

Dr R.W. Tomlinson, School of Geosciences, The Queen's University of Belfast.

Dr D.N. Wilcock, Professor of Environmental Studies, University of Ulster at Coleraine.

ACKNOWLEDGEMENTS

This book, Soil and Environment: Northern Ireland, is a soil survey memoir with a difference. It is essentially the final report of the DANI Soil Survey Project (1987 - 1997), which has been extended to include related environmental themes. For this, six contributors from Northern Ireland universities and the Geological Survey (NI) were invited to participate and their contributions have been appreciated greatly. These six academics are known as authorities in their respective specialisms, and so particular thanks are due to R A B Bazley, Nicholas L Betts, Margaret M Cruickshank, Joan E Moss, Roy W Tomlinson and David N Wilcock for making this publication more valuable and more interesting. Their full titles and affiliations are given in the list of contributors.

The DANI Soil Survey has been active for the past decade, but long before that in the early 1980s, senior figures in the DANI Science Service and in other parts of the Department, argued the case for a soil survey of the Province. The Soil Survey would not have come into existence without the enthusiasm and commitment of Kerry Garrett, supported by Chief Scientists Charlie Wright and Cecil McMurray, together with James Kerr of the former Drainage Division and Donagh O'Neill. Their influence was critical in the conceptual stages, while the financial assistance provided by DoE (NI) Environment and Heritage Service allowed this book and all the soil maps to be published in the late stages of the project. For this, particular thanks are due to Roy Ramsay of DoE (NI).

The work of the DANI Soil Survey has involved many staff, both permanent and temporary, over the period 1987 to 1997. The greatest share of the field work was the responsibility of scientific staff in the former Food and Agricultural Chemistry Division, namely Scott Gardiner, Hubert Logan and Alex Higgins, who were supported by field officers John Ferguson, Gilbert McBride, Chris McGinn and Raymond Stewart. These DANI staff each gave several years of service to the soil survey, and temporary, short-term service was given also by John Courtney, Donal Donnelly, Nicholas Coyle and Niall Maguire (in the west of the Province).

Within the same Division, soil analytical work was undertaken successively by Noel Quinn and by Peter Scullion and the team in the Soil Analytical Laboratory. Their work produced over 30,000 items of soil attribute data from soil samples collected by soil survey staff, and created an extremely valuable body of information about the inherent character and probable behaviour of the 308 soil series which were mapped. This information now forms a large part of the DANI Soil Attribute Database.

The staff of the Ordnance Survey of Northern Ireland (OSNI) have played a large part in the production of the 17 soil maps at 1:50,000 and the 1:250,000 soil map of the Province, as well as the compilation of the Soil Graphical and Textual Databases. Particular thanks for friendly and productive co-operation are due to the Director, Michael Brand, as well as to Gerry Mitchell, Kevin Donnan, Geoff Mahood, Henry Douglas, Catherine Miskelly and Lynn Porter. Many others at OSNI were involved and deserve thanks. The assistance of our printers, Nicholson and Bass, is gratefully acknowledged, especially that of Raymond Rainey.

Throughout the years of the Soil Survey, Dr M K Garrett has given me, as Project Leader, continuous support and encouragement. Kerry Garrett had the vision for the soil survey, and fought many battles on its behalf both before and during its life. Also, I would like to make special mention of Crawford Jordan who has recently joined with me to lead the new Land Research Group, and who gave both the soil survey and current land research programmes very considerable help with his expertise in preparing soil databases and using GIS techniques in derivative map production. Kevin Hamill has helped with many aspects of the Soil Survey Project, Roy Anderson provided many of the landscape photographs in this book, and Gill Alexander in Geosciences, Queen's University, undertook most of the cartography.

Finally, it is my very pleasant task to thank the many farmers and land owners who have facilitated this project by providing access to their land, and the local officers of DANI who assisted with this.

J G CRUICKSHANK
Editor and Main Author

FOREWORD

Sustainability has become the most influential concept guiding development as we approach the 21st Century and nothing is more fundamental to sustainable development than a knowledge of our soils and environment. This is the reason why publication of this book is so significant to the future of Northern Ireland. It marks completion of the first systematic study of the soils of Northern Ireland and coincides with the publication of soil maps of the Province at a scale of 1:50,000. This information is also available in a unique digital database together with a companion database on soil attributes, opening an extensive range of applications through the medium of GIS or Geographic Information Systems. A generalised map at a scale of 1:250,000 has also been produced. Publication of this book now places this information in the context of Northern Ireland's environment.

In Northern Ireland a diverse parent geology has been complicated by the effects of glaciation and climatic history to create a fantastic diversity of soils. Rational decisions about land use rely heavily on the interpretation of high quality soil information and this is particularly important in a region with such diversity. The outputs of the Soil Survey Project will have enduring value to all concerned with the planning process and with the use of land for agriculture, construction, conservation, recreation or industry. The information will also be invaluable to those concerned with research and education, environmental impact assessment and land valuation for investment appraisal. The value of soil information will extend further as derivative classifications are developed and the first examples of these are included at the end of this book.

The Soil Survey Project has illustrated what can be achieved by teamwork and cooperation. The survey and analytical work was undertaken by the Agricultural and Environmental Science Division of the Department of Agriculture for Northern Ireland, digital processing and map production work was undertaken by the Ordnance Survey of Northern Ireland and publication has been funded by the Environment and Heritage Service of the Department of Environment for Northern Ireland. It is a tribute to the leadership of Mr J G Cruickshank and the commitment of all concerned that this project was completed in only eight years. But this is a beginning - not an end. The challenge now is to harness this resource to benefit the future of Northern Ireland and its people by applying it to the many issues upon which the development of prosperity and quality of life depend.

M K Garrett
Deputy Chief Scientist
Department of Agriculture for Northern Ireland and Professor of Agricultural and Environmental Science
The Queen's University of Belfast

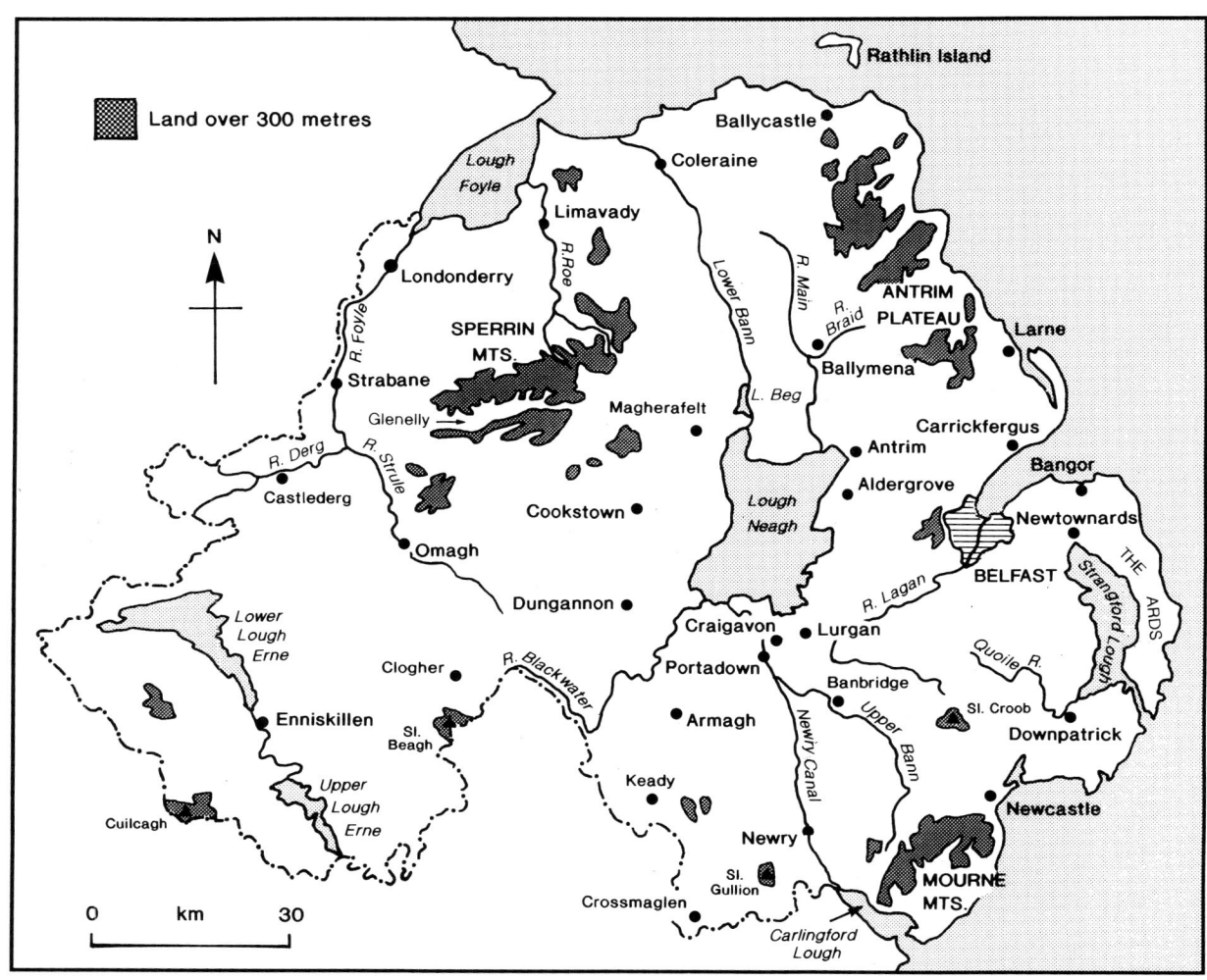

Figure 1.1: Place names of the main rivers, towns and mountains mentioned in the text.

CHAPTER 1

INTRODUCTION

THE SOIL-FORMING ENVIRONMENT IN NORTHERN IRELAND

The remarkable variety of soils in Northern Ireland reflects the varied character of the natural environment, which is the same as the soil-forming environment. All the elements of the soil-forming environment (so-called 'soil-forming factors') are expressed here over a wide range, and their interaction over the land area of Northern Ireland has produced an incredible diversity of soils. 'Soil' in the field or landscape, may be defined and mapped as soil series, where each map unit has the same soil profile on the same parent material. The DANI Soil Survey has identified 308 distinctively different soil series (each over 50 hectares in area) in the Province, and they have been developed from 97 soil parent materials. All this has happened in an area less than 14,000 km^2, and when compared with the area of England and Wales, which is almost 11 times greater, it is surprising that there are only about twice as many (around 700) soil series to be found in the latter.

However, the comparison with the number of soil series in England and Wales requires qualification. Although the soil classification adopted by the DANI Soil Survey in Northern Ireland has been the *Soil Classification for England and Wales (Higher Categories)* (Avery, 1980), significant changes were made for its application in Northern Ireland. Soil profile terminology is fully described in Chapter 2.

In an effort to achieve a more even distribution of areas of freely draining soil types, all the shallow, rocky soils or RANKERS (lithomorphic soils) were grouped together, comprising 8.8 per cent land area, as were all the BROWN EARTHS, comprising 13.1 per cent, and the PODZOLS, comprising 4.1 per cent (see Table 1.1). On the published Northern Ireland soil maps, each soil profile type in these groupings is shown with its appropriate abbreviated-label (eg BE or SBE, etc) and has a colour indicating the soil parent material. If these soil profile types are individually linked with parent material, they are counted as separate soil series, and overall total numbers of soil series can be found by addition.

The approach was reversed when classifying poorly drained soils or GLEYS. These were subdivided into three degrees of gleying (see Chapter 2 on soil classification), and each in turn, combined with its particular parent material, is regarded as a soil series. In theory, the number of soil series among gleys could be three times greater for the same area in Northern Ireland, compared with England and Wales. The reason for this subdivision is that all gleys (including alluvial soils) aggregate to 56 per cent, and in the early stages of soil survey (1987–1990), it was expected to be a higher overall percentage. In the effort to achieve a more useful distribution of the main types of soil, it was considered desirable to divide the gleys into three categories, reflecting degrees of gleying. The occurrence of the most common soil types in Northern Ireland, and comparison with England and Wales, and Scotland, is shown in Table 1.1. The differences among these three territories reflects the major differences in their soil-forming environments.

In the case of Northern Ireland, the remaining 4 per cent land area not shown in Table 1.1 is organic alluvium, lake deposits, and other specialised alluvial deposits, all of which are undivided gleys, and so should be added to the GLEY category.

This last point, made with reference to Table 1.1, implies that the possible total of all types and degrees of GLEYS in Northern Ireland should be 60 per cent of land area, a proportion far in excess of that in England, Wales or Scotland, and hence reflects a much wetter environment in the Province. Betts discusses the interaction between climate and soil in Chapter 4, and here, it should be noted only that the current range of climatic parameters is considerable in Northern Ireland. Mean annual rainfall ranges from less than 750 mm at places on the south-east coast, and in the centre around Lough Neagh, rising with altitude and in westerly locations to a maximum above 2,000 mm in west Tyrone. Conversely, potential evapotranspiration (PT) ranges from over 500 mm in the areas of lowest rainfall in the east, down to about 350 mm or less in westerly hill margins. Potential evapotranspiration (PT) exceeds rainfall for the 4 months of April, May, June and July in coastal, lowland areas of the east, where the annual surplus of moisture may be only about 200–300 mm. In contrast, areas above 250 m in the east and generally in the west, no month has PT greater than rainfall, and the annual surplus of moisture between the two will be at least 1000 mm. In these two opposite extremes, moisture from rainfall can be limiting on agriculture and an influence on soil/land use either by excessive amounts leading to flooding or by drought in summer, particularly the latter in easterly areas and on shallow soils. Similarly, there is a wide range of temperature conditions, as can be seen in the two month difference in possible length of growing season from high values of about 295 days in the lowland east to less than 235 days in hill margins everywhere, at the upper altitudinal limits for agriculture. These points will be developed further by Betts in Chapter 4, where appropriate maps and diagrams are used to illustrate the influence of climate on soil development, agricultural land quality, and current agricultural enterprises.

Soil-forming factors in Northern Ireland

For the last one hundred years, soil literature written by the pioneer pedologists in Russia (Dokuchaev and Sibirtzev in the 1890s) and supported soon afterwards in America (Hilgard, Shaler and Whitney – USDA Soil Survey founded in 1898), proposed that soil is an independently-organised natural body, developed by the interaction of five 'soil-forming factors' (climate, relief, parent material, plant and animal organisms), working over a period of time. Although the Russian and American soil research organisations worked separately and in isolation from each other, when the two came together in the 1920s to exchange ideas and information about soil development, they were generally in agreement. They agreed that individual soils changed in form and character over time, and also over space, the latter because of the spatial variation in the interaction among the five soil-forming factors. Both agreed on this from the evidence of soil mapping and soil survey organisations established in both Russia and America by the end of the 19th century. Soil is well-known as a spatially variable medium, and the mapping of soil differences has been an important method of recording soil development in relation to environment, throughout the world during the 20th century. Soil surveys have been established and completed in most of the countries of Europe, and in the English-speaking territories of the former British Commonwealth (for a full description of the history and methodology of pedology see Cruickshank's, Soil Geography, 1972, and Avery's, Soils of the British Isles, 1990). It is the main purpose of this publication on the soils of Northern Ireland to demonstrate the spatial variety of soils through the 17 soil maps (1:50,000) of the Province.

CLIMATE: In the traditional thinking of the early Russian and American pedologists, climate was regarded as the dominant soil-forming factor which controlled soil profile development. They proposed the concept of 'zonality' in which a soil profile was representative of its climatic zone. In recent times, this concept has been regarded as too strict and deterministic, and apart from that, as applying only to freely draining soils. In Northern Ireland, the freely draining soils, with a profile deeper than 40 cm, are only brown earths and podzols, together comprising about 18 per cent NI area. The remaining soils, in particular 60 per cent gleys and 14 per cent peat, are not considered to be 'zonal soils', and yet the moisture affecting their poor drainage, comes from climate. The reality is that the influence of climate on soil development is found in at least four main ways: chemical and physical weathering of mineral parent material, the glacial or fluvial or aeolian movement of parent materials, the development of the soil profile by internal soil processes, and lastly, the erosion or physical removal of soil. The relative importance of these four types of climatic input will vary by the location of the region, and will also vary by location within the region.

Most of the island of Ireland was glaciated several times during the Quaternary geological period (see Bazley, Chapter 3), and the most visible legacy of glaciation is the small hill or drumlin landform which was produced by ice-moulding in the last major

TABLE 1.1
Occurrence of the most common soil types, per cent area

Soil Type	England and Wales	Scotland	Northern Ireland
Rankers	7	10	9
Brown Soils	45	19	13
Podzols	5	24	4
Gleys	40	23	56
Peat	3	24	14

advance of ice, about 20,000 to 25,000 years ago, and which advanced from an ice centre in the north of the island to a southern limit through the present towns of Dundalk, Athlone and Limerick in the Republic of Ireland. The 'drumlin' with its distinctive Irish name, has had an important topographic influence on soil drainage patterns, on farm size, and even on field shape. Ice movement was responsible also for the movement of glacial till or boulder clay (soil parent material) outward from the ice centre in the area of Lough Neagh, so that we find north-east to south-west movement of glacial till in Fermanagh, and north-west to south-east movement of basalt till material across the Lagan valley, from Antrim to Down (see Chapter 8 on Soil Maps). Such movements of soil parent material do affect the appearance of the soil profile, particularly if red coloured materials are involved, as well as the physical and chemical properties at all levels of the soil profile. It must be said that the distance of such 'carry-over' is short, and is not likely to be more than 5 km (McCabe, 1969). The ice-carriage of stones and boulders is much greater, but the effect of those on soil profile development is minimal.

The weathering of mineral material to produce the 'Fine Earth' of soil (the fraction less than 2 mm in particle diameter) is controlled by climate, and has been operating from distant geological time through the period of the Quaternary glaciations, and up to the present. Physical weathering has been restricted to periods when freezing and thawing was active, but chemical weathering has been taking place over a much wider range of climatic conditions and a much longer period of time. The rates of chemical weathering reactions are increased by heat and moisture, and although rates would be a maximum in the humid tropics, the processes are working in the humid, temperate environments of Ireland today. They are responsible for the chemical change of mineral matter and possibly their most important role is in the production of clay minerals in soil.

The development of soil profiles is relatively recent in Ireland, starting about 10,000 years ago at the end of the Quaternary glaciations, and the following cold period of Late-glacial time. This is a short period compared with millions of years available for soil formation in areas such as the humid tropics, or in fact any area beyond the reach of the glaciations. That period of two million years, of the Quaternary or Pleistocene period, effectively stopped the processes of weathering and soil profile development. However, soil development in the subsequent Post-glacial period has been able to use mineral material produced in pre-glacial weathering. The 10,000 years of the Post-glacial is not a long period for soil formation as it requires at least 1,000 years in this environment to make even minimal change to the appearance of a soil horizon, in an ideal situation where acid, leachate solution is passing downward through the soil. The direct influence of climate on soil profile development is modified where the soil is partially or completely saturated, and processes of gleying are found. Climate, through excesses of moisture entering the soil from the ground surface, does influence the process of leaching which leads to the formation of brown earths, brown podzolics, podzols, and peat podzols, in that sequence of increasing leaching and acidification. For leaching to be effective beyond the first stage of brown earth formation, a porous, freely-draining material is required in association with a climate where annual rainfall exceeds potential evapotranspiration by three or four times. In Northern Ireland, that means usually an excess of about 1,000 mm of rainfall over PT, and that usually has to be above 250 m altitude. There can be exceptions to allow leaching at lower altitudes where the soil parent material is acid, below pH 4.5, sandy and very porous.

The variable impact of climate on soil profile development is best demonstrated by grouping or classifying all soils, deeper than 40 cm, into two types as shown in Figure 1.2. The vertical dimension may be taken to be 100 cm, and those soil profiles with free drainage are those most likely to be changed by climate, but the time scale is one of thousands of years. Where poor drainage occurs due to high and fluctuating groundwater table, mineral soils are naturally saturated. The end stage of that sequence is evident in the accumulation of nutrient-rich lowland peat, a process which has been happening since the

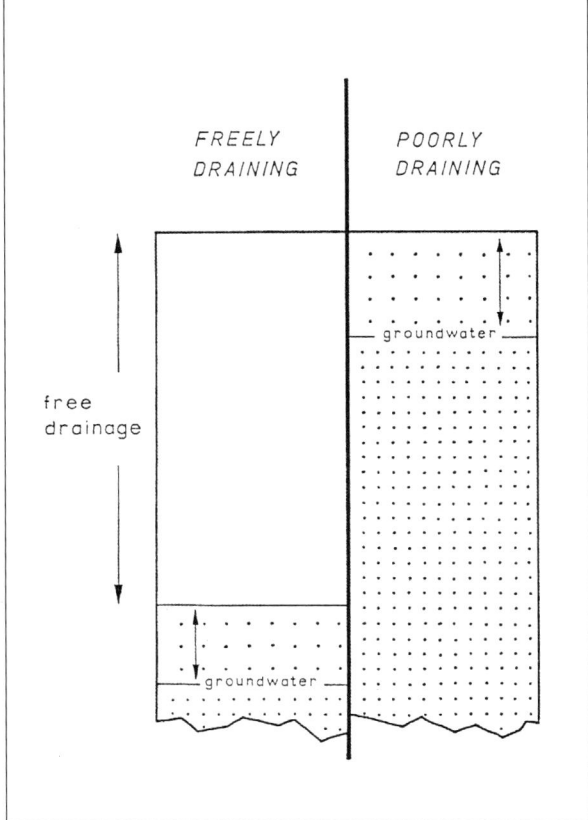

Figure 1.2: A simple two-fold classification of the soil profile, where the vertical dimension is about 100 cm.

end of the glaciations. In a contrast of time and chemistry, the blanket peat of the uplands is a product of surface or site wetness from the wetness of the atmosphere (mainly rainfall). Such peat is nutrient-deficient as it is fed only from the atmosphere, and in age, its initiation dates from an increase of climatic wetness about 3,000 years ago in sub-Atlantic time (although the range can be between 1,000 and 4,500 years BP).

Climate also has an impact on soil through soil erosion, but grassland agriculture has reduced this to a minimum in Northern Ireland. Grass cover protects the soil from possible erosion by running water, and clay-rich soils develop a structural resistance of their own. The few erosion incidents reported have been associated with sandy, arable soils during their exposed, vulnerable state, after ploughing in winter and spring, and in reclamation areas in uplands like the Mournes. Peat bog bursts have had an erosion impact on the peat and also on surrounding mineral soils, but the overall erosion problem in Northern Ireland is very small. Recent bog bursts (or slides) in Northern Ireland have been reviewed by Tomlinson and Gardiner, 1981, and by Cruickshank and Tomlinson, 1990.

RELIEF OR TOPOGRAPHY: The land area of Northern Ireland is 13,550 km^2 or 5,242 square miles, and while this is about the same size as Yorkshire, half the size of Belgium, one third of Denmark and only one sixth of the area of the island of Ireland, it contains a variety of relief and structure possibly unequalled in any other region of these islands. Relief reflects closely the underlying geology, and is explained partly by differential denudation or erosion (in the distant past) between hard and soft rocks, and partly by faulting of the main geological blocks.

The land surface of the Province is predominantly lowland, approximately 75 per cent of its total area being below 150 m or 500 ft, with most of it forming an extensive saucer-shaped lowland around Lough Neagh. From there, a coastal lowland extends along the north coast westward into the Foyle estuary; in the south, corridors of lowland connect to the basin of the Upper and Lower Lough Erne in County Fermanagh and also eastwards into the Lagan Valley, through Belfast and ultimately to the Ards peninsula beyond Strangford Lough. Much of the lowland is underlain by clay-rich glacial deposits of low permeability, and with stream gradients often being very gentle in lower courses, it is inevitable that poor land drainage restricts land use, unless artificial improvements have been made.

In Northern Ireland there are few mountains but often they have a striking appearance in the landscape by virtue of their isolated position. Only 6 percent of the total land area is above 300 m, or 1,000 ft, but the mountains are distributed in isolated blocks peripheral to the central lowland, giving them a dominance over the surrounding landscape that belies their areal extent. The location of the Mountains of Mourne and the Antrim plateau cliffs adjacent to the coastline allows these mountains to rise high above the shore, and to provide steep rocky slopes and shallow ranker soils.

There are four main uplands in Northern Ireland. Two are visible from Belfast, from where both near and distant skylines are made from young and tough igneous rocks. To the north, the skyline of three summits (Divis at 477 m, Black Mountain 375 m, and Cave Hill 358 m) is part of the Tertiary basalt plateau of County Antrim and beyond. From north Belfast, the extruded sheets of basalt lava extend some 80 km to the cliff coastline of the spectacular 'Giant's Causeway' in north Antrim, with the main plateau surface being between 300 and 400 m above sea level. From Belfast, south across 50 km, the distant skyline of the Mourne Mountains is built up of hard, Tertiary granites; in between, the lower upland of Slieve Croob in mid-Down was formed of much older 'Newry' granite. Towering above the coastal town of Newcastle, Slieve Donard at 847 m is the highest summit in the Mournes, indeed in the whole of Northern Ireland, and literally 'sweeps down to the sea'. The Mournes are small in area (only 187 km^2), but have fifteen summits above 600 m, and a further thirty above 300 m. As a result, over 80 per cent of the soils are shallow rankers, or peat podzols in mixes with rankers. The steeply-sloping topography restricts the areas of peat development. To the west of the Lough Neagh basin, the more extensive Sperrin Mountains, running about 30 km east to west, are formed from metamorphosed sediments, in the form of relatively resistant mica-schist rock. The main summits of the Sperrins are about 600 m, the highest being Sawel at 683 m, and only just stand out above rounded ridges and gentle slopes. Ice sheets of the Pleistocene swept over and smoothed the outlines of the Sperrins, while valley glaciers cut deep into the Mournes producing many more summits in a smaller area. The fourth upland forms the western Border of the Province; a table-land of hard Carboniferous limestone capped with grits, its cliff-line edge just over 300 m, makes a striking feature west of the Lough Erne basin in County Fermanagh. The highest peak is Cuilcagh at 670 m, rising above the area of the Marble Arch caves and an extensive, peat complex on a plateau at 300–350 m.

There are other mountains over 300 m in Northern Ireland but these are even more restricted in area. There is the isolated peak of Slieve Gullion at 574 m, with a few other hills in ring formation close by, in South Armagh. Again, these are formed from a mixture of hard, acid igneous rocks, produced from multiple volcanic activity. In Country Tyrone, there are rolling uplands of Old Red Sandstone where a few summits just break through the 300 m level, but are extensively peat covered. The upland structure of the

Province is also controlled by major fault lines, the most important of which is the Southern Uplands of Scotland fault which enters by Belfast Lough and fashions the hill slopes on the south of the city. The Scottish Highland Boundary fault plays a less obvious role in the landscape, being buried below the Antrim basalts, but it does delimit the south-eastern slopes of the Sperrins. Of greatest scale is the internal, and relatively recent Tertiary down-faulting of the basalts which takes those lava sheets down 800 m below sea level under Lough Neagh. The down-faulting of the central lowlands contributes in some measure to the relative prominence of the surrounding uplands, and has much affected the drainage pattern.

The most notable feature of the drainage pattern in the Province is that so much appears to feed inward and pass through Lough Neagh. This inward, or centripetal, drainage is now represented by the catchment of the River Bann and its tributaries, which drains 39 per cent of the land area of the Province, and which may have been more extensive in the past, when the basalt was more extensive. The system began to develop quite early in the Tertiary (about 55 million years ago in the Eocene period), and just after the main period of basaltic vulcanicity. First, the Lough Neagh depression started to be formed by downwarping and faulting while the rivers of the time carried down debris that was deposited, in places to a thickness of around 300 m, as the Lough Neagh Clays. These deposits survived to become a 'soft-rock', protected by being in a large series of depressions around the Lough, and are now an important soil parent material.

This outline of main structures of the Province has indicated how relief is interconnected with geology, through relative resistances of rocks, through igneous activity and major faulting. The interconnections between relief and climate are just as important in explaining the distribution of some soil profile types, and specifically that of peat. The latter covers 14 per cent of the Province and provides a good example of topography encouraging the accumulation of peat, given surface wetness of the site. Lowland, nutrient-rich peat (also called basin or fen peat) is fed initially by ground water, but later, may develop a mossy dome and become a 'raised bog', with the upper, acid layers being fed only by atmospheric wetness. These bogs have been developing since the end of the glaciations, 10,000 years ago. The development of upland blanket peat, usually above 300 m in the east and above 150 m in the west, also requires a level surface and site wetness. Good examples can be found on the Antrim plateau, the Sperrins, and plateau surfaces in west Fermanagh. Further discussion of peat development is given by Tomlinson in Chapter 6, and also was reviewed by Hammond in 1979. Subdivision of peat was excluded from the DANI Soil Survey and detailed consideration of peat is not included here.

The classic influence of climate combining with relief is in the altitudinal-hydrological sequence of soil profile types on freely draining parent material (see Figure 1.3). This sequence is best demonstrated in the Scottish Highlands, where it is possible to find the complete sequence from brown earths to podzols from sea level in the east to the boundary of blanket peat at about 500 m above. In Northern Ireland, there are only

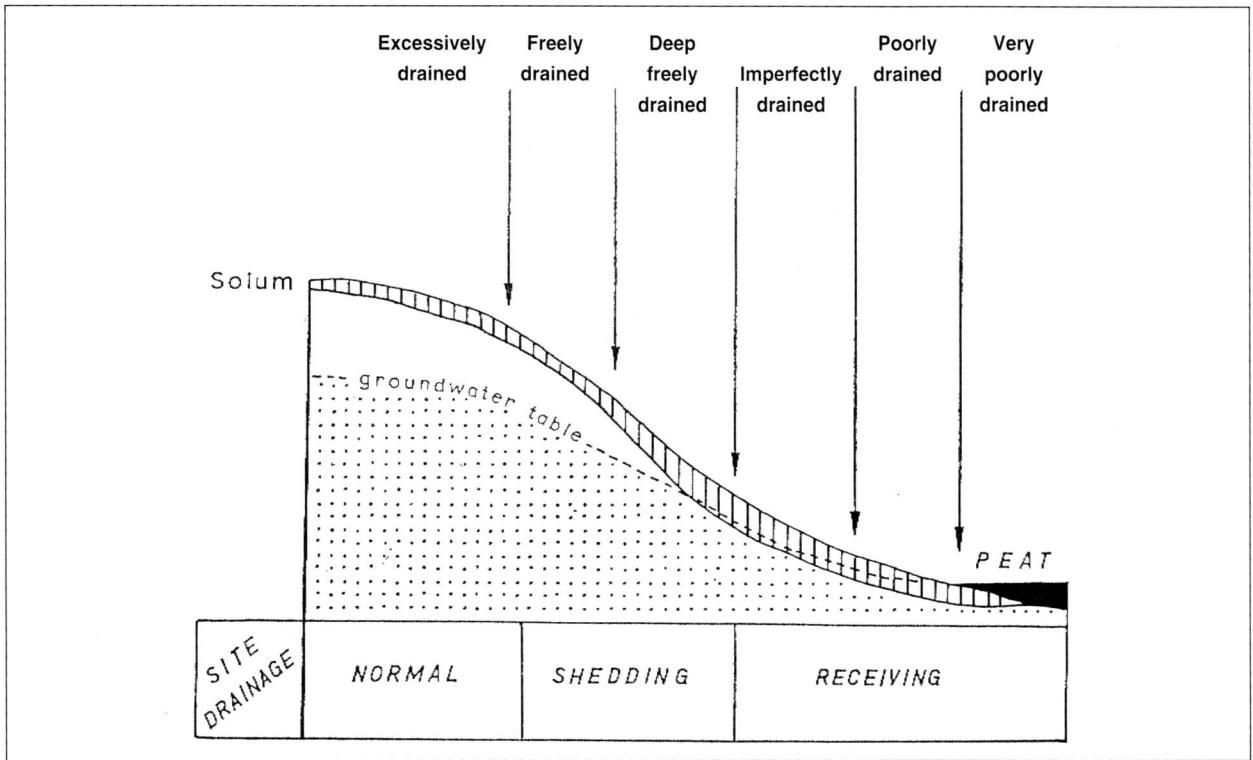

Figure 1.3: The altitudinal-hydrological sequence of soils along an ideal slope, from mountain to valley.

a few specific sites in the Mournes and the Sperrins where increasing acidification of freely draining soils follows increasing altitude. However, a general relationship with altitude is found for the sequence of freely draining soil profile types (18 per cent of area) in Northern Ireland.

The most common mineral soils in upland and mountain areas in the Province are the shallow <40 cm deep, RANKERS or LITHOMORPHIC SOILS, of steep and rocky slopes. These are found on almost all the main rock types. They include a variety of low organic content rankers (brown, gleyed and ferric rankers), as well as humic rankers in the climatic environment of blanket peat.

PARENT MATERIAL: It is sometimes said that the parent material of peat is water, but that apart, it is accepted that all the mineral and organo-mineral soils in Northern Ireland have developed from a very wide range of 97 parent materials. These fall into 5 main groups; 1) Rocks, and there are at least 20 common rock types in the Province; 2) Glacial till or boulder clay (unsorted, weathered mineral material moved by ice) derived from these rocks, and also including mixes of glacial tills; 3) Fluvio-glacial meltwater deposits of sand and gravel; 4) Lake deposits of stone free sand, silt or clay size materials, as well as diatomite; and 5) Marine alluvium of polder areas and other sites of coastal reclamation.

It is conventional to describe rocks by their genesis, or in the chronological order in which they were formed, and Bazley provides a detailed account in Chapter 3, but of particular relevance to soil development is the primary mineral composition and chemistry of rocks. Once again there is a remarkable range in the Province, from alkaline chalk, limestone and basic igneous rocks, through various sedimentary sandstones, shale, marl and mudstones of intermediate reaction, to acid and very acid igneous and metamorphic rocks, like granites, mica-schists and quartzite.

Northern Ireland happens to be a microcosm of the Earth's geology. Every period of the Earth's geological history (except a phase of rock formation in the Cambrian) is represented in the Province, and almost every known rock type can be found. There is a wide range of sedimentary rocks, as well as basic and acid igneous rocks. Metamorphic rocks, which are sediments altered by heat and pressure, are found in the ancient Dalradian mica-schist and quartzite rocks of the Sperrin Mountains. These rocks weather to produce soil material with a sandy texture and acid reaction. Similarly, adjacent granites in the Sperrins, as well as various types of granite in south Down and south Armagh are also sandy and acid, easily developing some type of acid, podzol soil.

The influence of parent rock type on soil texture in Northern Ireland can be demonstrated in the diagram, Figure 1.4, showing that the quartz content is a critical factor in determining the sand content, and through that, the texture of the derived soil. Conversely, clay textures are produced from rocks (such as basalt and slightly impure limestones) which have a high content of clay-forming primary minerals. The chemistry of soil also reflects that of the rock and glacial till from which the soil has developed. In particular, the

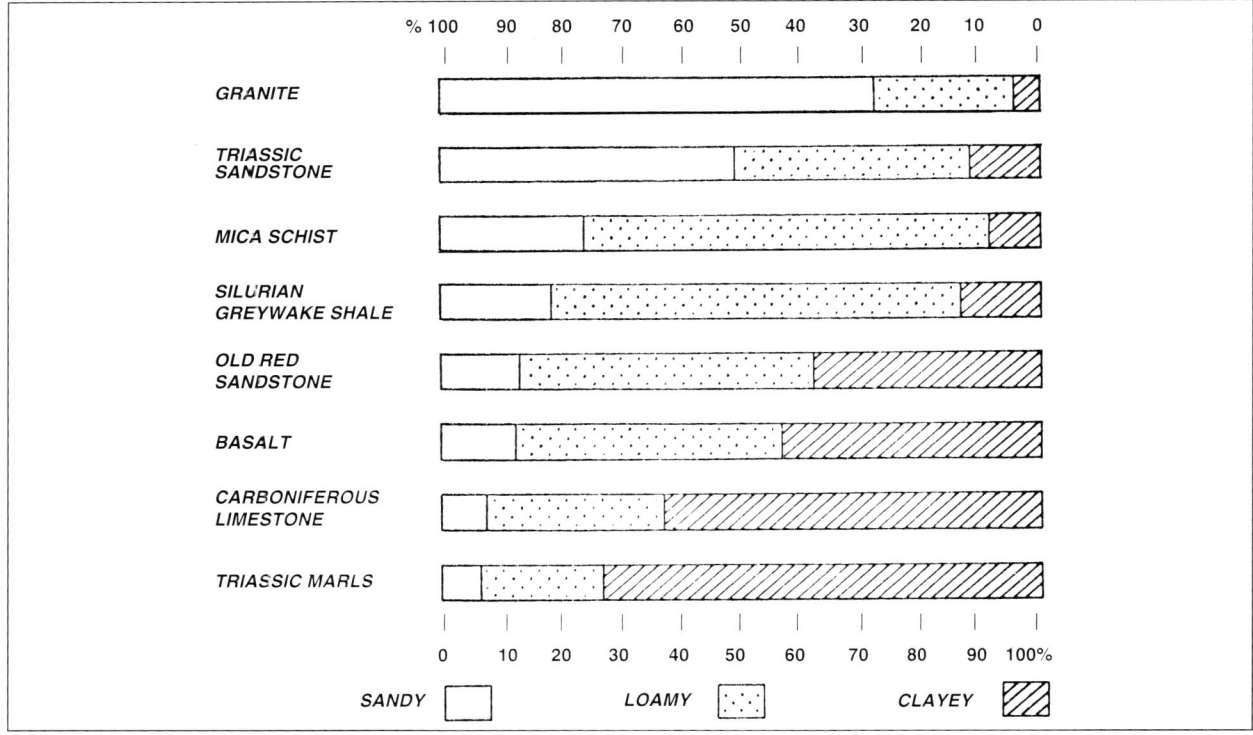

Figure 1.4: The influence of geology on soil texture – a Northern Ireland model first produced by McConaghy and McAllister in 1963 and closely reflecting the analytical results of the 1996 soil survey.

connection between soil and rock is close in their acidity, and in their abundance of the major plant nutrients coming from a geological source, such as calcium, phosphorus, potassium, magnesium and sodium. Soil acidity is important because it can control the availability to plants of soluble forms of these mineral plant nutrients.

Geology and relief have a close relationship, which is often expressed in the form of drainage basins. The igneous intrusion of rocks in the Ring Dyke complex, west of Newry, is a good example where resistant felsite rocks form a dyke-structure, containing basins of relatively less resistant granodiorites. And clearly, this is a general rule throughout the Province that hills and mountains, usually dominated by rankers, are a reflection of tough, resistant rocks, while valley systems and lowland basins are underlain by soft rocks, like sandstones, shales and marls (such as in the Foyle and Lagan valleys). Exceptions to this general rule are to be found where the same rock, such as the basalt in Antrim, has been topographically altered by geological faulting into upland and basin, as in the case of the Lough Neagh basin. Or, it may be the case that local weakness in the same rock type helped to initiate river valley erosion, leading ultimately to mature valleys and drainage catchments. This interconnection between rocks and topography produces a mosaic of miniature landscapes which gives the Province its distinctive environmental character, and is the topographic framework for soil development.

ORGANISMS (OR BIOTA): Plants/vegetation and animal/man organisms are the last two soil forming factors. It is difficult to quote widespread examples of the influence of plants/vegetation on soil profile development in Northern Ireland, particularly because most woodland vegetation was cleared so many centuries ago. To a limited extent there is vegetation influence on soil profile development in the relatively small areas of heather moor in the Sperrins (see Tomlinson in Land Cover, Chapter 6). However, the Province lacks extensive mountains and high-level uplands (only 6 per cent is above 300 m), where the interaction between climate and relief controls the nature of semi-natural vegetation and related soils, such as in the Highlands of Scotland.

Animal organisms include the impact of man, which has been, and continues to be, a significant influence on soil character, everywhere in the Province. It started with the clearance of woodland centuries ago, probably in two major pulses of activity initiated about 5,000 years BP and 1,400 year BP. At these times long before artificial drainage and in a generally wet environment, freely draining soils were selected for forest clearance. Experts are divided on whether such clearance accelerated the rate of soil leaching, by removing the intervening vegetation and allowing more potential leachate solution to enter the soil. Whether or not this is the case, only about 4 per cent of Northern Ireland soils show some degree of leaching, so this academic argument does not have large scale importance here. The development of humus-iron podzols in an oakwood on mica-schist material at Glenshesk, in north Antrim (D125337), is a localised example of forest clearance which does seem to have allowed the leaching of humus through the soil profile (Cruickshank and Cruickshank, 1981), as can be seen in Figure 1.5.

The history of man's impact on the 60 per cent of the soil landscape that is gleyed, plus 14 per cent in peat, starts much later. It is thought that the cutting of peat or turf for domestic fuel started after most of the forests had been cleared, and as the word 'turf' comes from the Norse torv, turf cutting probably started in Viking times, about 1,000 years ago. In wet and gleyed mineral soils, potato cultivation required the spade digging and building of 'lazy beds', raised above the level of the ground water table. Only in marginal hill land, never ploughed, does the topography of lines of raised lazy beds survive to this day. They were usually arranged vertically up and down the slope to improve surface run-off and field drainage, and incidentally, create channels which encourage soil erosion on bare soil surfaces. The legacy of lazy beds is often an accumulation of top soil at the base of the slope (McEntee and Smith, 1993), piled up on the upper side of cross-slope fences or hedge-banks. Until quite recently, it was the practice on many farms to re-spread this top soil back upslope.

The next event in man's impact on soil was the introduction of artificial drains into gleyed soil, which started just over two centuries ago. The first methods used were primitive drains constructed of large stones or grass sods or wood, and the first soils drained were in areas adjacent to well-drained land, and also soil only slightly gleyed and usually on large land holding units (demesnes and estates). The areas of worst drainage and clay-rich soils, such as most of Fermanagh, were not even enclosed until long after the east of the Province. Effective underdrainage, in the form of mole drains and gravel-filled tunnels, was not introduced until after 1960 and after 1970, respectively. Some of the early stone drains are still functioning, but their exact location is not known to land managers. Only since 1949 has land benefiting from underdrainage been recorded and locations of drains recorded by the Department of Agriculture (NI), and in that period, most of the installation has been in the three western counties. Wilcock reviews field underdrainage and considers its effect on agricultural land (in Chapter 5). Lastly, it should be noted that the benefits of underdrainage are mainly in improved soil fertility, including more efficient germination, better response to fertilizer application, increased yields and long-term sustained quality and flexibility of land use. Drainage does not change the visible morphology of gleys, and despite the

In conclusion, this introductory chapter should be seen as an attempt to evaluate the nature and the influence of the five soil-forming factors within the land area of Northern Ireland, and also the processes of soil development. The inter-action of the five soil-forming factors controls the rate and effectiveness of soil development processes, like primary stage weathering and organic decay, and secondary stage processes like leaching and gleying. The dimensions involved are time, a relatively short-time in this particular environment, and space which can be considered as a mosaic of miniature soil-forming environments across the Province.

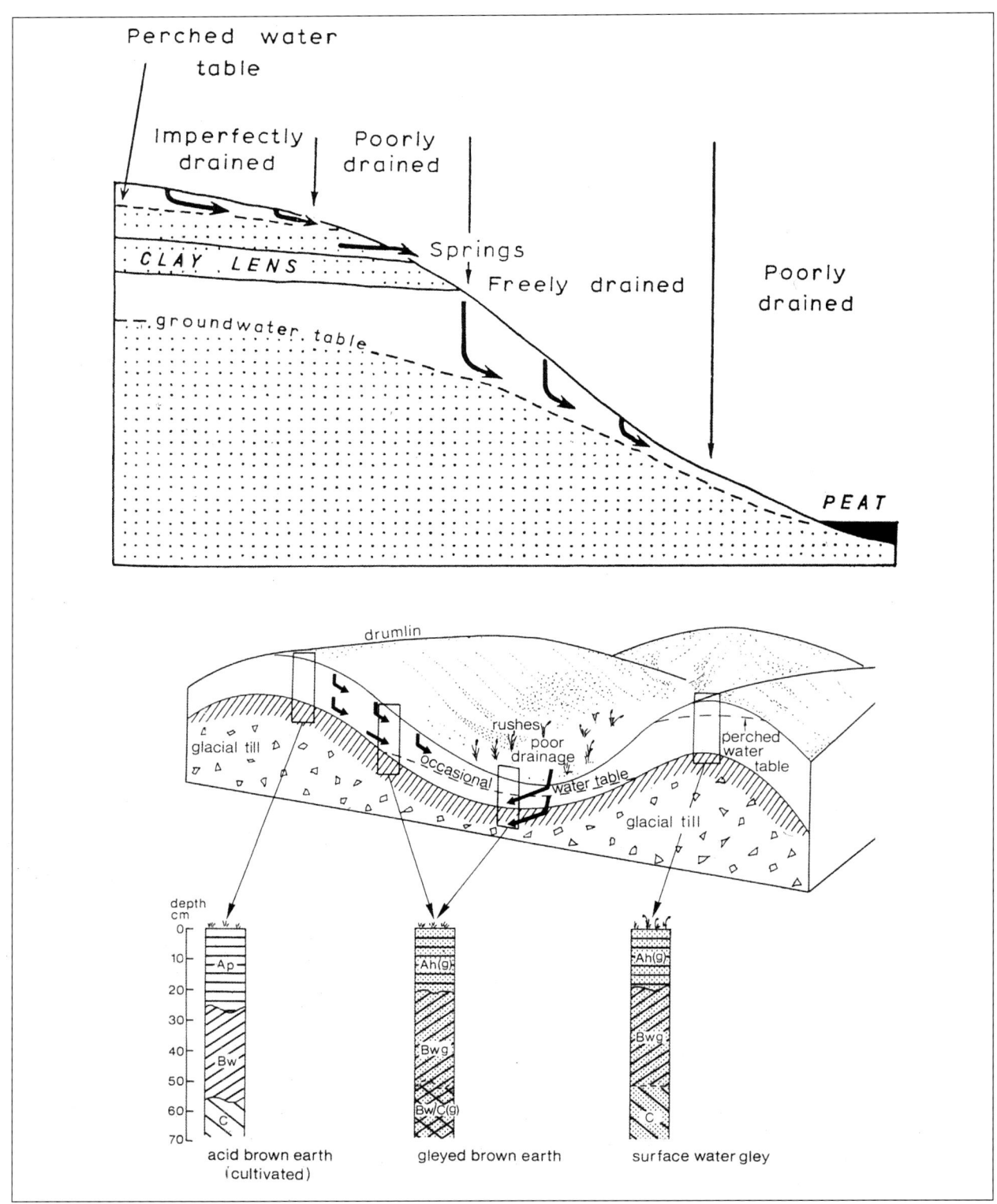

Figure 1.6: Perched or surface water gleying, and the relationship with ground water gleys.

CHAPTER 2

THE SOILS: THEIR NATURE AND VARIABILITY

Soil Classification

Many people think there is only one kind of soil, and that it is brown in colour, a mix of organic and mineral matter! Soil is much more than the surface layer of ploughed soil, the Ap horizon. The fact that this survey has identified and mapped at least 308 different soils on 97 parent materials within the land area of Northern Ireland, requires explanation for the reader to appreciate the complexity of soil profile development in this region. The explanation was started in the previous Chapter, and will now be developed in relation to the major soil types found in Northern Ireland and shown in Table 2.1.

Soil, as understood by pedologists who study the nature and distribution of soil, develops through layers or horizons into soil profiles. The latter are not produced by accumulation, except in alluvium and peat, but by internal processes of transfers and relocation, additions and losses, of both solid and dissolved substances, carried by soil solutions in many directions within the soil. Soil formation is a two stage process, and this re-arrangement of material within the profile follows after, and continues along with, the initial processes of weathering and organic matter decay. The early pedologists, at the end of the 19th century, described soil as an independently organised natural body, that is spatially variable, depending on the interaction of soil-forming factors. Soil forms in the three-dimensional skin of the Earth's surface, and the three dimensional soil unit has been called a pedon or pedo-unit. The two-dimensional section through the soil skin is the soil profile (see Figure 1.2). This sample unit of soil – the soil profile – is the object of soil order or soil classification, on which any soil survey depends.

There have been many international soil classifications proposed throughout the past century, but as Avery has described in his *Soils of the British Isles*, (Avery, 1990), the first soil classification for Britain and Ireland was being devised in the 1938–1940 period. In its simplest form, this was a primary division between freely draining soils (brown earths and podzols) and poorly draining soils (gleys), as demonstrated in Chapter 1, Figure 1.2. However, the soil classification used as the basis for soil survey in Northern Ireland was devised by Brian Avery in England in the 1960s, and published in different forms until its final refinement as the *Soil Classification for England and Wales (Higher Categories)*, (Avery, 1980). The original classification had 10 primary categories, called Major Soil Groups, but in this DANI Soil Survey, we have amalgamated the first three categories into one, covering all raw undeveloped and thin soils, and here called **RANKERS**. These are less than 40 cm in depth to the parent material, and have not developed horizons in the solum part of the profile (above the parent material). Most of the rankers (except rare gleyed rankers) are freely draining, and so this category joins the **BROWN SOILS** (or Brown Earths) and **PODZOLS** as the three categories of freely draining soils shown in the heading of the legend on every soil map, see Colour Plate.

Within the group of poorly drained soils, Avery's Major Soil Group, Pelosols, which are red in colour, slowly permeable clay-rich soils, have been incorporated in the gley soils because they are so rare in Northern Ireland. It is thought that red clayey Pelosols exist, developed on Triassic mudstone or marl till, north of Belfast Lough (Soil Map 15), and on limestone red till near Loughgall, County Armagh (Soil Map 19), but at best their identification was uncertain and polygon areas involved were very small. For that reason, the gleyed Pelosols, with gleyic features masked by red colour, have been included with the slightly gleyed SWG1 soils. Avery has two further Major Soil Groups called '**Surface water gleys**' and '**Ground water gleys**', and while the DANI Soil Survey retains the concept of two types of ground water table being responsible for gleying, it is accepted also that the difference between the two situations may be academic at many sites in the Province. We think it is of greater practical and applied importance to divide the gleys, which cover 60 per cent of land area, into three degrees of gleying. The definitions are given here in Box 1.

1:250,000 soil map of the Province, coastal sands are called sand rankers (**SR**).

In the Major Soil Group of **BROWN SOILS**, the DANI Soil Survey has identified shallow brown earths (SBE) which have 40-60 cm of solum horizons over the parent material, and the related brown earths, which are deeper. The two types are often intermixed in occurrence and together comprise 13.1 per cent land area. Brown earths have A, B and C (parent material) horizons, are freely draining, but have little visible differentiation between the horizons, which are normally brown or reddish-brown in colour throughout the profile. The middle or B horizon is weathered, but is only very slightly leached at most, and has not collected materials moved in from elsewhere in the profile. Surface Ap horizons are not humic, usually cultivated. These are the best agricultural soils of lowlands below 250 m, and mainly in the east of the Province. They are highly rated in any land quality valuation system. The variants of GBE gleyed brown earths, and CBE calcareous brown earths, are rare in the survey area. GBE might have been more extensive, but merge easily with G1 and SWG1 soils. CBE soils are difficult to identify in the field without a portable pH meter, and were mapped only around Middletown in west Armagh. Here, they were similar in morphology to grey-brown podzols, with clay-enriched B horizons (called Bt - t means texturally changed). These are not acid podzols, but are basic soils where clay particles have been physically moved by solution into the B horizon. Because the area is small, they have been included with brown earths as CBE, calcareous brown earths.

All of the **PODZOL** Major Soil Group comprises only 4.1 per cent land area in Northern Ireland although 6 Soil Groups occur. These are freely draining, acid and well-leached soils, with a visibly differentiated profile. SBP shallow brown podzolics, and BP brown podzolics, covering 2.7 per cent land area, are in the early stages of leaching, and as they are usually developed on acid, porous materials in lowland or middle altitude areas, they are within the agricultural land. Cultivation means that the surface Ap horizon is not humic, and incorporates the slightly bleached Ae or E (eluviated) horizon. Brown podzolics are identified by their rusty-orange coloured B horizon, enriched with iron and aluminium oxides. POD podzols are usually found on porous, acid parent materials above 250 m in a wet, climatic environment which encourages leaching. These later stage podzols collectively comprise only 1.4 per cent land area in the Province, partly because there is so little land above that elevation. Much of that is covered in peat or is in HR humic ranker. Podzols are very well-known, classical soil profiles, but here, they are usually outside the agricultural land on hill margins. They have a humic surface horizon, overlying the bleached and leached, grey-coloured E horizon, which in turn overlies the ferric oxide enriched Bs horizon. Podzols can develop a pan or hard crust of ferric oxide between the E and Bs horizons, and from that obstruction, gleying may subsequently develop in the E horizon to produce what is called a STP stagno-podzol or gleyed podzol. PP peat podzols develop where the surface humic horizon, on a podzol, builds-up to thickness between 25 and 45 cm. SP shallow podzols are rarely found, and such soils would be called PR podzolic rankers. All of these last mentioned soil profiles, POD, PP, STP and SP, comprise only 1.4 per cent land area together.

Rock types and soil parent materials

The immense range of rock types to be found in the political unit of Northern Ireland, which contributes to the variety of soils was noted in Chapter 1. It is useful to make this geological background the context in which to review soils and their properties. Soil profiles in the Province are still in the relatively early stages of development, although weathering dates back to time before the glaciations. Mineral soil property values reflect quite closely the character of the underlying rocks and the glacial till derived from them. Despite the fact that 97 soil parent materials have been counted in Northern Ireland, which makes it probably the most geologically diverse area of its size in Britain or Ireland, only about 10–12 rock types cover about 97 per cent of the land area, and they are parent materials (through weathered rock or related glacial till) to the soils in that part of the Province.

The major rocks and their approximate, percentage area is given in Table 2.2.

The percentage areas given in Table 2.2 are approximate and some are likely to be changed slightly, by redefinition of geological periods. The total area given is 97 per cent for these 7 major geological periods, and about 10–12 distinctive rock types. All these, except possibly Millstone Grit of the Carboniferous, have related glacial tills, which are also soil parent materials, and would be included in the 97 per cent NI area. However, the remaining 3 per cent NI land area contains rock such as rhyolite, Triassic mudstone or marl, dolerite, felsite, andesite, basic igneous rocks in the Sperrins, Cretaceous Chalk, Lough Neagh Clay, magnesian limestone and others, which easily raises the total of all rocks found in the Province to about 24 or 25, plus their related glacial tills. A wide range of transported materials and deposited materials such as alluvium, organic alluvium, marine alluvium, sand and gravel deposits, lake deposits and diatomite, as well as a long list of about 25 examples of glacial tills of a mixed rock origin, add to the list of mainly mineral soil parent materials. Organic material of peat (covering 14 per cent land area, and covering some of the rocks mentioned) completes the list totalling 97 parent materials. In the following review of the main soil series and their soil properties, most attention will be

Table 2.1
Summary of soil profile types and areas in Northern Ireland

		km^2	per cent land area of 13,550 km^2
1.	Humic Rankers 804 polygons	512.6	3.7
2.	All Other Rankers 1605 polygons	699.1	5.1
3.	Shallow Brown Earths 878 polygons	745.9	5.5
4.	Brown, Gleyed and Calcareous Earths 3428 polygons	1,036.6	7.6
5.	Shallow Brown Podzolics 124 polygons	103.9	0.8
6.	Brown Podzolic 527 polygons	196.3	1.9
7.	Podzols 412 polygons	270.8	1.4
8.	G1/SWG1/Pelosols 3558 polygons	2,411.1	17.8
9.	G2/SWG2 3353 polygons	3,286.6	24.2
10.	HG/SWHG (Humic Gleys) 1243 polygons	704.3	5.2
11.	G3/SWG3 794 polygons	453.8	3.3
12.	Alluvium (Mostly G2) 2607 polygons	726.5	1.5
13.	Organic Alluvium 1078 polygons	200.5	1.5
14.	Peat 3040 polygons	1,930.6	14.1
15.	Lake Deposits (Alluvium, Clay, Sand) 219 polygons	77.6	0.6
16.	Urban 478 polygons	556.4	4.1
17.	Disturbed 504 polygons	122.6	0.9
18.	Not Surveyed 106 polygons	7.5	0.1
19.	Marine Alluvium 7 polygons	28.5	0.2

Note: The figures of km^2 areas contain areas of included lakes. The total **land area** of Northern Ireland is 13,550 km^2 and the per cent land areas have been calculated on that figure. Consequently, the percentage total slightly exceeds 100. Province-wide, 24,765 soil polygons were mapped.

given to the soils of the top 10–12 rocks, and their related tills, as these are the most extensive in area, and hence are economically the most important soils.

As a first step, a comparison can be made between the areas covered by the 7 main geological periods and their related **mineral** soils, as shown in Table 2.3. It must be kept in mind that these 7 geological groupings are covered in different extent by superimposed deposits of sand, gravel, alluvium and peat.

The mineral soils, associated with the main rock types cover 59-60 per cent area of Northern Ireland. To that must be added a further 30 per cent area to include organic alluvium (1.5), alluvium (5.5), peat (14.1), sand (1.4), gravel (3.7) and urban areas (4.1). About 9-10 per cent area remains for all the soils of the minority rocks, and also till mixtures. Of these,

Table 2.4
Areas of the 25 major Soil Series in Northern Ireland

	Soil Series			Area in hectares	per cent NI area
1.	Basalt Till	–	Brown Earth	17,736	1.32
2.	Basalt Till	–	SWG1	55,510	4.13
3.	Basalt Till	–	SWG2	98,070	7.30
			SWHG	13,294	0.99
4.	Shale Till	–	Brown Earth	7,485	0.55
5.	Shale Till	–	SWG1	65,142	4.85
6.	Shale Till	–	SWG2	37,715	2.80
7.	Mica-Schist Till	–	Brown Earth	3,819	0.28
8.	Mica-Schist Till	–	SWG1	21,082	1.57
9.	Mica-Schist Till	–	SWG2	29,926	2.23
		–	SWHG	21,954	1.65
10.	Granite Rock	–	Brown Podzolic	12,001	0.90
11.	Granite Till	–	SWG1	10,972	0.81
12.	Lough Neagh Clay Till	–	SWG2	6,174	0.46
13.	Old Red Sandstone Till	–	Brown Earth	4,580	0.34
14.	Old Red Sandstone Till	–	SWG1	13,219	0.98
15.	Old Red Sandstone Till	–	SWG2	26,469	1.97
16.	Carb. Limestone Till	–	Brown Earth	2,312	0.17
17.	Carb. Limestone Till	–	SWG1	15,787	1.18
18.	Carb. Limestone Till	–	SWG2	13,227	0.98
19.	Calp Till	–	SWG1	2,644	0.19
20.	Calp Till	–	SWG2	22,646	1.69
21.	Calp Till	–	SWG3	20,701	1.54
22.	Carb. Sandstone Till	–	SWG1 and SWG2	15,628	1.16
23.	Yoredale Sandstone Till	–	SWG3	11,462	0.85
		–	SWHG	6,226	0.46
24.	Red Trias Sandstone Till	–	Brown Earth	7,360	0.54
25.	Red Trias Sandstone Till	–	SWG1	13,220	0.98
				576,361	42.87

Note that soil series numbers above are also used in the following soil analytical tables, and are followed by 4 other quite common soil series, numbers 26 to 29.

Basalt soils of the Antrim Plateau and adjacent areas

Soils derived from basic igneous rocks have attracted an exceptional degree of research interest and have been discussed in a series of published papers for Northern Ireland (Brown, 1954, Smith, 1957, McAleese and McConaghy, 1957–58). This parent rock, through the nature of its constituent minerals, exerts an almost unique influence on associated till, and ultimately on the soils developed. Because the parent material is unusual, the influence of other soil forming factors may appear to be reduced in the basalt region.

Tertiary basalt lavas covering 4,009 km^2 (29.5 per cent NI area) comprise just less than one third of the landscape of Northern Ireland. It is the largest continuous unit of basic igneous rock in the United Kingdom and Ireland, and for this reason alone, would merit special consideration. The basalt series are divided generally into the earlier, lower flows, which are usually below 305 m, and the upper flows, which in their eastern remnants form an elevated plateau surface. Within this landscape, soil forming situations range from sea level to over 600 m, with a mean annual precipitation varying from 750 to 1,750 mm. Many of the basalt soils could be classed as variations of brown earths, but gleying is common to different degrees, with SWG2 being the dominant soil. Brown earths may be found up to 305 m where 1,250 mm annual rainfall is recorded and podzolisation would be expected. But podzols are uncommon even in high rainfall areas, and the rare cases are related to a particularly high iron content in the parent material. Basalt soils are characterised by large cation exchange capacities, high levels of exchangeable calcium and magnesium cations, an abundance of sesquioxides of iron and aluminium from ferro-magnesian minerals, giving high totals, and a strongly bonded soil structure, created by these chemical properties. These properties also help to maintain a high base status in basalt soil, and keep pH values high.

Basalt soils are easily identified by their colour (Munsell colour 5 YR 4/4 Reddish brown) which may mask the presence of organic matter. Analyses nearly always show a higher organic matter content than is apparent in the field. A fluctuating watertable, causing gleying, can be identified by bright yellow-orange mottles of ferric iron against a pale purple-brown background of hydrated compounds. Structure is clearly developed into strongly cemented units of crumb or blocky shape, which can be seen as discrete and consolidated aggregates, resisting break-up when dry, even under physical attack by hammering with a pestle. Strong aggregation may appear to reduce the amount of clay measured, as complete dispersion is difficult.

As early as 1907, Kilroe observed the unusual chemical properties of basalt soils, noting that they are rich in calcium and phosphorus, but poor in potassium, being notably deficient in potash feldspars. More recently, research has been concerned with the problem of high cation exchange capacities in basalt soils being less related to the clay and humus content than to the form and nature of the clay minerals present. The primary mineral composition of basalt is dominated by labradorite, augite and olivine, to over 80 per cent by weight and volume. Clay minerals produced in chemical weathering are numerous, but olivine weathers early in the sequence to form vermiculite. The particle shape of vermiculite is often larger than clay size (ie > 2 μ) and hence contributes its high cation exchange property to the larger-size silt fraction. These have been referred to as the pseudo-aggregates of silt size which contain some clay minerals (Curtin and Smillie, 1981).

The cation exchange capacities of basalt soils are usually in the range of 35 to 55 milligram-equivalents per 100 grams of soil, which is regarded as large, and much larger than would be found in similar textures on other mineral parent materials. Normally the cation exchange capacity (CEC) is related to the clay and humus content of the soil, but in basalt soil the CEC may not change and may even increase as the clay and humus decreases with depth. It has been shown that material of sand size may contribute up to 30 per cent of the CEC, and the silt fraction up to 50 per cent. The clay fraction contributes up to 80 per cent in the surface horizon but often only 20 to 40 per cent in the deepest horizon. In this zone of initial weathering from the parent material, the clay minerals are in a large size category or are cemented into larger aggregates. The bonding of these aggregates is particularly effective because of the ferric and aluminium oxides and high content of exchangeable bases. Calcium usually dominates the composition of the CEC in any soil (to the extent of 80 to 90 per cent), but in some basalt soils magnesium can make up 40 per cent of the total cations. The high magnesium content is another diagnostic property of weathered basalt, and has been used to trace the direction of movement of glacial drift southward from the basalt edge.

Extremely high values for total concentrations of Chromium, Copper, Magnesium and Nickel have been recorded in basalt-derived topsoil samples, and also the highest levels of extractable Copper from the same soils. These data are reported in the Soil Geochemical Atlas of Northern Ireland in Chapter 9D.

3a SWG2 ON BASALT TILL - MODAL SOIL PROFILE (SOUTH ANTRIM - GLENAVY)

		Available mg kg^{-1}			meq 100 g^{-1}					mg kg^{-1}						
Horizon	pH	P	K	Mg	Mg	Ca	Na	K	CEC	Dithio Fe	Total P	% N	% C	% LOI	% Sand	% Silt	% Clay
Ap	6.2	27	84	354	3.18	18.41	0.28	0.27	41.73	24,550	1711	0.49	4.81	12.1	27.2	35.1	37.5
Bg	6.8	7	62	696	5.42	18.52	0.37	0.96	48.21	25,060	1243	0.33	2.93	7.9	31.9	33.2	34.6
C	7.1	5	111	378	3.44	12.03	0.29	0.32	42.52	27,784	1371	0.29	1.99	4.8	39.2	33.1	27.5

Landscape: This modal SWG2 basalt till soil is representing the till plain of south-west Antrim, around Dundrod - Glenavy - Aghalee, on Soil Map 14.

Soil Properties: Available P is adequate, and the same for Mg. Available K is low in all horizons, which is not uncommon for basalt soils. Exchangeable Mg and Ca are very high values, as is the case also for all following chemical elements. Soil textures are all clay loams.

Munsell colour of B horizon is 5 YR 5/4 Reddish Brown, or 7.5 YR 5/4 Brown.

3b SWG2   ON BASALT TILL - MODAL SOIL PROFILE (MID AND SOUTH-ANTRIM)

| Horizon | pH | Available mg kg⁻¹ | | | meq 100 g⁻¹ | | | | CEC | mg kg⁻¹ | | % N | % C | % LOI | % Sand | % Silt | % Clay |
		P	K	Mg	Mg	Ca	Na	K		Dithio Fe	Total P						
Ap	6.2	33	194	729	6.7	21.5	0.52	0.68	53.2	33,200	1873	0.42	4.62	10.2	27.8	37.5	34.7
Bg	6.9	2	29	1231	16.2	21.2	0.87	0.14	45.2	49,100	1034	0.14	1.32	6.1	36.7	42.8	20.5
C	7.1	2	58	1608	19.1	23.7	0.96	0.21	43.8	46,900	1162	0.02	0.18	4.5	34.9	35.3	29.8

Landscape: SWG2 soils dominate basalt till, especially in Glenwhirry and in Glenravel in mid-Antrim, on Soil Map 9.

Soil Series area: These SWG2 soils cover 98,070 ha or 7.3 per cent area in NI. If the SWHG (humic gley) related soils are added, the combined figure is 111,363 ha or 8.29 per cent NI. This makes the SWG2 + SWHG or basalt till the most extensive soil series in the Province.

Soil Properties: Available P and K are both adequate, and Mg is excessively high. Likewise, the values for all other chemical elements are extremely high, and the highest found in the Province. Textures are all clay loam.

Munsell colour of B horizon is 5 YR 5/4 Reddish brown.

Shale soils of mainly counties Down and Armagh

The soils developed from Silurian and Ordovician shale or greywacke, in the south-east of the Province, rank second in area to basalt soils, but have not attracted the same degree of research interest. However, these soils are distinctive in their own right, are located near to, and contain large centres of population, and are intensively managed for mixed farming. Because the shale soils are closely associated with the drumlin landscapes of counties Down and Armagh, the shale areas are highly variable internally. Shale soils are even highly variable within the average farm holding, as rocky rankers occur alongside peaty wet hollows, and drier drumlin hills, It is exceptional in the Province that brown rankers (which are very shallow, less than 40 cm deep brown earths, and including areas of rock outcrops) in the shale areas cover 29,075 ha or 2.17 per cent NI area, compared with 0.44 per cent in the larger basalt region, 0.26 per cent in mica-schist areas, and 0.54 per cent NI area on granite. In other words, the shale area has the highest proportion of any of the main parent material types in brown rankers. These very shallow lowland soils with rock outcrops, comprise a very high proportion of the landscape in east Co. Down, east of Saintfield and Crossgar, as well as in Lecale and in The Ards (see Soil Maps 20 and 21). There is also a large amount of land in shallow brown earth (up to 60 cm deep) on shale till. If the 20,173 ha or 1.50 per cent NI area of shallow brown earth is added to the area of brown ranker, the total reaches 49,248 ha or 3.67 per cent NI area. This is a very considerable area, and although technically two soil series, together they are both seriously affected by summer drought. This was well-demonstrated in the summer of 1995 when brown rankers were completely 'burned', and for that reason, must be considered to be in ALC class 4 within a soil landscape that would otherwise be regarded as Class 3A (eg on drumlins).

Shale soils are very stoney, containing many sharp, angular stones. And another distinctive physical feature, especially of the gleyed SWG1 and SWG2, is the hardening of the B horizon on drying-out, even in normal summers. Below the Ap horizon, the mineral subsoil becomes very hard and compact, probably because of the tightening structures in the clay minerals. On re-wetting, the moist soil loses the compaction. There is also a very characteristic texture of clay loam or sandy silt loam in all horizons of the shale soils, produced by a highly predictable composition of 40 per cent sand, 40 per cent silt and 20 per cent clay. If the clay is over 19 per cent, the texture is clay loam, and sandy loam if below 19 per cent. The very high silt content is unique to shale soils, and gives them their smooth feel. Values of all chemical element are average or just below average (in the soil range) in shale soils. Shale soils are very slightly acid, with pH 5.6-6.3. The CEC is usually moderate in size, and with about 50 per cent base saturation. Available P, K, and Mg are normally adequate in Ap, but reserves are low in the horizons below.

4. BROWN EARTH ON SHALE TILL - MODAL SOIL PROFILE

Horizon	pH	Available mg kg⁻¹			meq 100 g⁻¹					mg kg⁻¹		% N	% C	% LOI	% Sand	% Silt	% Clay
		P	K	Mg	Mg	Ca	Na	K	CEC	Dithio Fe	Total P						
Ap	5.9	21	134	141	1.23	12.12	0.04	0.54	25.3	8000	1794	0.37	3.99	9.4	35.8	38.6	25.6
Bw	6.1	6	74	88	0.58	6.69	0.04	0.34	21.5	4050	977	0.16	1.66	4.7	29.7	42.4	27.9
C	6.3	4	33	78	0.45	3.54	0.01	0.22	10.2	8200	660	0.04	0.25	2.8	50.8	34.2	14.6

Landscape: Brown earths and shallow brown earths are found extensively in mid- and north Co Down on ridges and till plains, but not often on till drumlins. See Soil Map 21.

Soil Series area: Brown earths and shallow brown earths on shale till together cover 27,658 ha, or 2.56 per cent of NI area. Shallow brown earths on shale rock in Co Down cover a further 17,000 ha, or 1.3 per cent area. Shallow soil and rankers are very common on shale in Co Down.

Soil Properties: Available P, K and Mg are adequate in the Ap, but reserves in B and C are low. Base saturation of CEC is less than 50 per cent, and pH is just slightly acid. Textures of A and B are clay loam, becoming sandy silt loam high in the C horizon. Silt is high in all horizons, which is typical of shale till soils.

Munsell colour of B horizon is 10 YR 5/8 Yellowish brown

5. SWG1 ON SHALE TILL - MODAL SOIL PROFILE

Horizon	pH	Available mg kg^{-1}			meq 100 g^{-1}					mg kg^{-1}						
		P	K	Mg	Mg	Ca	Na	K	CEC	Dithio Fe	Total P	% N	% C	% LOI	% Sand	% Silt	% Clay
Ap	5.6	32	173	163	1.32	9.28	0.26	0.86	25.07	3950	1307	0.34	3.71	10.3	40.1	42.2	17.7
Bg	6.0	28	259	117	0.91	4.24	0.17	0.82	20.51	6050	1346	0.21	2.22	6.5	40.5	41.2	18.3
C	6.1	5	162	202	1.42	5.04	0.13	0.63	12.12	9150	411	0.02	0.44	3.1	38.3	44.1	17.6

Landscape: Most of the SWG1 soils on shale till are on drumlins in Co Down and Co Armagh. See soil Maps 19 and 20.

Soil Series area: SWG1 on shale till covers 65,142 ha, or 4.85 per cent NI. It is the most extensive soil series on shale till, and the second most extensive in the whole Province.

Soil Properties: The most distinctive property of shale till soils are the high silt content at around 40 per cent, along with a similar content of mainly fine sand. If the clay content is over 19 per cent, the texture is clay loam, and sandy silt loam if below. Available P, K and Mg are all adequate. CEC is moderate and is less than 50 per cent base saturated.

Munsell colour of B horizon is 10 YR 6/2 Light brownish grey

6. SWG2 ON SHALE TILL - MODAL SOIL PROFILE

Horizon	pH	Available mg kg⁻¹			meq 100 g⁻¹					mg kg⁻¹						
		P	K	Mg	Mg	Ca	Na	K	CEC	Dithio Fe	Total P	% N	% C	% LOI	% Sand	% Silt	% Clay
Ap	6.3	23	60	161	0.98	9.66	0.18	0.25	25.3	7605	1818	0.42	4.64	12.2	35.8	43.1	21.1
Bg	6.2	6	22	31	0.33	4.13	0.18	0.05	12.6	4329	498	0.31	2.84	7.5	41.2	42.2	16.2
C	6.4	11	39	277	1.94	5.33	0.16	0.11	12.8	14070	1071	0.03	0.33	2.8	51.5	31.9	16.6

Landscape: Soils of lowland, around drumlins and on flat tops of drumlins in Co Down and Armagh, but it is most extensive in Mid-Armagh, on Soil Map 28.

Soil Series area: SWG2 covers 37,715 ha, or 2.8 per cent NI. Shale rock and till soils comprise 14.3 per cent of the Province, but shale/greywacke rock of Silurian - Ordovician times comprise 19 per cent of rocks.

Soil Properties: Available K is low, but P and Mg are adequate. CEC and exchangeable cations are moderate, but typical for shale till. Dithio Fe (Total iron) is also moderate to low, and soil texture is clay loam or sandy silt loam.

Munsell colour of B horizon is 10 YR 5/2 Greyish brown.

Mica-schist soils of the Sperrins and north-east Antrim

Mica-schist soils always have over 50 per cent of Fine Earth as Sand, and usually it exceeds 60 per cent. They have sandy loam textures, with clay always less than 10 per cent. Mica-schist soils develop from a range of slightly different metamorphic rocks (see Geology Chapter 3), including quartzite and some metamorphosed limestones. But, throughout the mica-schist region of the Sperrins, clay content is consistently low, and similar only to granite-derived soils, being less than 10 per cent of the Fine Earth. As clay is the mineral source of most nutrients, low clay is an indicator of low chemical fertility. Because the altitudinal range of mica-schist soils is from 100 m to 680 m above sea level, leaching is quite active in the upper, wetter part of the range and will further lower natural fertility levels.

It should be noted from Table 2.3 that close to half the total area of the mica-schist region is covered by superficial deposits of sand and gravel, alluvium, and blanket peat. The gravel deposits, mainly around Pomeroy, Omagh and Newtownstewart, are podzolised by leaching, and the acid, blanket peat has a particularly low altitude for its outer edge. Both these features further emphasize the acidity and low fertility of the mica-schist region of the Sperrins.

Rankers and shallow soils developing from weathered rock comprise only 14 per cent of all mica-schist soils, and among the mica-schist till soils, all the freely draining soils collectively comprise only just 11 per cent. Three soil series are co-dominant, SWG1, SWG2 and SWHG, covering almost 21,000 ha, 30,000 ha and 22,000 ha respectively, collectively accounting for 63 per cent of all mica-schist soils. This relative dominance can be seen on Soil Maps 7 and 12, and confirms the character of mica-schist soils to be wet, gleyed, and sometimes humic on the surface.

It is perhaps surprising that soils with a sandy loam texture should be gleyed on such a scale. The explanation lies in a combination of reasons including, a high silt content of between 28 and 39 per cent (only just below that of the shale soils), a tendency for B horizons to become hardened or indurated, and a wet climate in the Sperrins. However, mica-schist soils are improvable with good management, and lowland brown earths and SWG1 soils in the Foyle and Roe valleys, can be raised to moderately good quality. Liming and fertilizer treatment are necessary for good results, and the probability of these will decline with altitude and increasing humic cover in the High Sperrins. Drainage improvement depends on conventional techniques.

7. BROWN EARTH ON MICA-SCHIST TILL - MODAL SOIL PROFILE

Horizon	pH	Available mg kg⁻¹			meq 100 g⁻¹					mg kg⁻¹		% N	% C	% LOI	% Sand	% Silt	% Clay
		P	K	Mg	Mg	Ca	Na	K	CEC	Dithio Fe	Total P						
Ap	5.9	21	126	45	0.55	4.79	0.11	0.37	15.1	7,100	1131	0.25	2.99	7.49	69.5	19.2	10.9
Bw	6.1	12	76	20	0.16	1.86	0.08	0.22	7.8	5,600	637	0.07	1.08	3.81	64.9	28.1	6.4
C	6.2	10	59	15	0.11	1.12	0.05	0.16	2.6	5,400	595	0.04	0.48	2.57	72.6	23.6	3.4

Landscape: Brown earths on mica-schist till are found on well-drained, level areas in the lowlands of the River Foyle, and are part of the best quality land in the Province. See Soil Map 7.

Soil Series area: Brown earth on mica-schist till covers 3,818 ha, or 0.3 per cent area of NI. It is a minority soil of very limited area on mica-schist parent material, which overall covers 8.6 per cent NI area (excluding peat or gravel cover).

Soil Properties: Available P and K are adequate for most crops, but Mg is low or deficient. CEC is small, in these sandy loam textures, and is low in base saturation (20-40 per cent). Soils are slightly acid, but with good drainage, respond well to good management and fertilizer application. They tend to become leached, and thus change into brown podzolic profiles (as are found around Castlederg, in West Tyrone on Soil Map 7).

Munsell colour of B horizon is 2.5 YR 5/2 Weak red.

8. SWG1 ON MICA-SCHIST TILL - MODAL SOIL PROFILE

Horizon	pH	Available mg kg^{-1}			meq 100 g^{-1}					mg kg^{-1}		% N	% C	% LOI	% Sand	% Silt	% Clay
		P	K	Mg	Mg	Ca	Na	K	CEC	Dithio Fe	Total P						
Ap	5.8	21	53	68	0.63	7.8	0.16	0.14	22.5	7,800	1059	0.41	5.4	12.7	52.6	35.4	11.5
Bg	5.8	10	41	32	0.25	3.1	0.09	0.11	12.3	8,300	549	0.14	1.2	3.9	53.1	38.3	8.2
C	6.2	9	37	23	0.08	1.4	0.06	0.07	5.8	7,400	475	0.03	0.3	1.8	54.6	35.8	9.4

Landscape: Usually these SWG1 soils are only very slightly gleyed, and are found in lowland areas of the north-west, in the valleys of Rivers Foyle and Roe, on Soil Map 7.

Soil Series area: SWG1 on mica-schist till covers 21,082 ha or 1.6 per cent NI area.

Soil Properties: Available P, K and Mg are all low or even deficient for crops or grass. Soils are slightly acid. Low or small size CEC, of these sandy loam soils, is poorly base saturated (less than 40 per cent). Total iron values are also low. Chemical elements all indicate low fertility but with fairly good drainage, these SWG1 soils on mica-schist till respond well to liming and fertilizer application.

Munsell colour of B horizon is 10 YR 6/4 Light yellowish brown.

9. SWG2 ON MICA-SCHIST TILL - MODAL SOIL PROFILE

Horizon	pH	Available mg kg⁻¹			meq 100 g⁻¹					mg kg⁻¹		% N	% C	% LOI	% Sand	% Silt	% Clay
		P	K	Mg	Mg	Ca	Na	K	CEC	Dithio Fe	Total P						
Ap	5.9	12	31	49	0.32	4.4	0.09	0.08	14.2	5,200	725	0.18	2.81	7.28	65.1	28.1	6.2
Bg	6.1	7	30	40	0.27	5.8	0.07	0.06	17.8	7,100	563	0.17	1.49	4.59	59.2	34.5	5.7
C	6.3	7	21	38	0.19	2.1	0.06	0.03	5.9	2,100	461	0.03	0.31	2.82	63.5	27.8	7.9

Landscape: The SWG2 and humic gleys (SWHG) on mica-schist till are the dominant soils of the Sperrin Valley slopes and hillsides, merging into deeper peaty gleys and blanket peat upslope. See Soil Maps 12, 13 and 7.

Soil Series area: SWG2 on mica-schist till covers 29,926 ha, or 2.2 per cent NI area. If a further 21,954 ha, or 1.6 per cent NI, of SWHG is added to SWG2, the combined area of the two related soils is 3.8 per cent of NI area, making it the 4th most extensive soil series in the Province.

Soil Properties: Available P, K and Mg are all very low or deficient in all horizons. pH is slightly acid, in the range pH 5.8 - 6.3, which is common for mica-schist till. Total iron (Dithio Fe) and total P are very low. Chemical fertility is poor overall, and areas of SWG2 on mica-schist till can be downgraded further in land evaluation by climate, and sometimes by slope as well. Refer to Soil Maps 7, 12, and 13 for distribution.

Munsell colour of B horizon is 2.5 YR 5/2 Weak red, or 10 YR 6/4 Light yellowish brown.

Soils developed on granite or from granite till

Reference should be made to the geology chapter (Chapter 3 by Bazley) for more detail on the three granite areas of Northern Ireland, which are the Mourne Mountains, the Slieve Croob - Newry - Ring Dyke complex, and places in the eastern Sperrins (including the southern slopes of Slieve Gallion). The age or geological period of formation is not relevant to soil profile development, and it is clear that age of formation of this acid igneous rock is widely variable. It is also known that there are different types of granite, depending on mineral composition, over the three areas of occurrence, but these variations do not seem to influence soil profile development or soil property values. What is important is that this mineral material is rich in quartz, hard and resistant, so that the area of soils developed from weathered granite rock is more than twice as large as that of soils of the granite till. Rankers of all types comprise one third of all granite soils, which occur in 4.09 per cent area of Northern Ireland. Much of this area is relatively high in altitude, rocky and exposed as rock or weathered rock. Soil textures are sandy loams, and clay is always less than 15 per cent of Fine Earth. Podzolised humic soils are common on weathered granite (see Soil Map 20), and slightly gleyed SWG1 soils dominate on granite till. They are soils of low nutrient fertility, poor holding capacity for applied nutrients, but having good drainage.

Soils on the Devonian Old Red Sandstone rocks and till

Apart from one outlying unit adjacent to Cushendall in north-east Antrim, the Devonian Old Red Sandstone and associated ORS conglomerate rocks are found in one, continuous unit, covering west Tyrone and east Fermanagh to the shore of Lower Lough Erne. Their distinctive character is that they are <u>not</u> particularly sandy, and indeed, the gleyed soils usually have at least 20 per cent clay and about 40 per cent sand in the Fine Earth. Such soils mostly have clay loam textures, although others are sandy clay loam. The soil analytical tables show that these ORS soils are not chemically fertile and well-gleyed so that SWG2 comprises about half the area of Old Red Sandstone soils. The area of occurrence rises to over 300 m above sea level in the conglomerate area of south-east Tyrone, so the soils there can be leached or covered by blanket peat in parts. One third of the ORS parent rock area is covered by blanket peat and gravel.

10. BROWN PODZOLIC ON WEATHERED GRANITE - MODAL SOIL PROFILE

		Available mg kg^{-1}			meq 100 g^{-1}					mg kg^{-1}						
Horizon	pH	P	K	Mg	Mg	Ca	Na	K	CEC	Dithio Fe	Total P	% N	% C	% LOI	% Sand	% Silt	% Clay
Ap	6.0	16	88	37	0.51	8.83	0.12	0.34	22.2	9,967	1545	0.27	3.44	8.4	63.4	25.7	10.3
Bs	6.1	6	12	11	0.12	1.38	0.13	0.03	17.4	4,474	1182	0.07	0.81	2.1	70.8	24.2	4.7
C	5.9	3	10	4	0.02	0.35	0.06	0.02	7.6	2,811	1144	0.02	0.22	1.5	80.9	15.6	2.8

Landscape: Brown podzolics are found on the rocky hill slopes of Slieve Croob, Co. Down on Soil Map 20.

Soil Series area: Brown podzols and shallow brown podzols combined cover 12,000 ha, or 0.9 per cent NI, mostly on the slopes of Slieve Croob and the Mournes.

Properties: In granite soils, the size of the CEC is increased by humus present, and base saturation is less than 50 per cent. Clay values are extremely low, and within the sand fraction, coarse sand dominates (not shown above). Soils are very freely draining, and loamy sand in texture class. Leaching is active due to climate and free drainage in acid parent materials. These soils are usually located at the upper limits of farmed land, and have received regular liming. However, the P, K, Mg nutrients are low in this case, and also on similar soils elsewhere. Outside the improved land, these soils have pH in the range 4.5 - 5.5.

Munsell colour of B horizon is 7.5 YR 5/6 Strong brown.

11. SWG1 ON GRANITE TILL - MODAL SOIL PROFILE

Horizon	pH	Available mg kg⁻¹			meq 100 g⁻¹					mg kg⁻¹		% N	% C	% LOI	% Sand	% Silt	% Clay
		P	K	Mg	Mg	Ca	Na	K	CEC	Dithio Fe	Total P						
Ap	6.1	56	148	102	1.01	8.23	0.16	0.49	23.01	6,961	1378	0.39	4.43	10.6	61.5	22.5	15.5
Bg	6.4	3	169	34	0.34	2.71	0.13	0.51	12.75	1,380	743	0.08	0.97	2.9	68.3	20.5	10.8
C	6.5	6	71	322	2.61	4.56	0.15	0.19	12.81	1,682	877	0.02	0.21	1.5	66.1	18.8	14.8

Landscape: SWG1 soils are on drumlins and hill slopes around Rathfriland, west of Slieve Croob, on Soil Map 29.

Soil Series area: Granite soils cover 55,439 ha or 4.09 per cent of NI. Granite rock soils predominate in the group, and much is in rankers. Granite till soils comprise less than half the total, mainly around Rathfriland and Kilkeel. SWG1 soils cover 10,971 ha, which is 0.8 per cent NI area.

Soil Properties: The size of the CEC is boosted by humus present, and base saturation is less than half. pH is slightly acid, but the sites are in farmed land. In the Ap, available phosphorus is adequate for all crops, K adequate for grass/grazing, and Mg for most crops. Total iron is not high, and will not interfere with availability. The soils are very sandy (sandy loam), with slight gleying due to topography.

Munsell colour of B horizon is 7.5 YR 5/4 Brown

12. SWG2/G2 ON LOUGH NEAGH CLAY TILL - MODAL SOIL PROFILE

Horizon	pH	Available mg kg^{-1}			meq 100 g^{-1}					mg kg^{-1}		% N	% C	% LOI	% Sand	% Silt	% Clay
		P	K	Mg	Mg	Ca	Na	K	CEC	Dithio Fe	Total P						
Ap	5.8	11	36	77	0.84	8.63	0.03	0.12	23.78	8,332	830	0.29	3.67	9.3	52.7	28.1	19.2
Bg	7.0	2	69	470	4.06	9.19	0.05	0.21	19.9	15,296	933	0.04	0.49	4.3	35.2	29.4	35.2
C	7.2	2	71	642	5.26	10.41	0.08	0.23	21.47	17,772	2275	0.02	0.25	2.7	34.7	30.5	34.5

Landscape: Lough Neagh Clay till is located around the southern end of Lough Neagh, where once in the geological past (at the end of the Tertiary), the lake was much larger and basalt-material deposition took place. The SWG2/G2 cannot be confidently separated, and this is the only soil series of this till. See Soil Map 19 for Lough Neagh Clay and intermixed peat.

Soil Series area: The series covers 6,174 ha (or 0.46 per cent NI) of weakly drumlinised topography, where peat infilling the lowland is the majority soil.

Soil Properties: Values for iron and magnesium are high, being similar to basalt soil, from which this parent material was derived. Ap horizon may be deep in places (up to 50 cm) because of a very long period of improvement, but is here very low in available P and K in all horizons. The Ap is a sandy clay loam, and has probably been produced by late-glacial sandy meltwater deposits. The true till below (ie in the B and C horizons) has a high clay content (about 35 per cent) and is difficult to drain. Deep, sandy Ap horizons create a platform of better drainage, over clayey parent material.

Munsell colour of B horizon is 10 YR 5/4 Yellowish brown, or 10 YR 5/3 Brown.

13. BROWN EARTH ON OLD RED SANDSTONE TILL - MODAL SOIL PROFILE

Horizon	pH	Available mg kg⁻¹			meq 100 g⁻¹					mg kg⁻¹		%N	%C	%LOI	%Sand	%Silt	%Clay
		P	K	Mg	Mg	Ca	Na	K	CEC	Dithio Fe	Total P						
Ap	6.2	14	343	123	1.23	8.33	0.15	1.06	23.5	10,600	1223	0.34	3.73	10.2	40.9	38.2	20.7
Bw	6.2	10	152	25	0.25	3.14	0.12	0.52	13.4	10,400	556	0.11	1.32	4.2	49.5	40.8	9.6
C	6.4	13	31	24	0.08	1.65	0.11	0.08	7.4	8,300	403	0.03	0.34	2.8	71.4	20.7	7.3

Landscape: Slightly drumlinised, lowland (100-110 m above sea level) near to Ballinamallard, Co Fermanagh, on Soil Map 17.

Soil Series area: Brown earths in ORS till cover 4,580 ha, or 0.34 per cent of NI area. Collectively, all ORS soils cover 4.03 per cent of the Province area.

Soil Properties: Soil textures in the lower horizons are sandy loam, and with steeper slopes, drainage is good or free in these brown earths. Soils are slightly acid, and slightly leached, but respond to liming. Available P is low in all horizons, but K and Mg are adequate for most crops. High Ex.Ca in the Ap (as in the other ORS soils) is due to liming.

Munsell colour of B horizon is 5 YR 4/4 Reddish brown, or 2.5 YR 4/4 Reddish brown.

14. SWG1 ON OLD RED SANDSTONE TILL - MODAL SOIL PROFILE

Horizon	pH	Available mg kg^{-1}			meq 100 g^{-1}					mg kg^{-1}						
		P	K	Mg	Mg	Ca	Na	K	CEC	Dithio Fe	Total P	% N	% C	% LOI	% Sand	% Silt	% Clay
A	5.7	15	65	49	0.49	6.01	0.13	0.21	22.5	11,800	1509	0.36	4.30	11.52	34.5	42.1	22.8
Bg	5.4	10	81	33	0.33	3.55	0.09	0.24	20.2	8,600	712	0.12	1.65	4.13	38.9	40.2	20.5
C	5.6	3	75	36	0.25	3.11	0.15	0.17	8.7	5,400	872	0.04	0.42	2.12	35.4	38.8	25.7

Landscape: Soils are found on hill slopes west of Tempo in Co Fermanagh, or about 180-200 m above sea level, on Soil Map 18.

Soil Series area: SWG1 on ORS till covers 13,219 ha, or 0.98 per cent NI, mostly around Fintona and to the east of Enniskillen.

Soil Properties: Even the slightly gleyed SWG1 on ORS till, has a clay loam texture. Clay is relatively low at 20-25 per cent, but is supported by 40 per cent silt. Available P, K and Mg are all low in all horizons, too low for most crops. The horizons are all slightly or moderately acid in pH 5.4-5.7 range, and surface horizon tends to be humic or carbon-rich (perhaps due to higher altitude). Total iron values are relatively low. The ORS till soils respond to liming and NPK fertilizer application.

Munsell colour of B horizon is 2.5 YR 4/2 Weak red.

15. SWG2 ON OLD RED SANDSTONE TILL - MODAL SOIL PROFILE

| Horizon | pH | Available mg kg^{-1} | | | meq 100 g^{-1} | | | | CEC | mg kg^{-1} | | %N | %C | %LOI | %Sand | %Silt | %Clay |
		P	K	Mg	Mg	Ca	Na	K		Dithio Fe	Total P						
Ap	5.9	40	171	147	1.56	6.41	0.19	0.60	16.3	6,100	1211	0.35	4.54	11.90	43.5	37.1	19.3
Bg	6.1	5	94	137	1.10	3.12	0.15	0.34	9.05	5,200	329	0.09	1.36	4.51	42.5	36.9	20.5
C	5.7	11	71	246	1.73	4.83	0.14	0.18	8.12	8,100	1058	0.03	0.22	2.68	36.9	37.8	25.1

Landscape: Sites are in lowland, drumlinised topography, at 120-140 m above sea level, near Irvinestown, Co Fermanagh on Soil Maps 17 and 18. Annual precipitation about 1200 mm.

Soil Series area: Old Red Sandstone soils cover 4.03 per cent of the Province, mainly in Tyrone and Fermanagh (see Soil Map 17 and 18). Approximately half of that total (2.16 per cent NI) is in SWG2 and SWHG, covering 29,062 ha. It is the most extensive soil series in the west of the Province.

Soil Properties: The Old Red Sandstone soils are frequently clay loam in texture (as above), and usually have less that 50 per cent sand. They are not sandy, and the dominant soil in the group is this SWG2 with poor drainage. CEC is quite small, and base saturation is less than 50 per cent. Soils, are slightly acid, and respond to liming. P, K and Mg are just adequate in Ap. Soils are only moderately fertile.

Munsell colour of B horizon is 5 YR 5/2 Reddish grey.

Soils of the Carboniferous rocks

On a solid geology map (see Chapter 3 Geology), the Carboniferous rocks cover about 20 per cent of the Province, but the related soils cover only 8.5 per cent of the NI area. This means that a very large proportion, nearly 58 per cent, is covered by alluvium, organic alluvium, peat and water, in the area of the two Lough Ernes in Co. Fermanagh, as well as blanket peat and gravel in east Fermanagh and Tyrone.

The Carboniferous period was long and changeable for rock formation. The rocks are all sedimentary, but were laid down in changing conditions of air and water, so that the range includes very pure limestone through to shales, and various sandstones. At the base of the sequence lies Carboniferous Sandstone, otherwise known as basal clastics, composed of coarse grey sandstone, grits and some stoney conglomerate. Carboniferous Sandstone till is found in the Roe Valley and along the shores of Lough Foyle, around Draperstown, Cookstown, Omagh and in south Tyrone, in a total of 10 separate areas. All the Carboniferous Sandstone soils collectively cover only 2.39 per cent NI area. The soils (shown on Table 22) are mostly SWG1 or SWG2, usually sandy loam texture, acid with pH 5.4 to 5.8, and with low chemical fertility. However, drainage can be improved by conventional methods, and the soils respond well to fertilizer treatment.

The Carboniferous Limestone group belong to Lower and Upper Limestone series, occurring mainly in Fermanagh. The related soils are highly calcareous with very high pH values, around pH 7.0, high Ex. Ca but not very impressive values for available P, K and Mg. Brown earths are the least extensive soil series, and have an exceptional pH range from 7.0 to 8.0, being very close to underlying pure limestone. The dominant gleys (SWG1 and SWG2) can have a clay content around 30 per cent in the C horizon, which makes them just suitable for mole drainage or gravel-filled tunnels. These drainage techniques have been developed in Fermanagh, both by DANI and commercial organisations. Much attention has been given to soil conditions (usually clay mineral type and amount) that may prolong the useful life of mole drainage, or cause its failure (Spoor, G. *et al.*, 1982). Much field research into soil drainage was carried out by D G O'Neill in the 1970s and early 1980s at Enniskillen Agricultural College (see DANI Ann. Rep. Res. and Technical Work). Despite their calcareous character, these limestone soils have low nutrient status and some drainage problems.

The so-called CALP soils are soils developed from a range of rocks (impure limestone, shales and sandstones) from the middle series of the Carboniferous. CALP is a name that gives no clear impression of the character of the soils, which cover 3.62 per cent NI area. They are the most extensive group of Carboniferous soils, found mainly in south Fermanagh and south Tyrone. They are often quite sandy in surface horizons, with up to 55-60 per cent sand in sandy clay loam textures, but become very clayey in lower B and C horizons. Clay content is frequently 35-40 per cent of Fine Earth, and makes CALP soils suitable for mole drainage. Despite high pH and high base status, these soils have low nutrient status and until the advent of moling and gravel-filled tunnels, had a very serious drainage problem. The introduction of these new drainage techniques since 1970 has transformed the value and productivity of the CALP or yellow clay soils of Fermanagh. Surprisingly, the red clay limestone soils between Loughgall and Armagh City also belong to the same Carboniferous series.

The last notable group of soils of the Carboniferous belong to the highest sandstone rocks in the sequence. We have called them Yoredale Sandstone (a pale yellow, brown sandstone), as Yoredale was a term used in early geological memoirs. The soils are strongly gleyed, usually SWG3, and found high up on Slieve Beagh (see Soil Map 27). The entire group of soils covers 2.06 per cent NI area. Its character is acid (pH 5.1 to 5.7), nutrient deficient in all main nutrients, and down the profile, changes from sandy loam in surface horizons to clay loam at 50 cm depth. These soils have been extensively planted in forest, but several researchers have pointed to problems of plant nutrition in the early stages of tree growth (Adams, S.N. *et al.*, 1970, Savill and Dickson, 1975), in research based at Lisnaskea Forest on SWG3 Yoredale Sandstone till. Millstone grit does occur on Cuilcagh in west Fermanagh, but the only soils are rankers.

Around Ballycastle and Coalisland, Namurian coal measures are included in this late phase, but are very localised in their impact and do not seem to affect the soil parent material.

16a. BROWN EARTH ON CARBONIFEROUS LIMESTONE TILL - MODAL SOIL PROFILE - TYPE A

		Available mg kg^{-1}			meq 100 g^{-1}				mg kg^{-1}						
Horizon	pH	P	K	Mg	Mg	Ca	Na	K	CEC	Dithio Fe	Total P	% N	% C	% LOI	% Sand	% Silt	% Clay
Ap	7.3	4	55	110	0.40	25.7	0.11	0.15	26.8	22,200	1080	0.36	3.75	10.1	39.4	32.3	28.3
Bw	7.8	4	23	124	0.10	18.2	0.10	0.03	18.2	6,700	280	0.08	1.13	5.5	74.1	17.5	8.4
C	8.0	3	24	76	0.01	17.7	0.01	0.04	17.6	5,800	240	0.02	0.25	8.2	76.8	14.9	8.1

Landscape: Soil is from the lowland drumlin landscape on Fermanagh - Monaghan Border, near Clones, shown on Soil Map 27.

Soil Series area: All variants of brown earth on limestone till cover only 4,904 ha, or 0.36 per cent NI, and the occurrence is always on isolated rises or drumlins in western areas.

Soil Properties: Highly calcareous brown earths have very high pH or pH 7.3 - 8.0, high base saturation and very high exchangeable calcium, but seriously deficient in available P and K. Because of their location in isolated spots along the Border, management has been neglected. High Loss on Ignition values are due to the presence of carbonates. Soil textures in B and C horizons are sandy loam.

Munsell colour of B horizon is 7.5 YR 4/6 Strong brown

16b. BROWN EARTH ON CARBONIFEROUS LIMESTONE TILL - MODAL SOIL PROFILE - TYPE B

| Horizon | pH | Available mg kg⁻¹ | | | meq 100 g⁻¹ | | | | | mg kg⁻¹ | | % N | % C | % LOI | % Sand | % Silt | % Clay |
		P	K	Mg	Mg	Ca	Na	K	CEC	Dithio Fe	Total P						
Ap	7.6	53	54	37	0.42	18.62	0.09	0.15	20.59	18,647	1771	0.38	4.23	11.10	53.2	35.1	11.7
Bw	8.1	21	34	27	0.27	15.01	0.05	0.07	15.54	11,626	824	0.09	0.88	4.40	60.1	24.6	15.3
C	8.2	11	55	46	0.43	16.17	0.06	0.08	16.82	14,743	729	0.05	0.62	9.41	59.4	25.6	14.9

Landscape: Shallow brown earths, with rock close in places, near Navan Fort, Armagh, on Soil Map 19.

Soil Series area: All brown earths together on limestone till cover 4,904 ha, or 0.36 per cent NI area.

Soil Properties: Available P is adequate, but K and Mg are deficient, possibly because of very high values for Ex. Ca and calcium carbonates (9.4 per cent LOI in C horizon). Soil textures are on sandy loam to loamy sand border in B and C horizons. Limestone rock is intimately mixed with soil in this case, as can be seen in very high pH 8.2.

Munsell colour of B horizon is 7.5 YR 4/6 Strong brown.

17. SWG1 ON LIMESTONE TILL - MODAL SOIL PROFILE

Horizon	pH	Available mg kg⁻¹			meq 100 g⁻¹					mg kg⁻¹		% N	% C	% LOI	% Sand	% Silt	% Clay
		P	K	Mg	Mg	Ca	Na	K	CEC	Dithio Fe	Total P						
Ap	6.2	25	164	137	1.23	10.21	0.22	0.82	15.7	7,400	731	0.35	3.82	9.61	40.9	34.3	24.5
Bg	6.4	2	44	121	0.74	6.42	0.12	0.11	7.2	13,500	700	0.04	0.34	3.31	33.9	33.5	32.5
C	7.1	2	73	179	1.32	8.51	0.13	0.19	9.9	16,000	349	0.03	0.31	4.34	40.5	31.6	27.5

Landscape: Occurs on lowlands of Fermanagh and West Tyrone in glacial till plains, particularly in the area of Fivemiletown to Lisnaskea. See Soil Maps 18 and 27.

Soil Series area: SWG1 on Carboniferous limestone till covers 15,787 ha, or 1.2 per cent NI area.

Soil Properties: Soil property values are similar to SWG2 on the same till. Available P, K and Mg are adequate for most crops. CEC size is quite small, but is well saturated with base ions. There is base saturation in C horizon with pH 7.1. Total P is low. Soil textures are clay loam, and there will be associated drainage problems. High Loss on Ignition values in C horizon denotes carbonates present.

Munsell colour of B horizon is 10 YR 4/3 Brown.

18. SWG2 ON LIMESTONE TILL - MODAL SOIL PROFILE

Horizon	pH	Available mg kg⁻¹			meq 100 g⁻¹					mg kg⁻¹		%N	%C	%LOI	%Sand	%Silt	%Clay
		P	K	Mg	Mg	Ca	Na	K	CEC	Dithio Fe	Total P						
Ap	6.2	25	143	150	1.32	8.18	0.13	0.66	25.8	5,500	965	0.36	4.3	10.1	40.1	41.3	18.5
Bg	6.5	5	107	112	0.91	7.82	0.11	0.49	21.5	17,700	560	0.15	2.4	7.2	41.2	37.3	21.3
C	6.8	2	81	122	1.07	8.68	0.19	0.51	20.8	20,200	369	0.05	1.2	5.8	31.3	36.2	32.2

Landscape: Occurs in poorly drained lowlands around the shores of both Lough Ernes in Co Fermanagh. See soil Maps 17, 26 and 27.

Soil Series area: SWG2 on Carboniferous limestone till covers 13,226 ha, or 0.99 per cent NI area.

Soil Properties: Available P, K and Mg are adequate for most crops in the Ap, but not in the horizons below. Ex. Ca is quite high, but CEC is only 50 per cent base saturated. Total P is low. Soil textures are clay loam, and high clay values in C horizon create drainage problems. Soils like this SWG2 will respond to mole or gravel - filled tunnel drainage.

Munsell colour of B horizon is 10 YR 6/2 Light brownish grey.

19. SWG1 ON CALP TILL - MODAL SOIL PROFILE

Horizon	pH	Available mg kg^{-1}			meq 100 g^{-1}					mg kg^{-1}						
		P	K	Mg	Mg	Ca	Na	K	CEC	Dithio Fe	Total P	% N	% C	% LOI	% Sand	% Silt	% Clay
Ap	5.6	20	102	87	0.87	7.46	0.09	0.33	23.6	9,100	939	0.28	3.25	7.91	60.1	20.2	19.5
Bg	6.2	2	119	100	0.91	9.52	0.09	0.42	20.7	15,600	370	0.06	0.88	4.42	48.9	21.6	28.3
C	6.8	2	72	132	1.27	10.74	0.11	0.29	19.6	14,800	364	0.03	0.46	4.14	41.6	32.1	26.1

Landscape: SWG1 on Calp till (till from a mix of shales, limestones and sandstones) is found in isolated areas of better drainage in lowland Fermanagh and Clogher Valley, on Soil Maps 18 and 27.

Soil Series area: The series covers only 0.2 per cent NI area, and is a small minority series of Calp till. The actual area covered is 2,645 ha.

Soil Properties: Available P, K and Mg are all just adequate. Total iron and total P are moderate to low. CEC is only around 50 per cent base saturated. pH of surface horizons is slightly acid. Soil textures are usually clay loam. This is a soil that does respond to drainage improvement and balanced fertilizer treatment.

Munsell colour of B horizon is 2.5 YR 6/2 Pale red.

20. SWG2 ON CALP TILL - MODAL SOIL PROFILE

Horizon	pH	Available mg kg^{-1}			meq 100 g^{-1}					mg kg^{-1}		%N	%C	%LOI	%Sand	%Silt	%Clay
		P	K	Mg	Mg	Ca	Na	K	CEC	Dithio Fe	Total P						
Ap	5.8	13	76	155	1.41	15.7	0.36	0.36	23.6	10,400	282	0.42	4.4	11.6	42.1	29.2	28.7
Bg	6.8	3	80	140	1.16	20.7	0.28	0.20	24.1	27,900	472	0.12	1.2	6.6	29.3	28.5	41.1
C	7.4	4	65	119	1.10	21.5	0.24	0.15	22.8	19,700	332	0.06	2.7	12.7	31.9	36.4	31.7

Landscape: SWG2 is the dominant soil series of Calp till, and is found in weakly drumlinised lowland on both sides of Upper Lough Erne, in Co Fermanagh. See Soil Maps 26 and 27.

Soil Series area: Calp till covers 3.62 per cent area NI, and SWG2 covers 22,646 ha or 1.69 per cent area NI.

Soil Properties: Available P, K and Mg are low for all crops. Total P is also very low. Low or moderate pH 5.8 in Ap reflects poor management. Soils such as this SWG2 with 30-40 per cent clay (textures are clay loam or clay) have C horizon distinguished by high pH 7.4. Higher Ex. Calcium and per cent LOI are both high due to carbonates present in C horizon.

Munsell colour of B horizon is 10 YR 6/2 Light grey.

21. SWG3 ON CALP TILL - MODAL SOIL PROFILE

Horizon	pH	Available mg kg⁻¹			meq 100 g⁻¹					mg kg⁻¹						
		P	K	Mg	Mg	Ca	Na	K	CEC	Dithio Fe	Total P	% N	% C	% LOI	% Sand	% Silt	% Clay
Ap	5.8	20	105	163	1.73	15.5	0.42	0.51	44.8	7,200	1586	0.41	4.80	12.8	20.8	43.7	35.2
Bg	6.5	2	67	188	1.23	11.5	0.11	0.15	24.4	38,800	381	0.08	1.10	7.3	19.1	40.3	40.4
C	7.1	2	110	257	1.48	17.1	0.12	0.23	28.8	40,800	301	0.04	0.42	7.8	10.1	34.9	54.9

Landscape: SWG3 on Calp till is found extensively on low ground in South Fermanagh, west of Upper Lough Erne, around Kinawley and the Arney River. It is intermixed with lowland peat. See Soil Map 26.

Soil Series area: SWG3 on Calp till covers 20,701 ha, or 1.54 per cent area of NI. It is almost the same area and in the same places as SWG2 of the same till. Combined they cover 3.2 per cent NI area.

Soil Properties: Available P and K are very low or deficient, but Mg is satisfactory in these much neglected soils. Very high clay content (35-55 per cent clay) and clay texture of soils in low-lying sites, leads to seriously bad soil drainage, which is almost impossible to improve. These soils are of poorest quality for agriculture in the lowlands. High clay and organic matter produces large CEC, but still low base saturation and slightly acid pH below Ap. Carbonates present must explain high per cent LOI in C horizon.

Munsell colour of B horizon is 10 YR 5/4 Yellowish brown.

22a. SWG1 ON CARBONIFEROUS SANDSTONE TILL - MODAL SOIL PROFILE

		Available mg kg⁻¹		meq 100 g⁻¹					mg kg⁻¹						
Horizon	pH	P	K	Mg	Ca	Na	K	CEC	Dithio Fe	Total P	% N	% C	% LOI	% Sand	% Silt	% Clay
Ap	5.8	6	35	73	0.49	6.3	0.11	0.12	11.2	12,500	547	0.23	3.63	4.98	55.2	32.7	11.9
Bg	5.9	4	23	55	0.25	2.9	0.12	0.12	3.6	12,300	148	0.06	0.62	3.33	56.1	33.6	10.1
C	5.4	1	43	104	0.58	3.1	0.1	0.11	7.2	22,000	156	0.04	0.32	2.72	50.1	34.9	14.3

Landscape: SWG1 on Carboniferous sandstone till is found in better drained lowland areas of Roe Valley, and the area west of Omagh. See Soil Maps 7 and 12.

Soil Series area: This SWG1 is a minority soil series covering only 8372 ha or 0.62 per cent NI. It is very similar in property values to SWG2, and is regarded here as a variant.

Soil Properties: Available P, K, and Mg are all very low or deficient. Soil horizons are slightly acid, becoming acid in C horizon. Total P is low. CEC is small, because of low clay content, and about 50 per cent base saturation leads to slightly acid pH values. Soil textures are sandy loam, and soil drainage conditions can be improved by conventional methods.

Munsell colour of B horizon is 10 YR 5/3 Brown or 5 YR 6/1 Grey.

22b. SWG2 ON CARBONIFEROUS SANDSTONE TILL - MODAL SOIL PROFILE

		Available mg kg⁻¹			meq 100 g⁻¹					mg kg⁻¹						
Horizon	pH	P	K	Mg	Mg	Ca	Na	K	CEC	Dithio Fe	Total P	% N	% C	% LOI	% Sand	% Silt	% Clay
Ap	5.9	13	45	48	0.51	4.49	0.11	0.15	16.5	7,800	890	0.25	3.35	10.21	61.9	26.4	11.3
Bg	5.9	5	40	19	0.17	2.54	0.11	0.13	10.2	8,200	633	0.05	0.51	3.04	58.4	26.1	15.4
C	5.8	5	48	81	0.68	1.75	0.08	0.11	9.8	10,200	620	0.03	0.29	3.55	46.4	30.9	22.1

Landscape: SWG2 Carboniferous sandstone till is found in several lowland areas on the west of the Province, and sometimes in low drumlins such as in the Roe Valley and to the north west of Omagh. See Soil Maps 7 and 12.

Soil Series area: SWG2 on Carboniferous sandstone till covers 15,628 ha, or 1.2 per cent NI area. Combined with adjacent SWHG soils, the two cover 1.27 per cent NI area, just over half the total Carboniferous sandstone till area.

Soil Properties: Available P, K and Mg are very low or deficient. CEC and exchangeable cations are all low. pH is slightly acid due to less than 40 per cent base saturation of CEC. Total P is also quite low. Soil textures are sandy loam, or sandy silt loam in the C horizon. Drainage problems are not severe, and can be improved by conventional methods.

Munsell colour of B horizon is 5 YR 6/1 Grey.

23. SWG3 ON YOREDALE SANDSTONE TILL - MODAL SOIL PROFILE

		Available mg kg⁻¹				meq 100 g⁻¹				mg kg⁻¹						
Horizon	pH	P	K	Mg	Mg	Ca	Na	K	CEC	Dithio Fe	Total P	% N	% C	% LOI	% Sand	% Silt	% Clay
Ap	5.7	15	94	168	1.4	11.5	0.26	0.23	33.9	1,900	790	0.61	14.92	29.1	75.6	15.2	9.1
Bg	5.1	2	12	15	0.8	1.5	0.25	0.03	15.1	1,200	280	0.22	2.26	5.1	65.2	21.4	13.3
C	5.2	1	56	78	0.6	4.5	0.26	0.01	17.2	10,600	590	0.08	0.82	4.2	42.3	24.2	33.4

Landscape: These SWG3 soils are developed on Carboniferous sandstone till at 100-150 m on Slieve Beagh upland, east of Lisnaskea in Fermanagh. The area is on Soil Map 27.

Soil Series area: The SWG3, together with the adjacent SWHG (Peaty or humic gleys), cover 17,688.5 ha or 1.3 per cent NI.

Soil Properties: The soils are usually very wet and humic. They are slightly acid to acid, and below the surface humus, have low base saturation of a small CEC. Available P and K are low. These are poor, infertile soils with a very sandy texture at the surface, becoming sandy or clay loam at depth in C horizon. They are ALC Class 4.

Munsell colour of B horizon is 7.5 YR 6/2 Pinkish grey.

Soils of the New Red Sandstone or Sherwood Sandstone (here called Red Trias Sandstone)

Soils of Red Trias Sandstone occur on three lowland landscapes in the east of the Province, namely 1) around Cookstown, 2) Dungannon to Armagh, and 3) the Lagan Valley eastwards through to Comber. These areas are found on Soil maps 13, 19, 20 and 21. Because the rock is a soft rock found in lowland areas, rankers are almost unknown, being about 500 ha in all. The dominant soil profile is slightly gleyed SWG1, which comprises about half the total area of all Red Trias Sandstone Soils. Brown earths are ranked second in extent. These soils are sandy, usually with sandy loam or sandy clay loam texture, and moderately good or free soil drainage. They are found in relatively dry, lowland areas, so leaching is minimal. They are soils with optimum pH in the range 6.5-6.9 and good nutrient status. They are fertile soils which constitute good quality agricultural land, usually in Class 2 ALC.

24. BROWN EARTH ON RED TRIAS SANDSTONE TILL - MODAL SOIL PROFILE

Horizon	pH	Available mg kg⁻¹			meq 100 g⁻¹					mg kg⁻¹		% N	% C	% LOI	% Sand	% Silt	% Clay
		P	K	Mg	Mg	Ca	Na	K	CEC	Dithio Fe	Total P						
Ap	6.4	61	221	143	0.8	9.5	0.25	0.56	17.4	10,300	950	0.18	2.48	6.1	62.1	23.9	12.4
Bw	6.7	11	73	164	0.9	9.7	0.31	0.15	17.8	12,400	380	0.06	0.76	3.5	62.9	24.1	12.5
C	6.9	7	67	403	3.1	10.7	0.32	0.13	15.9	14,700	130	0.02	0.25	2.8	62.1	24.2	12.7

Landscape: Red Trias Sandstone till is found in weakly drumlinised lowland landscapes in the east, in three districts, 1) around Cookstown, 2) Dungannon to Armagh, and 3) the Lagan Valley eastwards through to Comber. Brown earths provide very good quality land (Class 2) around Cookstown and Comber.

Soil Series area: Brown earths cover 7,360 ha, or 0.55 per cent land area NI. The Red Trias Sandstone till covers 2 per cent NI area, and in addition is common in mixed tills in adjacent areas.

Soil Properties: pH is in optimum range, and available P, K and Mg, are all adequate for most crops. The CEC is fairly small in a sandy soil, but is 60-90 per cent base saturated. Total iron (Dithio Fe) is quite high in these red soils, and may complex with the phosphorus to take some out of circulation. These are freely draining soils with sandy loam texture, and a wide range of possible uses.

Munsell colour of B horizon is 2.5 YR 4/6 Red.

25. SWG1 ON RED TRIAS SANDSTONE TILL - MODAL SOIL PROFILE

		Available mg kg^{-1}			meq 100 g^{-1}					mg kg^{-1}					
Horizon	pH	P	K	Mg	Mg	Ca	Na	K	CEC	Dithio Fe	Total P	% N	% C	% LOI	% Sand	% Silt	% Clay
Ap	6.5	15	71	117	1.07	10.18	0.07	0.33	18.36	11,000	1140	0.19	2.25	6.3	67.5	17.4	14.9
Bg	6.8	18	89	130	2.62	5.14	0.14	0.18	14.25	13,000	560	0.02	0.29	2.5	66.9	20.1	12.5
C	6.9	8	74	500	3.54	6.28	0.12	0.92	16.85	18,000	590	0.01	0.11	2.5	61.2	18.8	19.6

Landscape: Occurs mainly on lowland drumlins between Dungannon and Armagh, in an area of low annual rainfall of about 800 mm, on Soil Map 19.

Soil Series area: SWG1 covers 13,221 ha, or 0.98 per cent NI area. It is the dominant soil series of the Red Trias Sandstone till which covers 26,800 ha, or 2 per cent NI area. If all the mixes with other tills are included, the overall till area (related to Red Trias Sandstone) is 3.12 per cent NI area.

Soil Properties: Available K is low, but P and Mg are adequate for most crops. pH is in the optimum range for soils from 6.5 to 6.9, and a moderate sized CEC is at least 60 per cent saturated. Total iron (Dithio Fe) becomes quite high in the C horizon, but total P is low. The soil is only slightly gleyed or SWG1, has a sandy loam texture, and is capable of a wide range of use.

Munsell colour of B horizon is 2.5 YR 5/4 Reddish brown.

26. SWG1 ON RED TRIAS MARL TILL FROM MARINE SOURCE, MIXED WITH SHALE TILL - EAST ARDS, CO DOWN

| Horizon | pH | Available mg kg⁻¹ | | | meq 100 g⁻¹ | | | | | mg kg⁻¹ | | % N | % C | % LOI | % Sand | % Silt | % Clay |
		P	K	Mg	Mg	Ca	Na	K	CEC	Dithio Fe	Total P						
Ap	5.5	13	97	185	1.75	6.95	0.17	0.37	14.9	6,500	736	0.29	3.13	8.23	54.4	27.9	17.7
Bg	6.8	1	155	517	4.45	7.25	0.28	0.46	18.8	17,400	344	0.08	0.75	5.01	25.5	34.2	40.3
Bg/C	7.6	1	135	783	5.55	8.25	0.35	0.43	15.9	18,300	500	0.07	0.55	4.12	25.2	34.5	40.3

Landscape: Level till plain, close to sea level, on east coast of The Ards, Co Down, on Soil Map 21.

Soil Series area: SWG1 and SWG2 together cover 8,531 ha, or 0.66 per cent NI area.

Soil Properties: Below the top soil, most soils on the mixed-marl till have over 40 per cent clay, and are also quite high in silt. The texture is clay, and the topography is fairly level. Available P, K and Mg are barely adequate, because these soils are not intensively managed. Consequently, surface pH at 5.5 and base saturation are low. Total P is quite low at all sites. These are difficult soils to manage, and are classed as 3B in Agricultural Land Classification (ALC).

Munsell colour of B horizon is 2.5 YR 4/4 Reddish brown.

27. SWG2 ON TRIAS MARL TILL, NEAR LARNE AND CARRICKFERGUS, CO ANTRIM

Horizon	pH	Available mg kg⁻¹			meq 100 g⁻¹					mg kg⁻¹						
		P	K	Mg	Mg	Ca	Na	K	CEC	Dithio Fe	Total P	% N	% C	% LOI	% Sand	% Silt	% Clay
Ap	6.2	16	74	482	5.88	27.25	0.12	0.23	44.84	7,700	930	0.35	3.59	13.29	35.3	30.4	34.2
Bg upper	7.1	1	94	640	9.46	29.44	0.35	0.28	41.94	19,800	620	0.05	0.52	8.39	27.4	34.7	37.8
Bg lower	7.6	1	92	730	13.58	27.25	0.49	0.29	47.03	20,400	810	0.03	0.27	7.66	34.3	36.4	29.1
C	7.9	1	102	610	11.52	32.44	0.47	0.31	49.92	21,800	720	0.02	0.21	7.09	36.2	34.4	29.3

Landscape: Mostly found on fairly level till plains on north side of Belfast Lough and Antrim east coast, on Soil Maps 9 and 15.

Soil Series area: SWG1 and SWG2 together on Trias Marl cover 8,531 ha, or 0.66 per of NI area.

Soil Properties: Soil property values are similar to sites on the eastern lowland of The Ards, which are also developed on Triassic Marl till. Textures are on the clay loam to clay boundary. pH covers a wide range, up to 7.9 at depth. Likewise, Ex. Ca and Ex. Mg are very high, as also are CEC and base saturation. Loss on Ignition values are very high due to carbonates. Although in a low rainfall area of about 900-1000 mm, these soils are difficult to manage, and are in Class 3B.

Munsell colour of B horizon is 2.5 YR 4/4 Reddish brown.

Soil profile codes are explained in Chapter 2

1. BE on basalt
2. SWG1 on basalt till
3. SWG2 on basalt till
4. BE on shale till
5. SWG1 on shale till
6. SWG2 on shale till

CP 1

7. *SWHG on shale till*

8. *Humus-iron podzol on mica-schist*

9. *SWG1 on mica-schist till*

10. *SWG2 on mica-schist till*

11. *BE on ORS till*

12. *SWG1 on ORS till*

13. *SWG2 on ORS till*

14. *BE on mixed Carb. Sandstone and Red Trias Sandstone till*

15. *SWG1 on Carb. Sandstone till*

16. *SWG2 on Calp till*

17. *BE on Carb. limestone till*

18. *SWG2 on Carb. limestone till*

CP 3

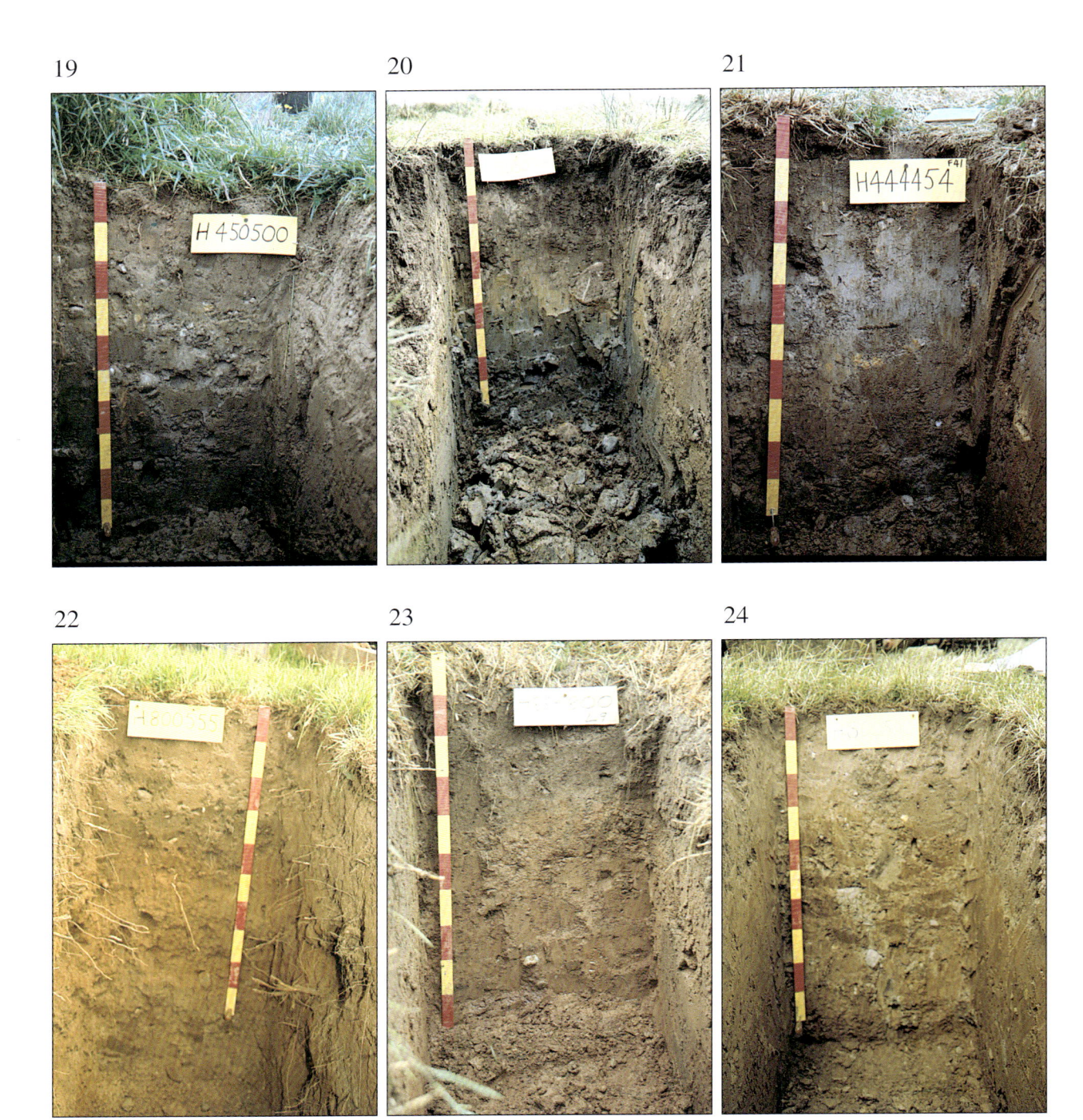

19. *SWG1 on Clogher Valley limestone till*

20. *SWG2 on Yoredale Sandstone till*

21. *SWG3 on Yoredale Sandstone till*

22. *BE on very sandy Red Trias Sandstone till*

23. *BE on Red Trias Sandstone till*

24. *SWG1 on Red Trias Sandstone till*

25. SWG1/SWG2 on Red Trias Sandstone till

26. SWG1 on Marl till

27. SWG2 on Marl till

28. BP on Granite

29. BE on gravel

30. BP on gravel

31. G1 in marine alluvium (Myroe)
32. G2 in estuarine alluvium
33. Deep Ap on SWG2 shale till
34. G2 in lake clay
35. G1 in diatomite
36. SWG2 on mixed basalt and Lough Neagh Clay till

37 *SWG1 on Marl till*

38 *SWG2 on Carb. Sandstone till*

39 *Brown podzolic on shale till*

40 *Deep humic gley on granite*

41 *Deep Ap-SWG2 on shale and granite mix till*

42 *SWG2 on Carb. limestone till*

43 *SWG2 on shale till in central Armagh*

44 *SWG2 on mixed Lough Neagh Clay till*

45 *Brown podzolic on shale*

46 *Gleyed brown earth (GBE) on chalky till*

47 *Peaty Ap in SWG1 on mica-schist till*

48 *Brown earth on mixed shale and Red Trias till*

28. CALCAREOUS BROWN EARTH ON CARBONIFEROUS LIMESTONE RED TILL AT LOUGHGALL, CO. ARMAGH

Horizon	pH	Available mg kg^{-1}			meq 100 g^{-1}					mg kg^{-1}		% N	% C	% LOI	% Sand	% Silt	% Clay
		P	K	Mg	Mg	Ca	Na	K	CEC	Dithio Fe	Total P						
Ap	7.0	64	102	85	0.80	15.12	0.09	0.31	24.42	16,500	770	0.27	2.92	8.5	50.6	25.1	24.3
Bw	7.8	3	99	94	0.89	15.08	0.06	0.30	16.88	15,200	841	0.09	1.28	5.6	42.4	14.4	43.2
C	8.3	3	91	78	0.64	16.96	0.06	0.22	18.21	16,800	1009	0.06	0.60	7.7	43.3	26.5	30.1

Landscape: This unusual clay-rich red soil is found on drumlins in Co Armagh, from Loughgall to Armagh City and Navan. Centuries ago it was selected for several large estates. It is a soil landscape in stark contrast to the Lough Neagh Clay till, adjacent to the north. See Soil Map 19.

Soil Series area: Red limestone (Carboniferous) till covers 3,005 ha, or 0.22 per cent NI area.

Soil Properties: Available P is adequate in the Ap, but K and Mg are low. pH is extremely high in B and C horizons, reflecting the high Ex. Calcium values and high total of exchangeable bases. CEC is base saturated. High per cent LOI reflects presence of carbonates. The strong red colour of the soil is not reflected in total iron values. Soil textures are clay loam and clay (clay is 43.2 per cent) in the B horizon. The soil may be classed possibly as a PELOSOL.

Munsell colour of B horizon is 2.5 YR 4/4 Reddish brown.

29. SWG1/2 ON RHYOLITE TILL, NEAR TARDREE HILL, CO ANTRIM (J200950)

		Available mg kg⁻¹			meq 100 g⁻¹					mg kg⁻¹						
Horizon	pH	P	K	Mg	Mg	Ca	Na	K	CEC	Dithio Fe	Total P	% N	% C	% LOI	% Sand	% Silt	% Clay
A	6.0	26	66	215	2.14	12.03	0.52	0.24	42.8	1,100	1150	0.57	6.45	14.2	57.0	29.2	13.7
Bg	6.1	3	206	650	5.76	11.48	0.52	0.85	24.8	5,300	392	0.05	0.51	4.4	57.2	32.1	10.1
C	5.9	5	905	950	7.32	10.98	0.57	0.91	17.2	7,550	334	0.02	0.14	2.4	71.5	23.8	3.7

Landscape: Tardree Hill, composed of volcanic rhyolite, reaches a height of about 250 m above sea level, and on the hill slopes, some rhyolite till is to be found. See Soil Map 9 and 14.

Soil Series area: Rhyolite soils cover only 1,450 ha, or 0.11 per cent NI area, in total. SWG1/2 on till covers only 742 ha.

Soil Properties: The parent material is acid igneous mineral, and pH is slightly acid. There may have been some mixing with the surrounding basalt materials as Ex Mg and Ex Ca are moderately high. Available P and Mg are adequate, but K is low. Soils from rhyolite are very sandy, with all horizons being sandy loam, and having a "cinder-like" consistency.

Munsell colour of B horizon is 10 YR 6/2 Light brownish grey.

CHAPTER 3

GEOLOGICAL BACKGROUND FOR SOIL DEVELOPMENT

DR R. A. B. BAZLEY, GEOLOGICAL SURVEY OF NORTHERN IRELAND

A glance at the geological map of Northern Ireland shows a country with a remarkable variety of rocks for its size. They range from Precambrian to Recent and, with the exception of the Cambrian, rocks from every other system are represented (see Figure 3.1). Just as their age varies so do the rock types, from igneous granites and basalts to sedimentary mudstones, sandstones and limestones. The oldest rocks are metamorphosed into schists and gneisses in complicated structural settings. The more recent Quaternary deposits are, of course, mostly unconsolidated sediments from the glacial episodes which were followed by, amongst other processes, river erosion and deposition, scree formation and peat growth. These superficial deposits cover much of the solid rock but although they complicate the background for soil development they are derived from the solid rock. So, a detailed understanding of the solid rock is important.

The solid rocks are complex as is shown by a study of the 1:250,000 geological map of Northern Ireland (GSNI 1997); even more detailed complexity is shown on the geological maps of about 1:10,000 scale that are available if the origin of soils within individual fields need to be considered. Even so, the solid rocks generally fall into four structural blocks (Figure 3.2) which give the best framework for their description. These natural blocks are 1) the metamorphic basement in the north west, 2) the Lower Palaeozoic rocks of the Down-Longford Massif in the south east, 3) a complex zone in the south west from Coalisland in south Tyrone to Fermanagh and 4) the volcanic rocks of the Antrim Plateau in the north east. Block 3 includes varied basement rocks and Upper Palaeozoic rocks, particularly Carboniferous limestones, sandstones and mudstones; it is commonly referred to as an extension of the Midland Valley of Scotland.

As mentioned above the structural complexity of the solid rocks is a fact, but from the soil development viewpoint, it is relatively unimportant how individual rock types found their way to their present day positions. It is where they lie today that is significant, or at least where they have lain during soil formation which in Northern Ireland really means since the retreat of the last ice sheet about 13,000 years ago.

It can simply be supposed that where a sandstone bed rock lies so a sandy soil will lie above. Similarly a limestone bedrock will provide a lime-rich soil. This may indeed be the case but in Northern Ireland the effects of the last glaciation have been fundamental. Ice swept across the country in several directions scraping away and moving the upper surface of the rock and dumping glacial moraine or drift over more than 80 per cent of the land surface. This glacial drift is largely till (boulder clay), but it includes sand and gravel formed in torrential meltwaters beneath and at the margins of the ice. One of the more remarkable facts is that the bulk of the glacial drift appears to be mostly derived from the rock found within just a few kilometres of the site; so although there is this thick cover, the soils still commonly reflect the character of the underlying rocks. Blocks of rock or pebbles that have travelled great distances within the ice, for instance the Donegal granites found in County Londonderry or the Scottish Ailsa Craig riebeckite microgranite along the Ards Peninsula of County Down, are well known but it is more common to find erratics derived from local rock outcrops.

So, over about 20 per cent of the country and without peat, identifying the soil will help identify the solid rock that lies directly beneath. Over about 80 per cent of the country where there are superficial deposits such as glacial till or river alluvium between the solid rock and the soil, it is more difficult. Indeed, because the solid rock is commonly deeply buried by other deposits, so a geological map is necessarily an interpretation of where expert geologists consider the rock types lie. This interpretation is based on detailed examinations of rocks where they are exposed at the surface, on the results of boreholes, on an understanding and extrapolation from nearby areas, and on geophysical and geochemical surveys as well as other remote sensing techniques such as satellite imagery. Nevertheless, even with all these modern techniques although there is tight control on the rock types in some areas, in many areas there is an element of uncertainty. So the soil surveyor must carefully consider how well understood the geology is at each location before deciding on the soil's derivation. It is hoped they will be few but there will be places where the soil does not directly reflect the solid rock that is considered to lie below; in these places perhaps the geological interpretation should be questioned. Also there will be places where the soil type cannot be explained by the underlying superficial deposits having come from a certain direction; again the interpretation should be questioned. The soil surveyor has an advantage over the geologist in always being able to take a sample and have proof of the soil type. Even so, the survey of the geology of Northern Ireland is as advanced as most places in the world with all the country having been surveyed at the detailed 6 inch to 1 mile (1:10,560) scale by the early 1900s and now mostly resurveyed at 1:10,000 scale since 1960.

Solid Geology

Taking the four natural blocks of geology from the oldest rocks to the youngest (Figure 3.1 and 3.2) the great variety of rock types is demonstrated. See also the geological map (GSNI 1997) for details.

BLOCK 1. Western part of County Londonderry, north west Co Tyrone and the northern part of Co Fermanagh – the metamorphic basement rocks.

The oldest rocks are the Moinean Lough Derg Group which cover an area of about 200sq km between Lough Erne and Lough Derg. They mostly comprise metamorphosed sandstones, termed psammites or granulites. They have been penetrated by numerous intrusions of basic igneous rocks, which when metamorphosed are called epidiorites, and are extensively veined by quartz and pegmatites (mostly coarse feldspar). However, most of this block is made up of Dalradian rocks which comprise grey or greenish grey mudstones and sandstones with relatively thin interbedded limestones and are called mica-schist on the soil maps. Being some of the oldest rocks they have been subjected to past folding and faulting with much fluid movement within the rocks. These fluids have at times been mineralised and in some zones deposited base metals such as iron, lead, zinc and copper. Associated with mineralisation may be other relatively rare minerals such as arsenic and uranium. These may be represented in local soils as relatively high levels of trace elements. The Dalradian rocks are also present in north-east County Antrim where they are of similar types. Generally the Dalradian rocks are considered poor aquifers with little potential for storing groundwater except at shallow depths within joints or weathered zones. Also, overall the Dalradian rocks tend to be lime-poor. Rather enigmatic on the edge of this block are the metamorphic rocks of the central inlier in County Tyrone. They include metamorphosed sediments as well as basic and acid igneous rocks, which appear as basic igneous rock and granite on the soil maps.

BLOCK 2. Co Down to Co Armagh block – Lower Palaeozoic rocks.

The Lower Palaeozoic rocks are mostly greywacke sandstones with subordinate slaty mudstones, and collectively called shale on the soil maps. The greywackes are much quarried as road and building stone. The presence of lead with silver in quartz veins in these rocks is notable and was the target for miners in the past. There are many old mine shafts and small tips in south Armagh and south Down. Overall the rocks tend to be lime-poor and are poor aquifers. They have been strongly compressed and, having very few pore spaces, carry virtually no groundwater except in joints or in weathered zones. These rocks have been intruded in many places by narrow igneous dykes mostly of dolerite. However, in south Down the massive intrusions of the late Lower Palaeozoic Newry granodiorite and the younger Tertiary granites of the Mourne Mountains are found. These striking pink and grey crystalline igneous rocks give rise to generally acid soils and because of their molten origin have interesting suites of trace elements that may be detected in soils.

BLOCK 3. Most of Co Fermanagh, south east Co Tyrone and north west Co Armagh – Devonian and Carboniferous rocks.

The Carboniferous sedimentary rocks mostly include a relatively high proportion of limestones or lime-cemented rocks. A new terminology has now been applied to most of these rocks in Northern Ireland, bringing them into line with modern international nomenclature. The formations are shown in detail on the 1:250,000 solid geology map of Northern Ireland (GSNI, 1997), a formation being a division of lithological (rock type) integrity which can be mapped at the surface. Hence the Ballyshannon Limestone Formation (previously termed the Lower Limestone) is a limestone that can be traced over a wide area. Similarly the Mullaghmore Sandstone (previously termed Calp Sandstone) and the Glenade Sandstone Formation (termed Yoredale Sandstone in the past) are mostly sandstone. The Devonian rocks (or Old Red Sandstone) and oldest Carboniferous rocks are mostly sandstones and conglomerates. Some occur as isolated patches outside the southwestern area, e.g. Carboniferous Sandstones in north County Londonderry and Devonian conglomerates near Cushendall. However most of the Carboniferous sedimentary rocks are calcareous mudstones and limestones with sandstones, commonly calcareous, developed in places. Thicker sandstones occur higher in the sequence and a significant area of red sandstones, pebble beds and mudstones is present

	SYSTEM	Lithostratigraphy	Thickness (metres)	Principal lithologies
CENOZOIC	Palaeogene (Tertiary)	Lough Neagh Group Antrim Lava Group	> 400m 770m	Sand, clay, lignite Basalt lava
MESOZOIC	Cretaceous	Ulster White Limestone (Chalk) Hibernian Greensand	120m 23m	White limestone with flint Sandstone, mudstone
MESOZOIC	Jurassic	Lias Group	270m	Mudstone, thin limestone
MESOZOIC	Triassic	Penarth Group Mercia Mudstone Group Sherwood Sandstone Group	15m 500m 700m	Siltstone, mudstone Mudstone, siltstone, salt (halite) Conglomerate, sandstone
UPPER PALAEOZOIC	Permian	Magnesian Limestone 'red-beds'	23m 300m > 1000m	Limestone. Marl, halite, sandstone, breccia, volcanic rocks
UPPER PALAEOZOIC	Carboniferous	Slievebane Group Leitrim Group (Millstone Grit) Omagh/Tyrone Groups	1100m 550m 4000m	Conglomerate, thin mudstone Sandstone, mudstone Limestone, sandstone, mudstone
UPPER PALAEOZOIC	Old Red Sandstone (Devonian)	Fintona Group Cross Slieve Group	2500m 1500m	Sandstone, siltstone, mudstone Volcanic rocks Conglomerate, sandstone
LOWER PALAEOZOIC	Silurian	Strangford Group	Over 1000m	Greywacke sandstone, siltstone, mudstone, shale/slate
LOWER PALAEOZOIC	Ordovician	Gilnahirk Group	Over 1000m	Greywacke sandstone, siltstone, mudstone, shale/slate
PRECAMBRIAN	Dalradian (Neo-Proterozoic)	Southern Highland Group Argyll Group	Over 1000m	Psammite, schist, pelite Psammite, epidiorite, limestone
PRECAMBRIAN	Moinian	Lough Derg Group Corvanaghan Formation	Over 1000m	Psammite, epidiorite Gneissose psammite, gneiss

Figure 3.1. The geological sequence in Northern Ireland. Most groups near or over 1000m thick crop out over substantial areas although they may be covered by superficial drift deposits.

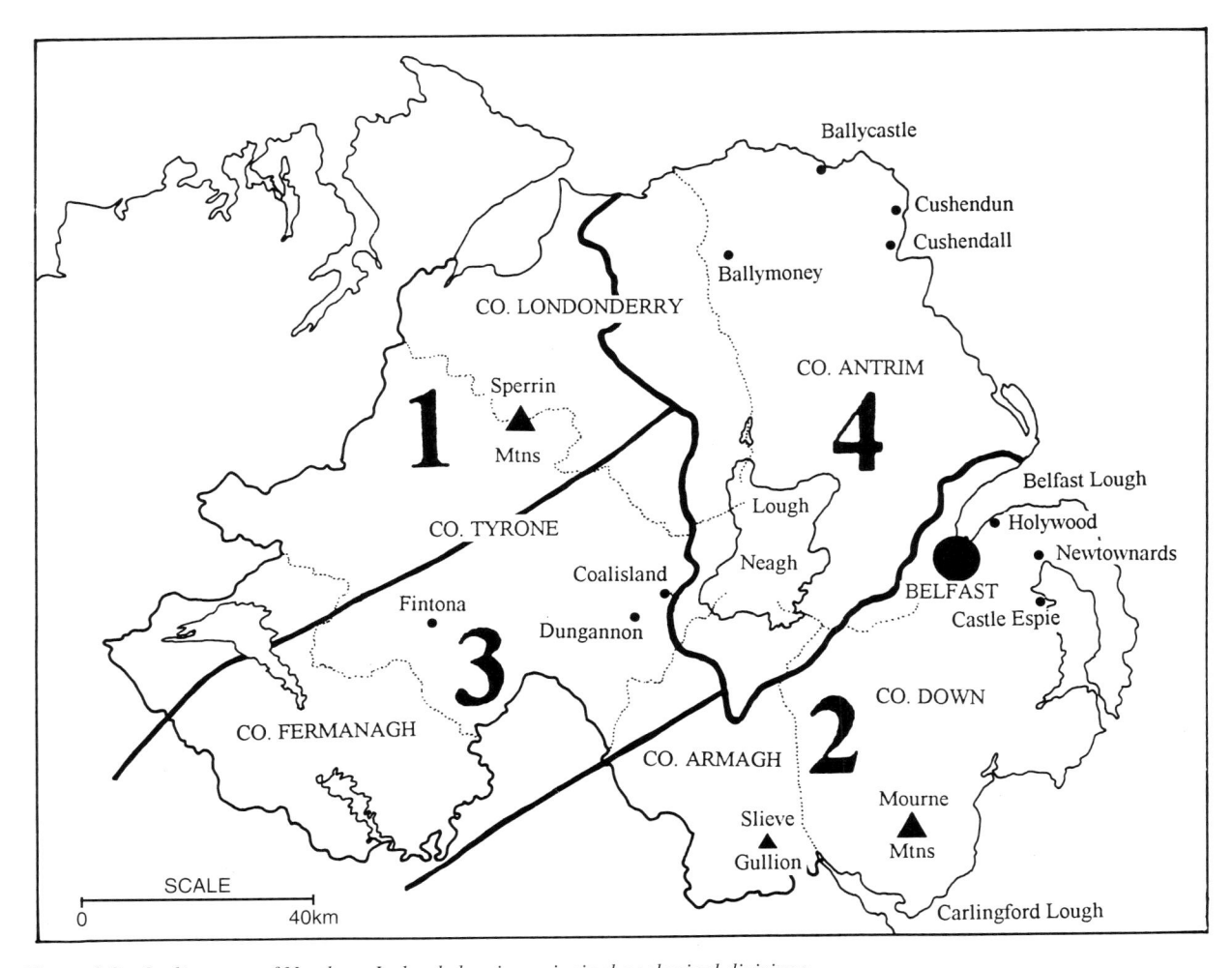

Figure 3.2. Outline map of Northern Ireland showing principal geological divisions.
Block 1 – Metamorphic basement;   Block 2 – Lower Palaeozoic rocks;   Block 3 – Devonian and Carboniferous rocks;
Block 4 – Antrim Plateau, mostly of Tertiary basalt lavas.

north of Tempo in County Fermanagh. It should be noted that names used on the soil maps for Carboniferous strata are the former names, in sequence upward from sandstones, Clogher Valley limestone, Calp mixed beds, upper limestone, Yoredale Sandstone and Millstone Grit.

Coal Measure strata occur in the Dungannon – Coalisland area, comprising dark grey mudstones with coals and thin sandstones. The coal is no longer mined, but there is a network of abandoned mines which can give rise to acid mine water drainage problems. Also, tips of waste material from the coal mining process are present in places. The influence of these rocks on soil is purely local.

Generally the Carboniferous Sandstones can act as groundwater aquifers, especially where any calcareous cement has been removed. The limestones can have karst surfaces, as on Cuilcagh Mountain, and where joints are widened by solution can carry water supplies. Cave systems and swallow holes are developed in places, especially south Fermanagh. Water from Carboniferous rock aquifers is typically lime-rich.

BLOCK 4. Co Antrim, eastern part of Co Londonderry and around Lough Neagh – the Palaeogene (Tertiary) volcanic lavas and Mesozoic rocks of county Antrim.

This area is the youngest geological entity in Northern Ireland. The Mesozoic rocks are mostly hidden by the later volcanic lavas, but Triassic rocks, especially the Sherwood Sandstone (red sandstone) and Mercia Mudstone (marl on soil maps) significantly crop out over a sizeable area of the Lagan Valley. The Sherwood Sandstone is a major groundwater aquifer, comprising thick reddish brown gypsiferous sandstones. The younger Mercia Mudstone includes substantial salt beds in the Carrickfergus to Larne area. Above, in places, come Liassic grey mudstones, followed by the Cretaceous Hibernian Greensand Formation and the Ulster White Limestone or Chalk. These all form mostly very narrow strips of land around the Antrim Plateau, the exception being an area of Chalk north of Larne where swallow holes are developed. The Ulster White Limestone is a hard white limestone that has a very low porosity and water flow is mostly through joints but the 'greensand' below is a significant aquifer and the source of many major springs. The Liassic mudstones, although of minor surface expression, do act as a major slip plane around the Antrim plateau. The overlying water bearing sandstones turn the top mudstone layers to weak clay which in slopes develop shear planes. Lias mudstones are at the base of a number of major ancient landslips with large areas of foundered ground notably below Cave Hill, south of Larne, and below Binevenagh in County Londonderry. Landslip areas are clearly marked on the geological maps. They mostly occurred at the end of the last glaciation when the ice melted leaving oversteepened slopes around the Antrim Plateau. However, the rocks within the big slips commonly remain unstable to the present and shallower mudflows can develop in response to heavy rainfall or imprudent excavation.

The Antrim Lava Group (of Palaeogene/Tertiary age), which covers most of County Antrim, part of County Londonderry, underlies much of Lough Neagh and extends to north County Armagh is mostly basalt lavas. Although divided into upper, middle and lower units the lavas are all of similar composition, to the extent that geochemically they are indistinguishable except perhaps for some of the trace elements. An interbasaltic bed of acid rhyolite lava is present in places, especially around Tardree Mountain, north of Antrim, but elsewhere it has limited outcrop. This bed includes the red to yellow bauxites and laterites that were once extensively mined in east Antrim from Lyles Hill to near Carnlough. The Antrim basalts, well known for the spectacular columnar jointing developed at the Giant's Causeway, comprise many lava flows piled one on top of another. Most flows have a slaggy top which commonly weathers to a reddish brown colour due to the iron content. Here and there, ancient volcanic vents from which some of the basalt lavas possibly erupted stand above the plateau, the most spectacular being Slemish Mountain. The basalts are not aquifers but water is in places present within joints and weathered zones in the lavas.

Above the basalts, under and around part of Lough Neagh, are Palaeogene clays and silts of the Lough Neagh Group. These clays are poorly consolidated and include brown coals (lignite) in places. Recently substantial areas of the Lough Neagh Group have been discovered around Ballymoney. They include thick lignites, but are not exposed at the surface.

Drift Geology – Superficial deposits

The variable sedimentary, metamorphic and igneous rocks described above represent a wide chemical mix which would cause significant variation in soil types. However, after the outpourings of volcanic lavas ended around 57 million years ago the landscape was subject to weathering and subaerial erosion until the Quaternary period which started about 2 million years ago. There followed a series of climatic fluctuations causing glacial episodes which may well have swept away much of the weathering debris of the previous 55 million years. In Northern Ireland the last of these glaciations is the only one for which much evidence remains. It reached its peak just over 25,000 years ago, involving ice moving across the country from a major ice centre over the central part of Northern Ireland as well as ice sweeping across from Scottish centres. The Scottish sourced ice was an early feature in the glaciation and coming across the North Channel picked up fine muds with shells. The resulting till is distinctive because it is highly calcareous. In particular it was deposited around the south side of Lough Foyle and in the coastal area of county Down. However relatively little of this calcareous till is at the surface because at a late stage the ice from the central part of Northern Ireland became dominant. The last major re-advance of ice took place about 25,000 years ago when ice moved generally northwards and southwards, depositing and moulding the till into drumlins. As the ice melted away, in late glacial times about 13,000 years ago, the flood waters deposited outwash sands and gravels.

There was a minor readvance of Scottish sourced ice at a late stage of the glaciation (slightly later than the drumlin re-advance) which impinged especially on the northern part of County Antrim, depositing a chalky till and an end moraine along its margin, called the Armoy Moraine. However most of the till now at or near the surface and covering possibly about 75 per cent of the country is from the 'Lough Neagh' ice centre. It varies according to the substrate but is commonly a stiff silty clay with locally derived boulders. It is not obviously calcareous in most places. Geologists used to term the deposit 'boulder clay' in the past but now avoid the term in favour of 'till' because in places, again often depending on the substrate, it is more a silt or fine sand with boulders rather than a clay. In its upper metre or so the till, which commonly reaches tens of metres in thickness, weathers to allow limited superficial water movement. It is not always in drumlin form, but in lowland Antrim, across Counties Down, Armagh and Fermanagh, these features are dominant. Indeed, the drumlin swarms of County Down (GSNI 1991) are classic examples of the 'basket of eggs' topography. The drumlin forms in the landscape cause impeded drainage and peat has commonly developed in inter-drumlin hollows, although modern drainage is now causing peat wastage.

The moraines and glacial sands and gravels are illustrated on the 1:250,000 Quaternary Edition of the geological map of Northern Ireland (GSNI 1991). Moraine is a mixed bag of lithologies including sand, gravel, boulders and till in a chaotic mixture and forming moundy ground. The Armoy Moraine is the best example. High level moraines and corries, relics of the last mountain glacier phase, are found on the high ground of the Mourne Mountains. The sands and gravels occur as outwash terraced deposits, notably on the south side of Lough Foyle and along the Faughan valley near Londonderry where kettle holes are features. Kettle holes are collapse depressions, now commonly filled with peat, which developed when blocks of ice, left within the sand and gravel, melted.

Other sands and gravels occur in typical ridge form, probably developed in tunnels within and beneath the ice. These commonly sharp ridges are termed eskers. Good examples of eskers are found around Lisburn, e.g. at Causeway End, at Muntober west of Cookstown and at Eskermore near Beragh. Other extensive areas of outwash material, termed kames, occur near Capecastle, around Pomeroy, east of Ballygawley and in the upper valley of the Ballinderry River, near Cookstown. The sand and gravel generally gives rise to well drained land and leached soils. Where these deposits fill old river valleys, such as the Foyle and Faughan rivers, they carry considerable quantities of groundwater.

Significant sands and moraines also occur in the Lagan Valley from Lisburn to the Malone Road, Belfast, in the Enlar Valley between Dundonald and Comber, along the Sixmilewater between Antrim and Ballyclare, north of Ballymena, along the lower Bann Valley, the Armoy moraine, throughout the valleys around the Sperrin Mountains, and from Tempo to north east Fivemiletown in County Fermanagh. Associated with the ice retreat of the last glaciation were also many glacial lakes, some small and temporary, others large and longer lasting, e.g. in the lower Bann valley south of Coleraine. In these lakes, clays, commonly varved, were deposited.

It was probably in late glacial times that the major landslips around the Antrim Plateau occurred and formed unstable ground that remains vulnerable to more superficial mud flow activity. Around 8,500 years ago there was a general rise in sea level of 20m or more which covered the peat deposited on the Lagan and Bann estuary shores. The resulting clay deposits, known locally as 'sleech' contain abundant marine shells and are also found in Larne Lough, Strangford Lough and Lough Foyle. It was probably at this time that the floor of the Bann valley was flooded to the south of Portglenone and this may have ponded back the precursor of Lough Neagh in which diatomite was deposited on top of the inundated peat. Many of the raised beach sandy deposits date from this episode.

During the last 10,000 years the rivers have matured to form flood plains with quite extensive alluvium in places. Notable are the large expanses of lake alluvium around Upper Lough Erne, near Lisnaskea, and along the Lower Bann River. Blanket peat bogs have also developed over much of the higher ground, with the most extensive areas in the Sperrin Mountains, but the Antrim Plateau also has some fine examples. Blown sand occupies large areas around Magilligan Point and Dundrum Bay mainly in the form of stable dunes.

Given this very varied relatively recent Quaternary history, and the concealing of over 75 per cent of the solid rocks, it is not surprising that in a few areas, the soils do not obviously match the known geology. However, overall there is generally good agreement which emphasises the quality of the work of past geology and soil surveyors. Where there are mismatches perhaps the interpretation of the geology needs to be checked but more commonly the cause is likely to be some presently unknown happening in Quaternary times or human activities. The latter are probably not to be underestimated because landfill of sorts must have happened earlier than just the last hundred years. Carriage in of foreign materials, such as seaweed and seashells inland from the shore, was quite common in an attempt to improve land. In any event the soils are all ultimately derived from the bedrock and an understanding of the rocks and their history is essential if the soils themselves are to be fully appreciated.

References

Wilson, H.E., 1972 *Regional Geology of Northern Ireland*, HMSO, Belfast.

Geological Survey of Northern Ireland. 1991 Geological map of Northern Ireland, Quaternary edition, 1:250,000 scale.

Geological Survey of Northern Ireland. 1997 Geological map of Northern Ireland, solid geology edition, 1:250,000 scale.

CHAPTER 4

CLIMATE

DR. NICHOLAS L. BETTS
SCHOOL OF GEOSCIENCES, THE QUEEN'S UNIVERSITY OF BELFAST

Introduction

Soil is a product of a wide range of factors, primary of which are the characteristics of the geological parent material, past and present climate, landscape position, the activity of living organisms and time. Climate is not associated solely with the atmosphere, for atmospheric influences extend into the soil along fissures and pores, and the effects are felt throughout the body of the soil. Discussion of soil-climate relationships emphasises the considerable importance of moisture and temperature. In terms of soil development, atmospheric temperature determines the speed of chemical reactions and to some degree, the amount of biological activity in the soil. Moisture is important since the balance of precipitation and evapotranspiration determines the amount of water available to percolate through the soil and become involved in the physical, chemical and biochemical processes occurring within the soil. Indirectly, the climate also determines the natural vegetation or crops which may be grown, and thereby influences the nature of the organic component of the soil. An interaction also exists between the soil and the atmosphere through the biochemical reactions occurring in soils, producing substantial quantities of gases prominent in the global warming scenario.

SYNOPTIC CLIMATOLOGY OF NORTHERN IRELAND

The general circulation

The climate of Northern Ireland owes much to a mid-latitude oceanic position on the western side of a land mass, the maritime influence being enhanced by heat transfers from the neighbouring relatively warm surface waters of the North Atlantic Drift to the overlying atmosphere (Perry and Walker, 1977).

The upper atmospheric pattern consists of a warm ridge extending north-eastwards from the Azores high and a cold trough changing in position between a mid-Atlantic location and longitudes 30–40°E. On occasions when a strong zonal flow dominates the Atlantic sector, there is little evidence of this trough.

Generally, if the axis of the trough is located in mid-Atlantic, Northern Ireland experiences mild and usually wet weather. When located over or to the east of the country, colder, showery conditions prevail. With a decrease in wavelength, accompanied by increasing amplitude of the upper waves, the warm ridge extending over Northern Ireland and on towards Scandinavia may become enlarged, producing a 'blocking' situation with increased meridional airflow and weather which is often drier than normal. During this blocking of the westerlies, the upper jet stream still exists, but meanders around the high pressure ridge and low pressure trough, with its surface expression, the polar front, following in similar fashion. As a result, cyclonic activity occurs along tracks of north-south alignment, rather than along the more usual west-east track. Boucher (1975) and Lamb (1972a) have outlined case studies showing the relationship of surface weather to the middle troposphere westerly flow.

The fluctuating upper airflow pattern shifts between two extreme stages: a 'high index' circulation is associated with strong zonal movement (Figure 4.1a) while a 'low index' situation is characterised by meridional airflow (Figure 4.1b). The climate of an individual period of the year is usually determined by the relative dominance of one of these two forms of circulation. Overall, years when a 'high index' predominates are warmer and wetter than average, whilst 'low index' circulations tend to produce greater climatic variability, dependent upon the location of the upper trough.

De la Mothe (1968) has indicated that the upper airflow pattern follows a fairly common sequence at certain times of the year, producing a recurrence of distinctive features of the climate. Often, the upper westerly airflow is in a 'high index' phase during December and January, with associated depressions at the surface progressing rapidly eastward accompanied by strong winds and appreciable frontal rainfall over Northern Ireland. A gradual shortening of the upper wavelength then follows during the spring, and in association, a blocking pattern becomes prevalent.

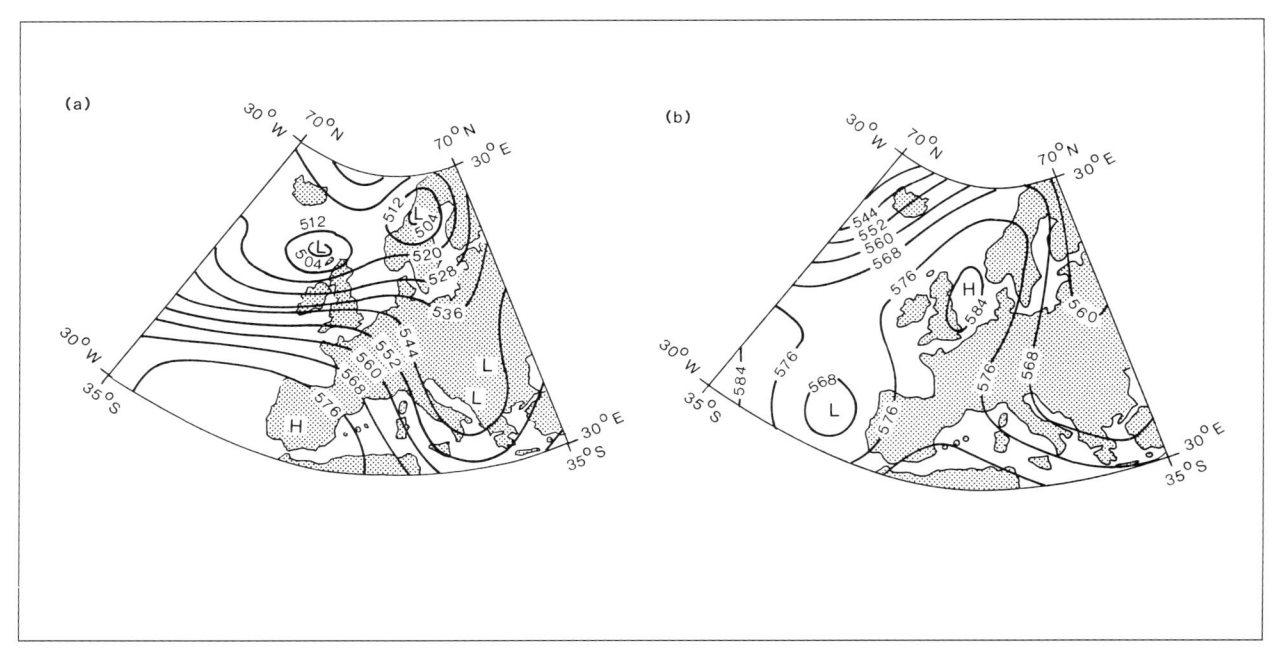

Figure 4.1(a) 'High-index zonal circulation at 500 mb over the Atlantic and Western Europe'. (Derived from Europäischer Wetterbericht, 00 00 hrs, 17 December 1979).

Figure 4.1(b) 'Low-index meridional circulation at 500 mb over the Atlantic and Western Europe'. (Derived from Europäischer Wetterbericht, 00 00 hrs, 21 August 1976).

This is associated with an extension of the continental anticyclone towards Northern Ireland which prevents the advance of depressions, and accounts for a greater frequency of dry spells in spring than at other seasons. Towards late June, there occurs a renewed increase of the wavelength pattern aloft and a return of the westerly airflow, although associated depressions are often quite shallow and progress less rapidly than in winter and may sometimes remain almost stationary for several days. In September, the progressive pattern is often interrupted by a spell of anticyclonic weather. From mid-October on, however, progressive stormy weather is the most prevalent situation, although long spells of a particular weather type are less common at this period of the year (Lamb, 1950). With the variability of weather in oceanic middle latitudes, several of the above mentioned features may not occur during the course of an individual year.

A useful approach to the analysis of weather characteristics is the daily categorisation of the circulation pattern over an area. The circulation pattern over the British Isles has been classified daily from 1861 to date using the criteria adopted by Lamb (1950; 1972b), and a modified catalogue for Northern Ireland has been produced by Betts (1989). These classifications overcome some of the inadequacies of the more traditional air mass approach, for they are based upon the more general movement of pressure systems and associated airflow over the region. Seven types are defined: anticyclonic (A), cyclonic (C), northerly (N), easterly (E), southerly (S), westerly (W), and north-westerly (NW). Hybrids of two or three types are recognised, the latter cases sharing winds from neighbouring quadrants. An unclassifiable grouping is also included. The synoptic patterns producing the seven types and their associated weather characteristics are well documented by Lamb (1972b) and Perry (1976). Sweeney (1985) and Betts (1989) have indicated the marked seasonal variations in the average frequency of each of Lamb's weather types over Ireland. Particularly evident is the December/January peak of westerliness and a mid-summer peak of cyclonicity. Recently, Mayes (1996) has identified changes in the spatial pattern of monthly and seasonal airflow fluctuations over the British Isles between 1941–70 and 1961–90. Such changes, especially in terms of westerly and cyclonic types, have implications in respect of climatic change.

Meteorological network and records

Armagh Observatory possesses the longest continuous series of meteorological records in Northern Ireland dating back to 1794, although weather diaries exist from earlier periods. One such diary is that of Thomas Neve from Ballyneilmore, County Londonderry, for the years 1711–25 (Dixon, 1959). The lack of standardisation of meteorological instruments and general observational procedures requires caution in the analysis of early records, although by the 1880s a fairly homogeneous station network had been achieved (Rohan, 1986).

Expansion of the observational network was slow during the first half of this century, but with the establishment of the Belfast Meteorological Office in 1961 (now the Belfast Climate Office) to develop and maintain standards set by the World Meteorological Organisation, there followed a rapid development of both rainfall and climatological stations. The intensification of the network in the 1960s provided a significant increase in the number of upland rainfall stations (those above 150 m O.D.) and climatological stations. Since 1980, the policy of the Belfast Climate Office has been to improve the quality of instrumentation rather than to greatly expand the network. In particular, the quality of upland coverage has been maintained to satisfy the needs of both the Water Service and Forest Service. Furthermore, with the continued introduction of automatic weather stations in upland areas, the general meteorological coverage of Northern Ireland is satisfactory.

The variability in length of records at climatological stations in Northern Ireland makes it difficult to ensure a common time base for the description of mean conditions relating to all aspects of individual climatic elements. In the following account, figures have been derived for the standard climatological period 1961–90 where this was possible, but occasionally averages have been obtained from data relating to a shorter period. All time references are based on the 24-hour system, and all times are GMT.

WIND

Wind direction

Figure 4.2 presents 12-point wind roses for four stations, selected to give a broad impression of wind distribution over Northern Ireland. Aldergrove (1957–93), Ballykelly (1957–71), Kilkeel (1964–93) and Castle Archdale (1964–93) data are from continuous recording equipment. Clearly the predominant wind directions are generally between 200°–220° and 260°–280° (west), although the surface airflow is modified by relief features. Glassey and Durbin (1971) for example, suggest that winds at Ballykelly may be funnelled along the Foyle Valley thus producing maximum frequencies in the range 230°–250°. Similarly, the Roe valley provides a natural path for southerly winds (170°–190°). The least frequent winds at Ballykelly are from 50°–70°, and at Kilkeel from 350°–010°, a possible reflection of the shelter effect provided by the North Derry hills and the Mournes respectively. At Castle Archdale high frequencies of winds in the ranges 140°–160° and 290°–310° suggest a channelling effect through the Lough Erne lowland.

Analysis of wind directions throughout the year reveals a marked spring maximum of northerlies. Winds from 140°–160° have a well-defined summer maximum at each station, particularly Aldergrove and Castle Archdale. There is a distinct winter maximum for winds between 170° and 220°, and a pronounced summer maximum for those between 260° and 340°.

Wind speed

Close proximity to the paths of depressions over the Atlantic means that Northern Ireland generally experiences stronger winds than southern areas of the British Isles. The effects of relief may result in the values of mean wind speed at an individual station only being representative of that location, but generally, mean wind speed decreases with distance inland. This pattern is due to land exerting a greater frictional force upon the surface wind than the ocean. The annual average wind speed ranges from more than 6.7 m s^{-1} on the North Antrim coast to under 4.1 m s^{-1} at sheltered inland sites. June to September is the period of lowest mean wind speed, while highest velocities are most frequent between November and March. The wind regime, however, displays considerable variation from year to year, dependent upon prevailing synoptic pressure distributions.

Wind speed, like wind direction, varies

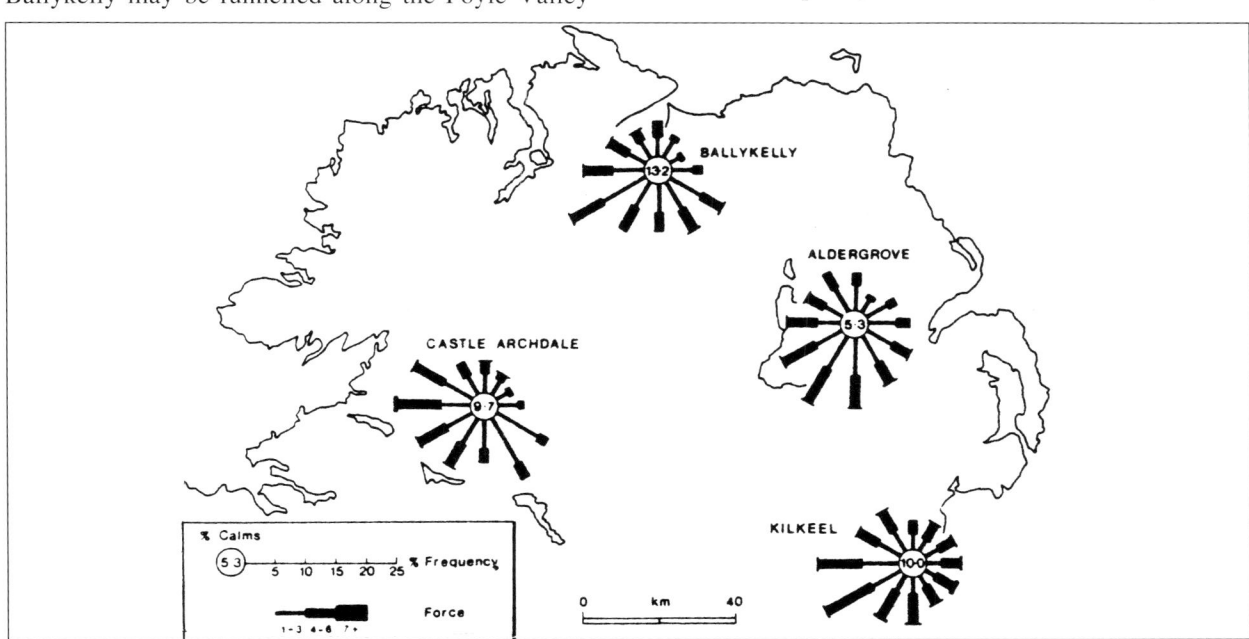

Figure 4.2 Annual percentage frequency of the force (Beaufort scale) and direction of wind at selected stations.

continuously, and although it is unusual for the mean wind speed to attain force 8 (17.2–20.7 m s^{-1}) inland, gusts exceeding force 9 (over 24.4 m s^{-1}) are not uncommon. Table 4.1 presents the highest hourly mean wind speeds and the highest gust speeds recorded at selected stations with at least 14 years of record. Despite the sheltering effect of the mountains of Sligo, Leitrim and Donegal, severe gales are occasionally experienced throughout Northern Ireland. The greatest intensities are reached along the Londonderry and Antrim coasts, although the County Down coastline experiences storms associated with the very strong south-east to east winds. Indeed, Figure 4.2 clearly shows that the strongest winds do not blow solely from the south-westerly quadrant. The highest hourly mean speeds so far recorded in Northern Ireland at low altitudes have not exceeded 28.8 m s^{-1}, whereas recordings in excess of 33.4 m s^{-1} have been monitored over north-west Scotland, and south-west England. Upland sites in Northern Ireland have probably experienced hourly mean speeds of such force, but the paucity of high-level stations and relatively short length of records prevent meaningful discussion of wind speeds at high altitude over the country.

At most stations in Northern Ireland, gusts exceeding 32.7 m s^{-1} have been recorded during each of the months of the winter half-year (October-March). Hardman *et al* (1973) updated the work of Shellard (1962, 1965) and produced estimates of both the hourly mean wind speeds and gust speeds for recurrence periods of 10, 20, 50 and 100 years. Table 4.2 presents data for Aldergrove, the only station in Northern Ireland with records of sufficient length to allow such analysis, although the record length used is only 43 years.

A 'day of gale' is one during which the mean wind speed reaches 17.2 m s^{-1} for a period of a least 10 consecutive minutes at a standard height of 10 m above the ground. The greatest frequencies of days with gales are to be found along the exposed coasts of north Antrim and County Down, averaging 18 each year. This contrasts with inland, low-lying sites: Aldergrove averages only 4 days each year, and Moneydig (near Garvagh, County Londonderry) and Armagh, 6 and 7 respectively. The frequency of gales increases with altitude, but even on the highest sites, the annual average never exceeds 40 days. The variation in frequency from year to year is of course considerable, but gales are most frequent during the period November to March. Generally, gales in summer are less severe than those occurring in winter. Nevertheless, events such as one gale in July 1978 during which hourly mean wind speeds of 19.6 m s^{-1} and a maximum gust of 30.9 m s^{-1} were recorded at Kilkeel, emphasise the possibility of intense storms at any season.

A sequence of very intense depressions have occurred over the North Atlantic sector in recent years (Burt, 1987, 1993; Rowe, 1990). Analysis of gales and windiness however, indicates that in general, recent storm events are within the range of 'random' fluctuations to be expected in the long-term wind climate of Northern Ireland (Betts, 1994).

SOLAR RADIATION, SUNSHINE AND CLOUD

Solar radiation

The source of energy behind atmospheric circulation on earth is solar radiation, and the small fraction of total solar radiation intercepted by the earth is scattered or absorbed within the earth's atmosphere, reflected back into space or absorbed by the earth's surface. Solar radiation received at the earth's surface therefore comes from many directions. The term global (total) solar radiation is used to denote the rate of receipt of solar energy on a plane surface, usually horizontal, coming from the whole hemisphere. It is made up of contributions directly from the sun, and also diffuse amounts from the sky.

Measurements of radiation, usually carried out with a thermopile instrument, are very few. In Northern Ireland, only Aldergrove has radiation records extending back to 1969. The marked difference between winter and summer global solar radiation receipt is caused by the geometrical aspects of the Earth's rotation and orbit. With the predominance of cloud cover throughout the year in Northern Ireland, the diffuse component normally accounts for over 60 per cent of the total radiation received at the surface.

Local variations in radiation received due to slope gradient and aspect have been examined by McEntee (1976). In mid-summer, the effects of aspect and slopes of up to 20° are relatively minor, but early and late in the year (October-March) south facing slopes of 20° receive up to 137 per cent of the radiation received on a horizontal surface. North facing slopes of 20° receive as little as 76 per cent of the amount received on a horizontal surface. Clearly, implications exist in respect of soil energy fluxes and out of season growth in undulating terrain.

Sunshine

The duration of 'bright sunshine' is measured by a Campbell-Stokes type sunshine recorder. A glass sphere focuses the sun's rays on to a chemically treated card. The card is charred during bright sunshine and a record of sunshine duration is obtained.

Monthly averages of the mean hours per day of duration of bright sunshine for selected stations over the period 1961–90, are given in Table 4.3. With the constantly changing declination of the sun throughout the year, and a corresponding annual variation of day length, there exists a marked annual regime in average daily sunshine duration. Values show a significant rise in April and a similarly marked fall in September, with May generally being the sunniest month and June also experiencing significantly high amounts, a reflection of the high frequency of drier airflows at this time of the year. Relatively low July and August averages correspond to the comparative wetness of the summer season, although December is the dullest month.

Also apparent is the relatively small percentage of bright sunshine that reaches the instruments in Northern Ireland. Annual sunshine values are only about 28 per cent of possible amounts, ranging from a minimum of 17 per cent in December to a May maximum of 38 per cent.

Over the year as a whole, the duration of bright sunshine is between 3.1 and 3.9 hours a day, with coastal areas having the highest values. In upland areas the enhancement of cloudiness, particularly on windward slopes, results in lower sunshine amounts than on neighbouring lowlands. Estimates at the few upland stations possessing sunshine recorders suggest that less than 25 per cent of the possible annual bright sunshine is received.

Cloud

Cloud cover not only influences the receipt of bright sunshine by day, but it also controls night temperatures, for clear skies under favourable synoptic conditions are associated with lower night temperatures and the occurrence of phenomena such as dew, frost or fog. Observations are made of cloud type and the amount of cloud, which is measured in terms of oktas of the sky covered.

Table 4.4 provides monthly and annual averages of cloud amounts at Aldergrove for the period 1961–90. These values are fairly typical of Northern Ireland, the mean annual value of 5.8 oktas emphasising the cloudy conditions so typical of a mid-latitude maritime climate. July is the cloudiest month, closely followed by August, and a secondary maximum occurs from November to January. The clearest period is from March to May. Increased cloudiness over uplands gives values well in excess of those for Aldergrove, and since humidity also increases with altitude, the uplands tend to be very wet.

Barrett (1976) shows the cloudy nature of Northern Ireland, but even so, the incidence of heavily overcast skies (7-8 oktas cloud cover) is somewhat less than in Wales, northern England and Scotland, and only a little higher than in southern England. Days with relatively clear skies (0-2 oktas cloud cover) are experienced with a markedly lower frequency in Northern Ireland than in southern England due to the more frequent occurrence of anticyclonic weather and the general lack of relief in the latter area. However, Northern Ireland also compares unfavourably in relation to much of Wales, northern England and Scotland. Within Northern Ireland, the number of days experiencing little cloud cover increases eastward, and ranges from 27 days per year in the west, to over 46 days on the Down coast. This pattern is partly a reflection of topography.

Temperature

In Northern Ireland, the relatively high average temperature for such a northerly latitude, and the rareness of extremes, results from an oceanic position in the mid-latitude westerly wind belt, and exposure to the effects of the North Atlantic Drift. It is during departures from the normal zonal circulation over varying time periods that extremes of temperature are recorded.

The extreme oceanic regime of Northern Ireland is indicated by the application of Conrad's (1946) Continentality Index upon temperatures at selected stations. The formula, which takes account of the annual range of mean temperature and the geographical latitude, provides a minimum continentality value of 4.2 at Ballykelly, and a maximum value of 6.7 at Armagh. In comparison with values of more than 60 for continental interiors, it is evident that the degree of continentality in Northern Ireland is very small. Indeed, Kirkpatrick and Rushton (1990) indicate that the degree of oceanicity of the climate is similar in Northern Ireland, western Scotland and south-west Norway. This is exemplified in the close relationship in the floristic composition of dwarf shrub vegetation communities between the three areas (Kirkpatrick, 1988).

Air temperature recording throughout Northern Ireland varies with station type. Dry and wet bulb temperatures are taken hourly at Aldergrove, and at least daily at 09.00 hours at meteorological or climatological stations. These stations also record maximum and minimum temperatures, although the reference period varies with the individual station. Spatially, air temperature is not as variable as precipitation. Significant local variations of temperature may however occur, which can be of critical importance to agriculture.

In the measurement of soil temperature, soil thermometers are read *in situ*. At shallow depths, the thermometers are placed in bare soil, monitoring soil temperature at depths of 50, 100 and 200 mm. For deeper measurements, earth thermometers encased in steel tubes are installed at depths of 0.3 m and 1 m beneath a grass surface.

Mean air temperature

Mean air temperature fluctuates within narrow limits in Northern Ireland, with only a 10 per cent probability of a departure of more than 0.5°C from the average (Betts, 1982).

Table 4.5 presents temperature data (uncorrected

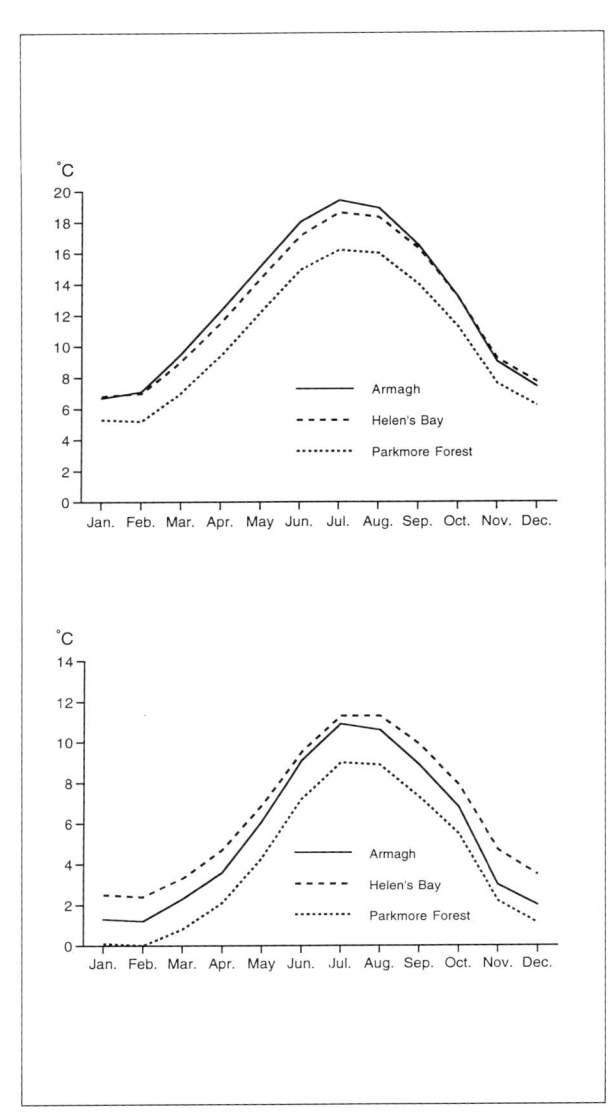

Figure 4.3(a) Mean daily maximum temperatures, 1961-90, for Armagh, Helen's Bay and Parkmore Forest.
Figure 4.3(b) Mean daily minimum temperatures, 1961-90, for Armagh, Helen's Bay and Parkmore Forest.

for altitude) at nine stations selected to give an adequate coverage of Northern Ireland for the period 1961–90, while Figure 4.3 illustrates the temperature differences between coastal, inland and upland stations.

In the free atmosphere, the lapse rate of mean air temperature corresponds to a decrease of over 0.6°C for every 100 m gain in altitude. Considerable spatial and temporal lapse rate variations exist however, depending on the prevailing synoptic situation. Generally, while lower daily maximum temperatures are probable on higher ground, there is a tendency for higher minimum values to be recorded there on clear nights with light winds or calm conditions. Under such circumstances negative lapse rates often develop (increase of temperature with height), as katabatic winds draining cold air down the hillslopes make adjacent lowland markedly colder than conditions prevailing at higher levels. Such a situation is frequently evident in the Lower Bann valley between the Sperrins and the Antrim Plateau (Betts, 1982).

Seasonal variations of air temperature

Consideration of temperature variations throughout the year reveals that highest mean temperatures in winter are found along the coast, reflecting higher mean daily maximum, and more importantly, higher mean daily minimum values (Figure 4.3). Due to the thermal properties of ocean surfaces, the waters surrounding Northern Ireland do not reach their lowest temperature until February, and for some coastal sites fully exposed to maritime influences, the lowest mean temperature is recorded in this month. January is generally the coldest month at most inland stations, although February is equally cold at some sites, another indicator of the maritime nature of the Irish climate. Furthermore, with a mean water temperature of 4°C in winter, Lough Neagh to some extent determines minimum temperatures in the adjacent area (Stewart and Gibson, 1987; Betts, 1989). Winter also sees the greatest fluctuation in mean temperatures, for temperature contrasts between air masses with westerly and easterly components are at a maximum at this time of year.

By spring the temperature pattern has changed considerably. The highest mean daily maximum temperatures during April are at Armagh (12.3°C). In contrast, at the coastal station of Helen's Bay daily maximum temperatures during April average 11.5°C. Occasionally persistent on-shore winds from the Irish Sea maintain very low temperatures along eastern coasts, and may be accompanied by coastal fog, the 'haar', in late spring and early summer. Mean daily minimum temperatures, however, are still warmer in coastal areas, as they are in winter, and the daily range of temperature inland may be very large in spring.

Coastal stations are generally cooler than inland sites throughout the summer months. During spells of anticyclonic weather, the sea breeze effect holds down maximum temperatures along the coast. The effect of a maritime and eastern location upon summer temperatures is well exemplified at Kilkeel, County Down. While Aldergrove, Armagh and even the less exposed County Londonderry coastal site of Ballykelly have recorded absolute maxima in excess of 27°C in at least 3 months of the year, at Kilkeel, absolute maxima are 24.2°C, 25°C and 24.7°C for June, July and August respectively. The pattern of mean daily minimum temperature, however, is the inverse of the maximum temperature distribution. The mean daily range is therefore at its greatest at this time of year, increasing inland with July values of 8.5°C at Armagh, in comparison with a range of 7.3°C at Helen's Bay.

By October the pattern of mean temperature

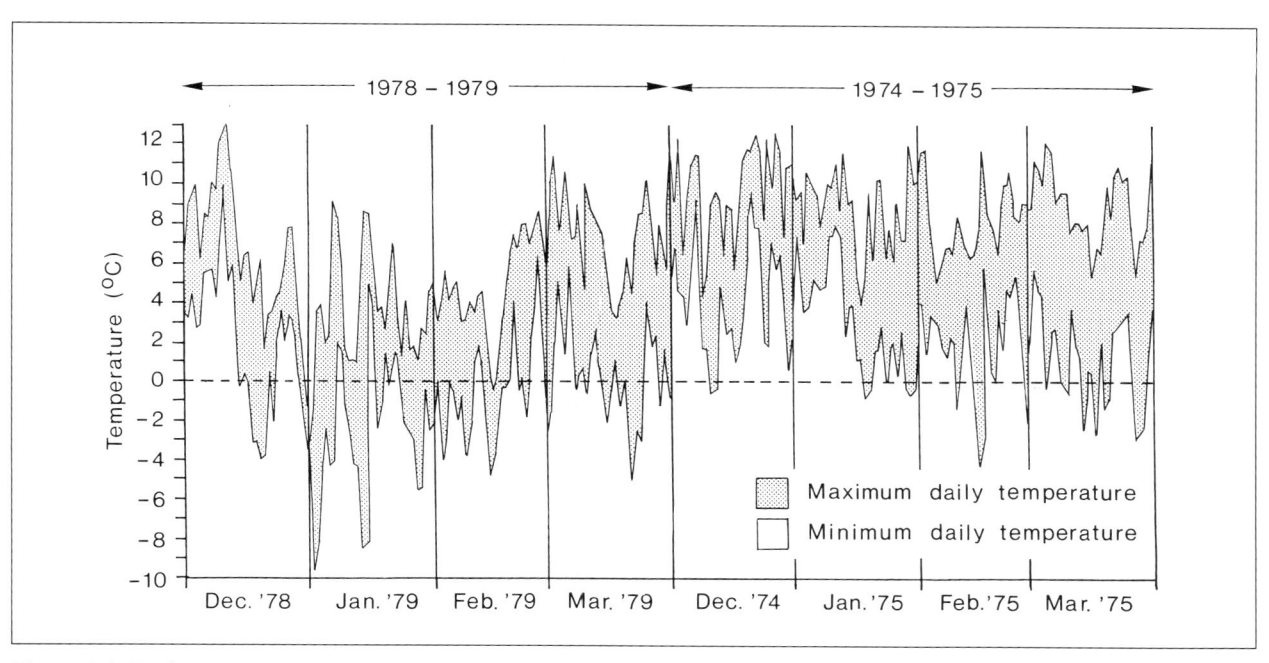

Figure 4.4 Daily maximum and minimum temperatures at Armagh during the winters of 1974/75 and 1978/79.

resembles that of January. Coastal areas enjoy higher mean temperatures than inland, due mainly to the significantly higher mean daily minimum temperatures at coastal sites. As at other seasons, the greatest daily range of temperature occurs at inland sites, but by autumn the range has fallen from the summer peak due to the decline in maximum rather than minimum temperatures.

Extremes of temperature

Table 4.6 presents the extreme maximum and minimum temperatures recorded for each calendar month at Armagh over the period 1901–93. Armagh was selected since it possesses the longest complete temperature record in Northern Ireland. Furthermore, having the maximum continentality index value, Armagh should provide representative examples of the temperature extremes experienced this century.

At Armagh, extreme maximum values have exceeded 30°C in only June and July. This contrasts with stations in central and southern England, some of which have experienced temperatures exceeding 35°C in June, July and August. Such high temperatures are usually associated with an inflow of tropical continental air accompanying anticyclonic conditions.

Temperatures below 0°C have been experienced at Armagh in all months other than July and August. Lowest minimum temperatures depend on local topographical influences such as well defined frost hollows and sheltered locations where cold air can collect. The lowest minimum recorded in Northern Ireland is –17.5°C at Magherally near Banbridge, County Down on 1 January 1979. Temperatures of this severity, particularly at lowland stations, are usually associated with a blocking anticyclone accompanied by a persistent easterly airflow.

In Northern Ireland, periods of extreme temperature are generally short-lived, but occasionally, persistence of certain circulation patterns may lend a particular character to an individual winter or summer. Owing to the influence of the relatively warm surrounding seas, few very cold days are experienced in winter. Occasionally, as in January 1982, January 1987 and December 1995, polar continental easterly and north-easterly airstreams can result in temperatures persisting below 0°C over a succession of days (Betts, 1982; 1987). Such airstreams in crossing the Irish Sea, however, ensure that temperatures are not of the severity experienced in Britain. Apart from the winter of 1962–63, those of 1978–79 and 1981–82 were the most severe in recent decades. During the winter of 1978–79, incursions of mild Atlantic air associated with the remnants of Atlantic depressions periodically interacted with easterly and north-easterly airflows – a situation that did not occur in 1962–63 (Jones, 1979). Figure 4.4 shows the pattern of daily maximum and minimum temperatures from December 1978 to March 1979, and for comparison, those of one of the warmest winters, 1974–75. The repercussions upon temperature of persistence of continental or maritime airflow are clearly evident.

In summer, anticyclonic circulation, particularly if accompanied by a continuous inflow of continental air from the south or south-east, will produce a hot, dry season. Enhanced anticyclonic activity over the British Isles during the well documented drought of 1975–76 (Miles, 1977; Murray, 1977; Ratcliffe, 1977; Royal Society, 1978; Doornkamp, Gregory and Burn, 1980), produced a remarkable persistence of above average daily maximum temperatures between June and August 1976 (Figure 4.5).

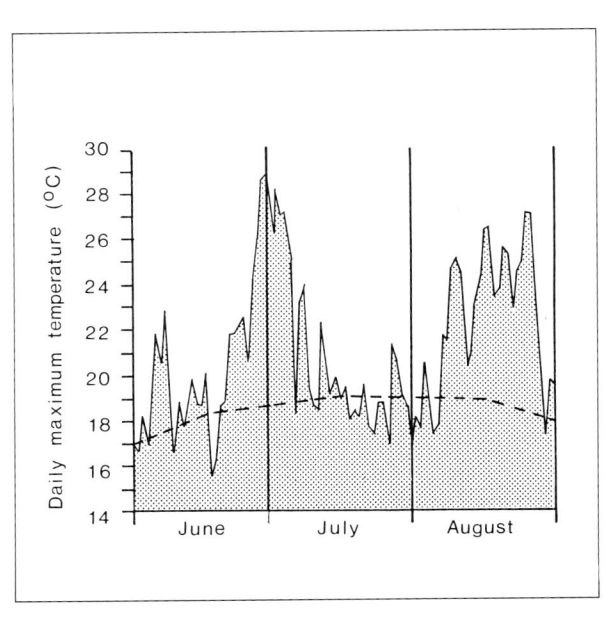

Figure 4.5 Daily maximum temperatures at Armagh, June-August 1976. The dotted line represents mean daily maximum temperatures.

Soil temperature

The temperature of the soil is primarily dependent upon the amount of energy received by radiation from the sun, a factor controlled by solar elevation and prevailing weather conditions. The thermal response of the soil to incident radiation depends upon the composition, arrangement and cohesion of the soil particles. Soil temperature displays a marked diurnal and annual cycle of variation. The diurnal amplitude depends on the time of year, being greatest in summer and least in winter, and on cloudiness, being more pronounced under clear conditions. The annual cycle of temperature displays a peak in June and a trough in January.

The top 10 cm experiences the greatest daily amplitude and annual range of temperature. An annual thermal phase lag becomes apparent with depth, the annual maximum and minimum temperatures occurring in later months. Soil moisture content is also an important determinant of temperature patterns. The extreme is reached in blanket peats, when due to the low heat conductivity of immobile water, maximum surface temperatures are recorded in late summer, and maximum sub-surface temperatures at 1 m are not reached until December (Collins and Cummins, 1996). The nature of vegetation cover and landuse will also modify diurnal and annual thermal patterns of soil (Jones and Brereton, 1986).

Frost

Frost is recorded as 'air frost' when the air temperature falls below 0°C. Since 1961, 'ground frost' statistics have been referred to as the "number of days with grass minimum temperature below 0°C". Ground frost is so much influenced by topography, soils and vegetation cover, that statistics of this element are indicative only of the site at which the recordings are taken. Connaughton (1975) established the mean date of last ground frost during the period 1944–68 at Aldergrove as 6 June, and 17 September as the mean date of occurrence of first ground frost in autumn.

Spatial variation of air frost is less pronounced, but it is still difficult to generalise about the dates of the last spring frost, or the first frost, in autumn. Connaughton (1969) and Glassey (1967) have both examined this problem for Ireland and Northern Ireland respectively. Glassey showed that no station in his selected network was entirely frost-free throughout any one winter in the period 1936–65, but dates of 31 December for the first, and 10 January for the last frost have occurred at some coastal locations. Aldergrove has recorded an air frost in June on four occasions. Other stations to record a June air frost include Hillsborough and Moneydig, each on one occasion.

Collins and Cummins (1996) indicate only a 10 per cent probability of air frost occurring in coastal areas of County Down after 28 April, whereas a 50 per cent probability of air frost exists in the Lough Neagh basin after 1 May. The probability of air frost before 1 November is less than 5 per cent along the Down coast, whereas most inland areas have a 50 per cent possibility of air frost occurrence before this date.

Freeze-thaw effects upon soil water aid structural improvement, causing a granulating action on soil clods. If however the soil is very dry during the winter, disintegration of clods will not occur. Stresses caused by freeze-thaw action upon soil water may cause compact soil to shatter, thereby alleviating the detrimental effects of excessive soil compaction. In Ireland however, winter frost penetration is normally restricted to no more than 10 cm. Compaction existing below this depth will thereby persist unless remedial measures are undertaken (Collins et al, 1986).

Growing season

The daily, seasonal and annual variations in soil temperature are important influences upon a crop's response to its environment. With different crops having specific critical temperature requirements, the concept of the growing season must be used only with reference to a specific crop. The grass growing season for example, approximates to the period when soil temperature at 10 cm depth is consistently above 6°C, for this is considered the critical temperature threshold for grass growth (Keane, 1986). Since at the commencement of reasonable growth there exists a tendency for the soil temperature to slightly exceed air temperature, in Northern Ireland the growing season is most often taken as that period when the mean daily air temperature exceeds 5.6°C. Such threshold means are not perfect, but are useful in comparing the

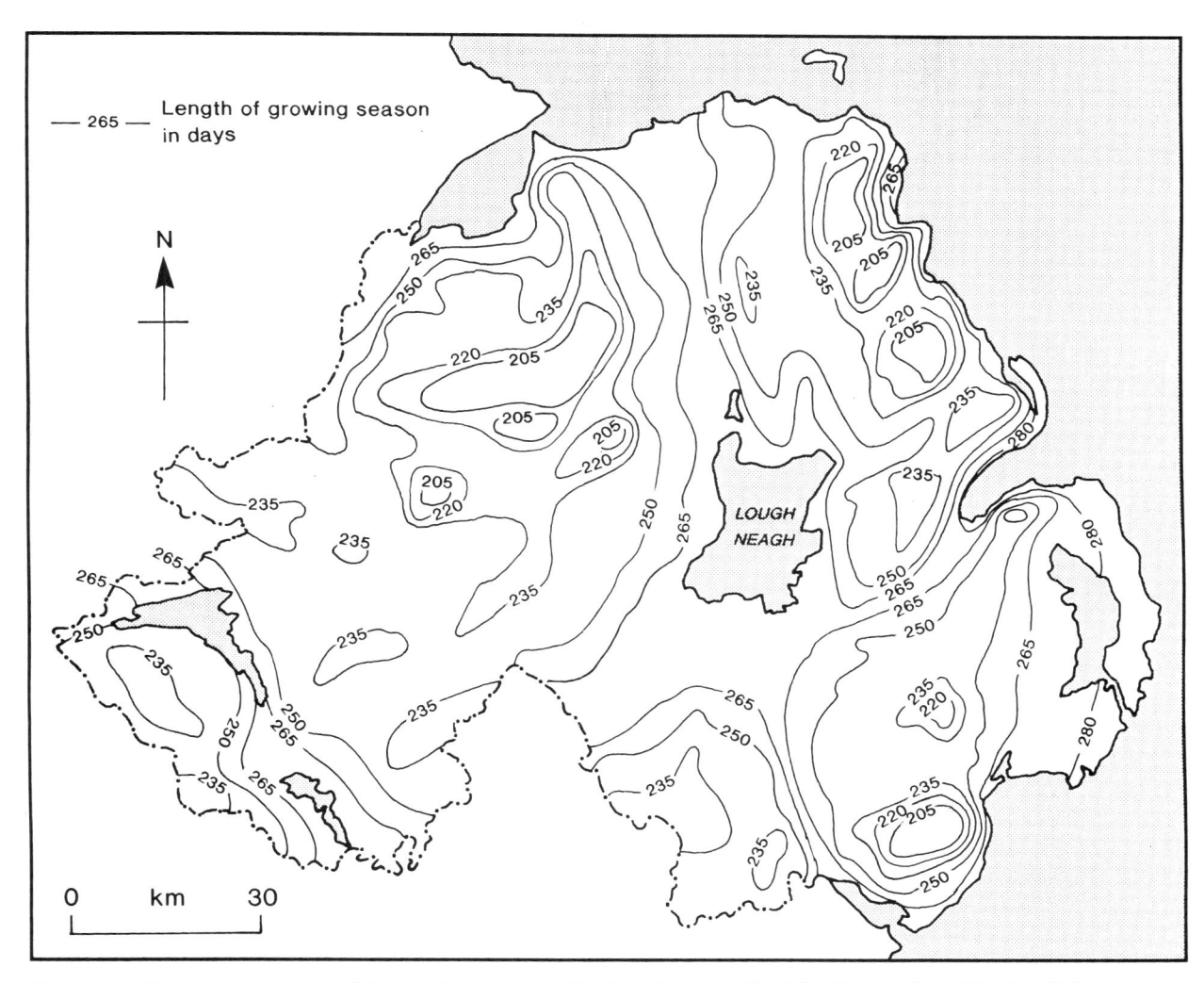

Figure 4.6 The average duration of the growing season in Northern Ireland, 1961-90. (Derived from Woods, 1995).

effectiveness of various places from the standpoint of general farming. Using this threshold of 5.6°C, Figure 6 illustrates the average duration of the growing season in Northern Ireland for the standard climatological period 1961–90. Excluding local factors, it is apparent that length of growing season is governed principally by altitudinal and maritime influences. In general, the length of the growing season decreases by about 20–30 days with every 100 m of altitude. A maximum of 235 days a year (late March to early November) are available over the uplands and fewer than 205 days in the highest areas. The season lengthens to more than 265 days (mid-March to early December) in the central lowlands. Values in excess of 280 days (early March to mid December) are found around Belfast Lough, east Down and the Ards peninsula (Woods, 1995).

Despite the somewhat arbitrary nature of the periods selected, Figure 4.7 shows the effect of climatic fluctuations upon the average length of the growing season throughout Northern Ireland, with the 1961–70 decade having much shorter growing seasons than either the decade which preceded it or the two following it. This was principally due to the cold Decembers of the 1960s which caused the growing seasons to end relatively early. December mildness, particularly in the early 1970s, is reflected in the recovery of the average length of growing season for the period 1971–80 (Table 4.7). The 1980s saw a continuation of this extension of growing season length. Such changes are responses to fluctuations of temperature, precipitation and atmospheric circulation as outlined by Lamb (1977, 1995).

To measure intensity rather than length of growing season, a unit termed accumulated temperature is often adopted. Intensity is represented by the accumulation of degree-days above a growth threshold air temperature appropriate to a specific crop. In respect of pasture grasses, the most extensively grown crop in Northern Ireland, a start date of 1 February is normally employed, and degree-days above 5.6°C after that date accumulated. Lowland throughout Northern Ireland experiences mean annual values of 1600 degree-days, although this compares unfavourably with southernmost regions of the island of Ireland, where values exceed 1900 degree-days (Collins and Cummins, 1996). Nevertheless, with the limit of cereal cultivation in the British Isles at 1500

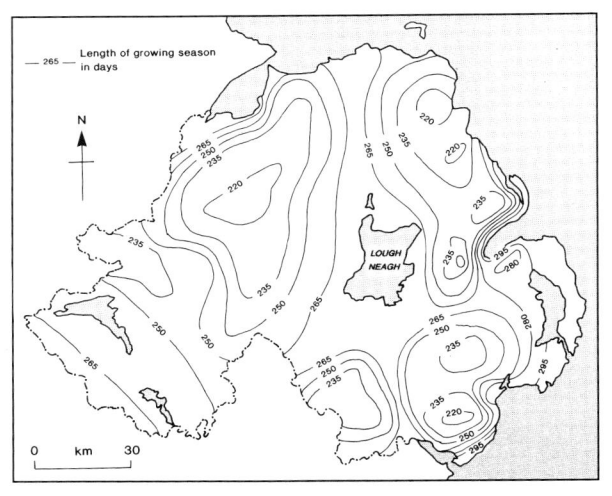

Figure 4.7(a)

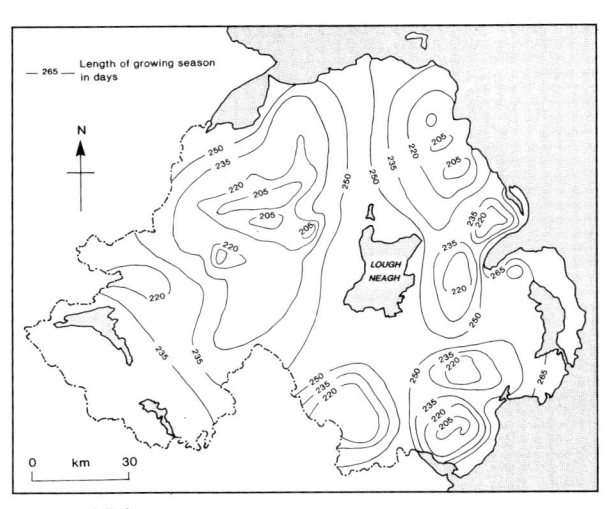

Figure 4.7(b)

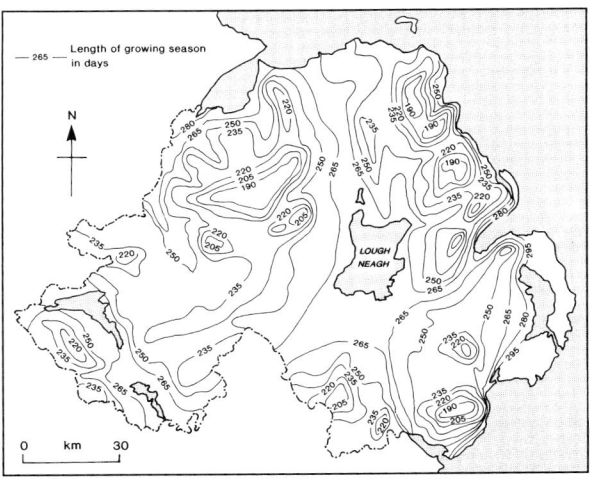

Figure 4.7(c)

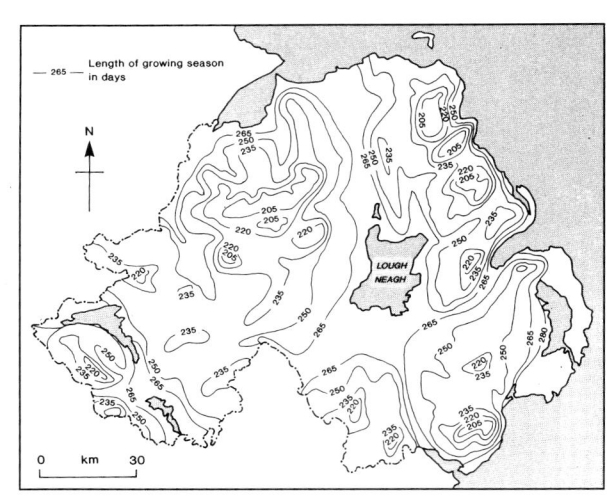

Figure 4.7(d)

Figure 4.7 The average duration of the growing season in Northern Ireland : (a) 1951-60; (b) 1961-70; (c) 1971-80; (d) 1981-90. (Derived from Bailie, 1980; Woods, 1995).

degree-days of accumulated temperature each year above 5.6°C (Stephens, 1963), the Province is favourably placed. Increased precipitation, humidity and wind speed, accompanied by lower temperatures, limit cultivation much above 210 m O.D. which marks the lower limit of rough grazing, although isolated field crops under favourable conditions may occur up to 300 m O.D.

As used above, the degree-day concept can be of limited value as it is based on assumptions that only one significant base temperature is operative throughout the life cycle of the plant, that maximum and minimum temperatures are of equal importance for plant growth, and that plant reaction to temperature is linear over the entire temperature range. One response to such limitations is the use of Ontario heat units, a degree-day accumulation developed in Canada that takes into account daily maximum and minimum temperatures separately (Collins and Cummins, 1996). Fitzgerald (1992) has used this parameter in assessment of the suitability of certain locations in the island of Ireland for new crops, and identification of favourable sites for the early harvesting of crops.

Certain indices of the growing season may emphasise soil moisture requirements. Pasture production is particularly dependent upon a suitable moisture regime, and Hurst and Smith (1967) have defined the grass growing season in Britain as the number of days between April and September inclusive when the soil moisture deficit does not exceed 50.8 mm. This value represents the estimated division between unhindered and drought-retarded growth. In an average year, the number of days lost to grass production varies from less than 10 in Northern Ireland, to more than 40 days per year in south-east England. This reflects the infrequent occurrence of long periods of moisture deficiency during the summer in Northern Ireland, which will be discussed later. Indeed, perhaps a more appropriate measure of the grazing season in the Province is one which considers general conditions, whereby animals and machines must be able to traverse the land surface without seemingly damaging the soil structure. A formula devised by Smith (1976) combining rainfall and air temperature (though not taking account of local soil properties), is often used to determine the grazing season. Much of lowland Ulster has 200–225 grazing days in the season, while in upland areas this is reduced to less than 150 days (Keane, 1988).

Precipitation

Precipitation is one of the most variable of climatological elements, both temporally and spatially. The latter variation is induced particularly by local effects of topography upon rain-bearing airstreams. Northern Ireland possesses a dense precipitation recording network, and with a number of stations having rainfall observations dating into the nineteenth century, precipitation is therefore perhaps the best documented of climatic elements in the Province.

Precipitation in Northern Ireland is principally in the form of rain or drizzle. Snowfall and hail are infrequent. Precipitation is caused by three principal mechanisms: frontal activity within extra-tropical cyclones, convection, and orographic ascent of moist air. These three mechanisms frequently operate simultaneously, though the cyclonic component contributes most rainfall, particularly in the west.

Annual and monthly amounts

Figure 4.8 shows the mean annual rainfall over Northern Ireland for the standard period 1961–90. A general decrease in rainfall totals from west to east is apparent, although upland areas with their heavier rainfall complicate this pattern. Furthermore, Perry (1972) and Logue (1978) have suggested the existence of north-south contrasts of rainfall regime under different synoptic patterns for Ireland as a whole. The dominant feature of the precipitation distribution in Figure 8 however, is the increase of precipitation with altitude. This orographic enhancement of precipitation is the result of complex processes, produced not simply by the upland acting as a barrier to moist airstreams, but by the hills functioning as high level

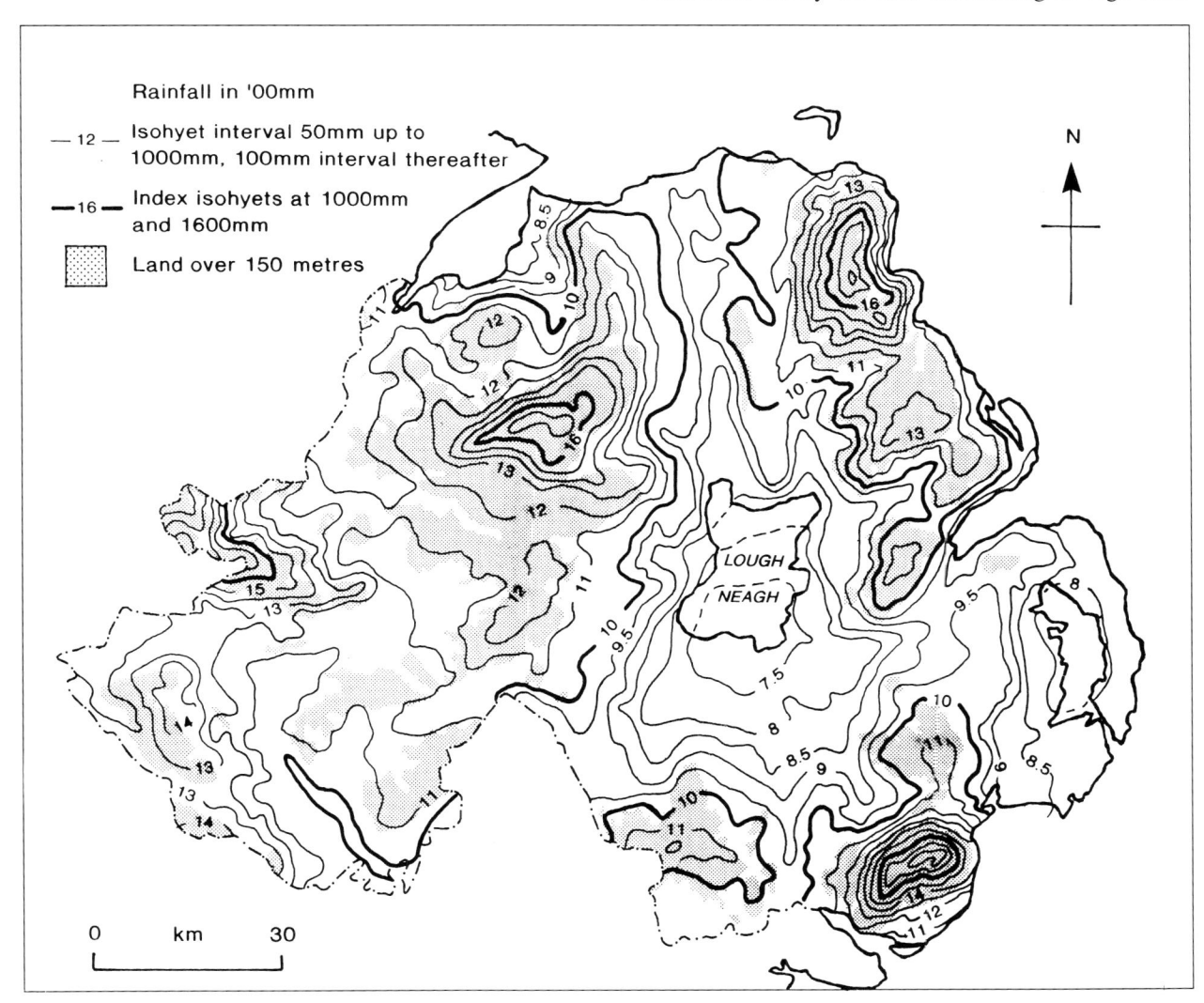

Figure 4.8 Mean annual rainfall in mm, 1961-90. (Reproduced by permission of The Met. Office).

heat sources on clear days thereby encouraging the development of convective clouds and the shower activity associated with them (Sumner, 1988). A mechanism for precipitation enhancement was advanced by Bergeron (1965), proposing that raindrops from pre-existing (seeder) clouds aloft wash out small droplets within low-level (feeder) clouds formed by ascent over hills. The amount of orographic enhancement is determined therefore by the rates at which pre-existing precipitation washes out the feeder cloud, and at which the feeder cloud is replenished by condensation.

In a study of meso-scale precipitation within a frontal depression, Betts (1990) identified orographic enhancement of up to 6 mm hr^{-1} in the Mournes, comparing rain gauge traces for the upland site of Trassey (215 m O.D.) and neighbouring coastal site, Murlough (12 m O.D.) (Figure 4.9). This is indicative of the importance of hill slope; a steeper gradient particularly in association with increasing wind speed, promotes more vigorously the vertical component of air motion, thereby resulting in higher condensation rates and greater precipitation enhancement.

Betts (1989, 1992) has also identified orographic precipitation enhancement over the Antrim Plateau, resulting from thermal instability caused by significant lateral and vertical motions in the lowest layers of airstreams forced through the North Channel.

These processes explain the very high precipitation experienced in the Sperrins, Mournes, Antrim Plateau, and in West Tyrone, with annual amounts exceeding 1600 mm on the summits. Downwind of high ground however, air may become subsident and therefore warmer and drier. Such pronounced rain-shadow effects are evident in the lee of major upland in Northern Ireland (Figure 4.8). Over lowland around Lough Foyle, annual precipitation is generally below 900 mm and in the Magilligan area, less than 850 mm. The driest areas are the Upper Bann – Lough Neagh lowlands, where annual totals are less than 800 mm and in places below 750 mm. Much of the drumlin country of east Down receives less than 900 mm, and over some areas of the Ards peninsula, annual totals are below 800 mm. These driest areas of Northern Ireland experience a wet climate compared to eastern England, where average annual rainfall is 500-600 mm. The Province is however, afforded some protection from moisture laden prevailing south-west airflows by upland areas to the west and south in the Republic of Ireland. This results in the wettest areas of Northern Ireland experiencing lower precipitation totals than the uplands of the Republic of Ireland, where annual amounts exceed 3000 mm. Similarly, the exposed uplands of Britain are considerably wetter than any part of Northern Ireland.

Table 4.8 presents for the period 1961–90 monthly and annual averages of rainfall for fifteen stations selected to give a representative indication of mean conditions in Northern Ireland. A period of marked minimum rainfall is evident between February and July, when the westerly circulation is less pronounced. A fairly marked rise in precipitation is apparent in August and this wetness persists until February. An agricultural implication of a dry period commencing in February is that the soil is provided an opportunity to dry out after the wet autumn and winter period; drying will allow soil temperatures to increase more rapidly than if the soil remained wet, and hence agricultural activities may proceed unhindered. Away from the lowlands, particularly in the west, a winter maximum of rainfall is more pronounced, associated with the most intense depressions at this season.

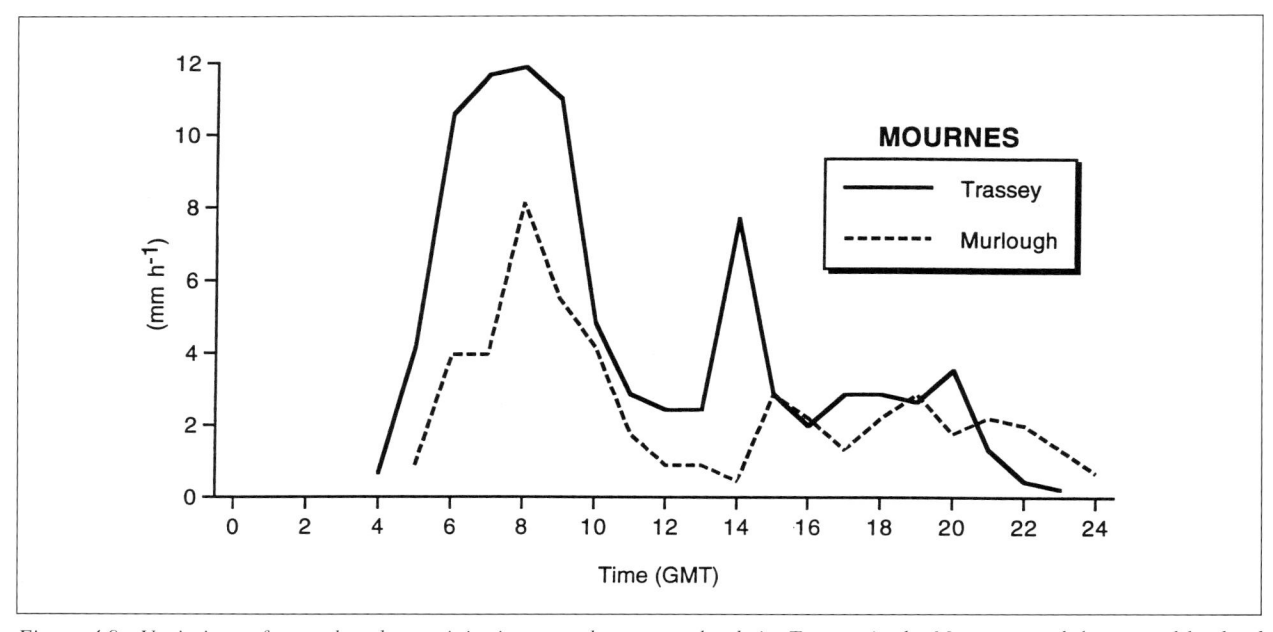

Figure 4.9 Variations of mean hourly precipitation rates between upland site Trassey in the Mournes, and the coastal lowland site of Murlough, 21-22 October 1987. Data points plotted at end of hour to which they refer.

Upland stations have in excess of 100 mm in each month between August and January, and some sites especially in the west, experience this amount in all months. Within the period August to January, few of the months are significantly wetter or drier than others, although with a few exceptions, January is the wettest month in the uplands. The general enhancement of autumn/winter rainfall implies the operation of an external control. Betts (1989) and Sweeney (1989) have indicated the roles of the Atlantic Ocean and Irish Sea as important influences upon the precipitation regime through the transference of latent energy to the atmosphere and to cyclonic systems at these seasons.

Rainfall variability

Adoption of standard climatological reference bases such as the current 1961–90 period, may be somewhat misleading in terms of monthly and annual averages of rainfall. Some variation exists from year to year as a response to the ever changing atmospheric circulation, and in particular to the location, intensity and duration of the main centres of cyclone and anticyclone activity relative to the British Isles. This variability is contrary to the popular belief in the reliability of Northern Ireland's rainfall. Examination of rainfall amounts in individual years reveals that an exceptional number of Atlantic depressions crossed the country in 1954 and produced annual rainfall totals 115 per cent of the long-term average, with 9 months having above average amounts. February (149 per cent), March (130 per cent), May (153 per cent), October (164 per cent) and November (142 per cent) were the wettest months relative to long-term average conditions.

In contrast, the anomalous high surface pressure during much of 1975 and 1976 mentioned earlier, produced a prolonged rainfall deficiency in Northern Ireland, and the monthly rainfall sequence is shown in Figure 4.10. Rainfall during 1975 was 75 per cent of the long-term annual average, with only January and September experiencing above average amounts, and May, June and December receiving below 40% of their mean. Rainfall was more variable during the winter and spring of 1976, when January, March and May were wet months, but the 3 month period June to August received only 48 per cent of the standard normal period operative at that time, 1941–70. The drought ended with the heavy rains of September and October 1976.

Droughts of the intensity of 1975–76 are exceedingly rare, but dry periods are not unusual in Northern Ireland. An 'absolute drought' is defined as a period of at least 15 consecutive days, none of which is credited 0.2 mm or more of rain. The period March to August is the most prone, and rarely would more than two spells of drought occur in a single year. A 'partial drought' refers to a period of at least 29 days during which the mean daily rainfall does not exceed 0.2 mm. This situation can be of importance to agriculture as the total rainfall for the 29 days does not exceed 5.8 mm. Such occurrences are much rarer than that of absolute drought. A period of at least 15 days, on none of which 1 mm or more of rain falls, is termed a 'dry spell'. Dry spells occur in most years, often twice a year, especially from March to October, and 1972 experienced four such spells (Betts, 1982).

Daily rainfall

The character of rainfall is not fully revealed by annual and monthly totals, but consideration of daily amounts affords more detail. Two indices of daily rainfall commonly used in this connection are 'rain days' with falls of 0.2 mm or more, and 'wet days' with falls of 1 mm or more. The average number of rain days increases from about 195 on the east Down coast, to over 250 days in the west, north-west and upland areas. Wet days range from 150 in the east to 200 days in the west and uplands.

Disadvantages of the rainfall-day as the arbitrary period for studies of heavy rainfalls have been stressed by Bleasdale (1963, 1970), but it is a convenient time unit. Amounts in excess of 100 mm in one rainfall-day have occurred on a number of occasions at stations in the Mournes, Sperrins and Antrim Plateau, but falls in excess of 125 mm are rare in upland areas, and absent altogether from lowland stations. The infrequent occurrence of heavy falls is due to relatively low relief and the low frequency of severe convectional activity in summer. It is not possible to present a comprehensive list of heavy rainfall occurrences, although Logue (1995) has statistically analysed extreme rainfall events in Ireland. Prior and Betts (1974), Houghton and Ó Cinnéide (1977) and Betts (1989, 1990, 1992) have discussed individual events.

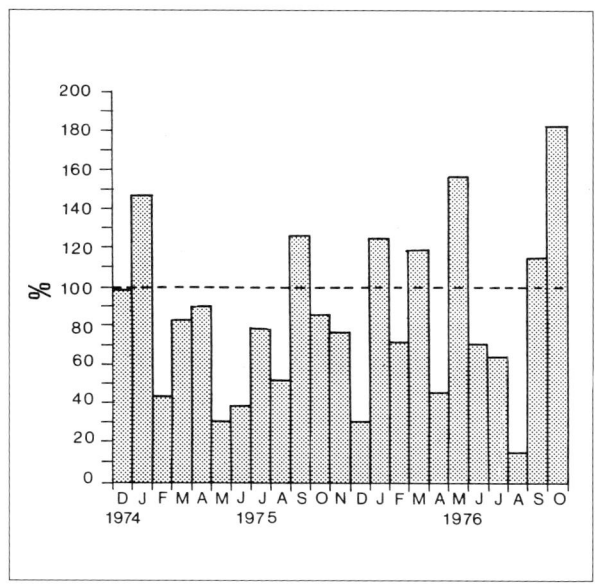

Figure 4.10 Monthly rainfall as a percentage of 1941-70 mean from December 1974 to October 1976 over Northern Ireland.

Rainfall duration and intensity

To appreciate rainfall duration and intensity characteristics in Northern Ireland, it is useful to compare values at Aldergrove with those in Great Britain. Aldergrove, for example, has 722 hours rainfall per annum in comparison with 506 hours at Camden Square, London. Consistently higher rainfall duration throughout the year reflects the closer proximity of Aldergrove, than London, to depression tracks. Aldergrove also has a mean rainfall intensity of only 1.12 mm per hour in comparison with a national average of 1.3 mm hr^{-1} (Atkinson and Smithson, 1976), a figure which supports the general perception that Ireland's rainfall is of low intensity and of long duration.

Important in connection with water management and drainage schemes is the relationship between depth, duration and frequency of heavy rainfall. The Flood Studies Report (Meteorological Office, 1975) provides data to allow the calculation of rainfall amounts for a range of duration and return periods at any location in Northern Ireland (see also Jackson, 1977a). Figure 4.11 presents this information at Annalong (130 m) and Armagh (62 m) for durations of 1 hour, 24 hours, 48 hours and 25 days with return periods of up to 50 years. The similarity of values of 1-hour falls over various return periods at the two sites suggests that during short-period heavy falls often localised in character, the orographic effect is not as significant as convective uplift. In contrast with longer durations, rainfall amounts are markedly greater at the upland site of Annalong. This is in accordance with the fact that areas experiencing high annual precipitation, receive a great proportion of these amounts from prolonged heavy events associated with large-scale weather systems, and these falls usually occur over uplands for the reasons given earlier.

Snowfall

Although heavy falls of snow are relatively rare in Northern Ireland, the economic consequences of such occurrences can be great. The synoptic patterns most often associated with snowfalls are those producing cold airstreams from northerly or easterly quadrants. Westerly and south-westerly airflows in winter give temperatures well above 0°C even on the highest ground. Snowfalls in the form of instability showers occur over the north and east coasts exposed to onshore winds, although rarely is any great depth of cover attained. The heaviest snowfalls occur with 'polar lows' and with slow moving warm fronts, but with the passage of the frontal system across the area, and the influx of warmer air, a rapid thaw often results.

The average number of days with snow lying on the ground is lowest on the East Down coast, the Lough Foyle lowlands, and the valleys of the Roe and Lower Bann all having less than 10 days. Inland stations under 150 m O.D. experience between 20–30 days of snow each year, although the Armagh region averages only 18 days. Over the Antrim Plateau, Sperrins and Mournes, snowfall frequency exceeds 30 days per year (Jackson, 1977b).

Water balance

Water vapour, the source of precipitation, is received by the atmosphere through evaporation from the oceans, inland water bodies, and most land surfaces, and through transpiration from plants. These processes are described by the term evapotranspiration, and together with precipitation, form part of a continuous transfer of moisture and heat energy between the atmosphere and the earth's surface. Globally, there exists a balance between these transfer processes, but locally, imbalance between precipitation and evapotranspiration results in periods of water deficit when the latter exceeds the former, and surplus when the reverse occurs. Prolonged deficit or surplus can have severe repercussions for water resource management, even in Northern Ireland (Betts, 1978a, 1978b, 1984). Furthermore, soil moisture content is of importance to agricultural operations at specific periods of the year. Normally, plant growth is unlikely to be severely constrained by lack of soil moisture, except for shallow-rooted grasses and horticultural crops in the east of the Province. In a dry summer, however, even deep-rooted crops may suffer from soil deficits throughout Northern Ireland. In assessing water losses, evaporation and evapotranspiration measurements are taken at a number of locations throughout the country.

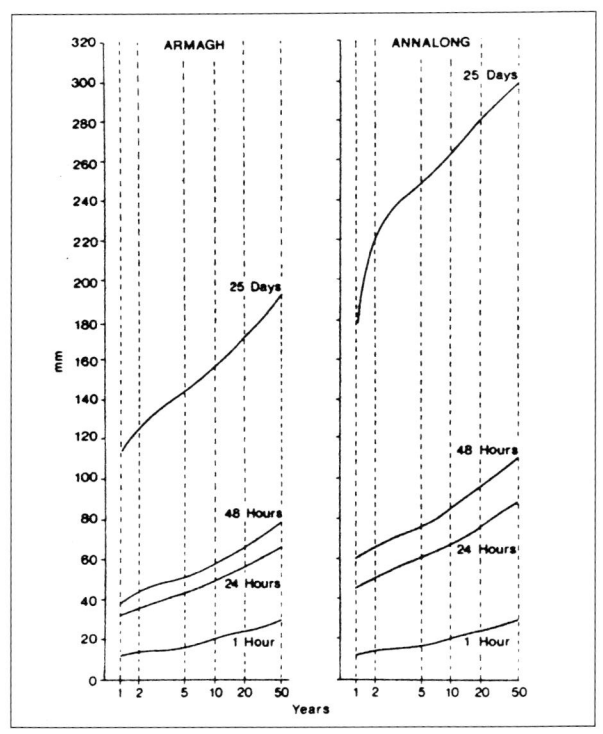

Figure 4.11 Rainfall in mm for a range of durations and return periods at Annalong and Armagh. (Data from the Met. Office, Belfast Climate Office).

A distinction is made between potential evapotranspiration (PE) and actual evapotranspiration (AE). PE is the maximum quantity of water loss that will occur from a moist surface under given climatological conditions. In contrast, water loss in the form of AE has a lower value, for this term allows for the occasional drying out of the soil surface, and under such conditions little or no evapotranspiration occurs. PE is a somewhat artificial concept, but being influenced by fewer factors than AE, it is less difficult to estimate. Data on PE are therefore easier to collect, and consideration will be restricted to this measure of evapotranspiration.

In the estimation of PE by The Met. Office, an amended version of Penman's formula (1962) is adopted (Grindley, 1972). The meteorological variables used in the Penman estimates of PE, (which Penman terms PT to distinguish his estimates from those based on Thornthwaite's formula), are mean air temperatures, mean deficit from the saturation vapour pressure, mean wind speed and mean sunshine duration. In Northern Ireland eighteen stations have PT data for fifteen years or more.

The influence of solar radiation and temperature upon evapotranspiration is reflected in the seasonal variation of PT rates at Aldergrove, an inland site, Murlough, a coastal site, and Lowtown, an upland site (Figure 4.12). At all three stations PT is considerably greater during summer than winter, with a maximum in June and a minimum in December. A smaller seasonal range is found at Murlough on the coast, with lower summer and higher winter values than at Aldergrove, a low lying inland station, although Lowtown, an upland station, has the smallest seasonal variation of the three. Annual PT values exceeding 560 mm along coasts decrease to less than 350 mm in upland areas. This spatial pattern reflects the longer duration of sunshine and higher wind speeds of coastal areas.

Jordan (1994, 1996) has used a Geographical Information System (GIS) to prepare maps of long-term averal annual, potential evapotranspiration (PT) and potential water balance (P-PT), based upon annual summaries derived from monthly records at 25 stations throughout Northern Ireland for the period 1969–94 (Figures 13a and 13b). Mean annual PT in Northern Ireland tends to be greatest in the east and on exposed coasts and lower in the west and upland areas (Figure 13a). Annual PT at Aldergrove (68 m O.D.) averages 504 mm, at the coastal site of Bangor (4 m O.D.) 568 mm, and 339 mm at Lowtown (213 m O.D.). These characteristics reflect again the availability of net radiation and the higher wind speeds in coastal areas. Conversely, reduced PT values in the upland areas are largely associated with low radiation availability.

In terms of the annual potential water balance (P-PT), Figure 4.13b clearly indicates that P greatly exceeds PT, especially in the west and north of the Province. The two areas where the differences are significantly reduced are again, the upper Bann area south of Lough Neagh, and parts of north Down and the Ards peninsula. Here, P exceeds PT by less than 350 mm, and in the Lurgan area 300 mm. In contrast, over much of the upland, P exceeds PT by more than 1000 mm annually.

Monthly data (Figure 4.14) show that the excess P over PT during winter is gradually reduced in spring, mainly due to the increase in PT, but also as a result of decreased rainfall. At most stations PT starts to exceed P in late spring, and the reverse trend occurs in early autumn. The shaded area in these graphs relates to the period of the year when precipitation is low and the soils are relatively dry. When PT exceeds P, vegetation must draw upon reserves of stored water in the soil, and a soil moisture deficit (SMD) is said to exist. SMDs build up to an average July value of 50 mm generally for Northern Ireland, although in the Lough Neagh basin and coastal areas in dry summer months, values will exceed 100 mm. Soils normally return to field capacity in September, and remain at field capacity or wetter until the following April or May, when SMDs again begin to develop (Wilcock, 1982). In individual years, however, there will be considerable variation from the average.

The moisture balance at altitude and in the west of Northern Ireland remains in surplus throughout the year (Figure 4.14). One exception however is Silent Valley in the Mournes, where a sheltered south facing aspect results in PT marginally exceeding P during July. In very dry summers, when sunshine is enhanced even in the upland areas, a moisture deficit will occur but there may still be adequate reserves whereas, in adjacent lowland, drought conditions prevail. A generalised pattern of the average number of months during the year when PT exceeds P is illustrated in Figure 4.15, and reinforces the west-east contrast in terms of wetness.

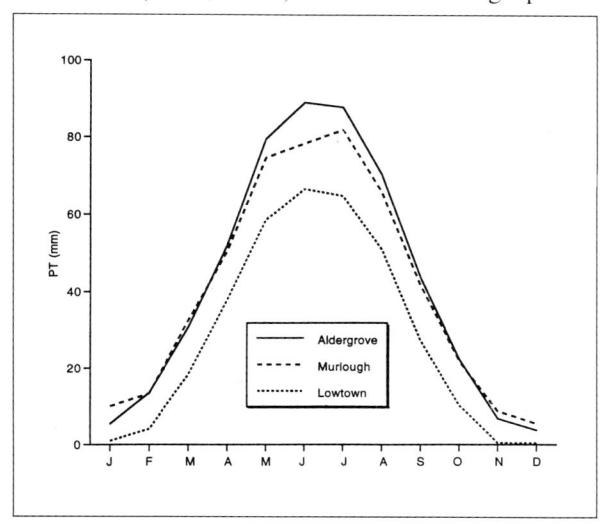

Figure 4.12 Mean monthly potential evapotranspiration (PT) at Aldergrove, Murlough and Lowtown, 1973-94.

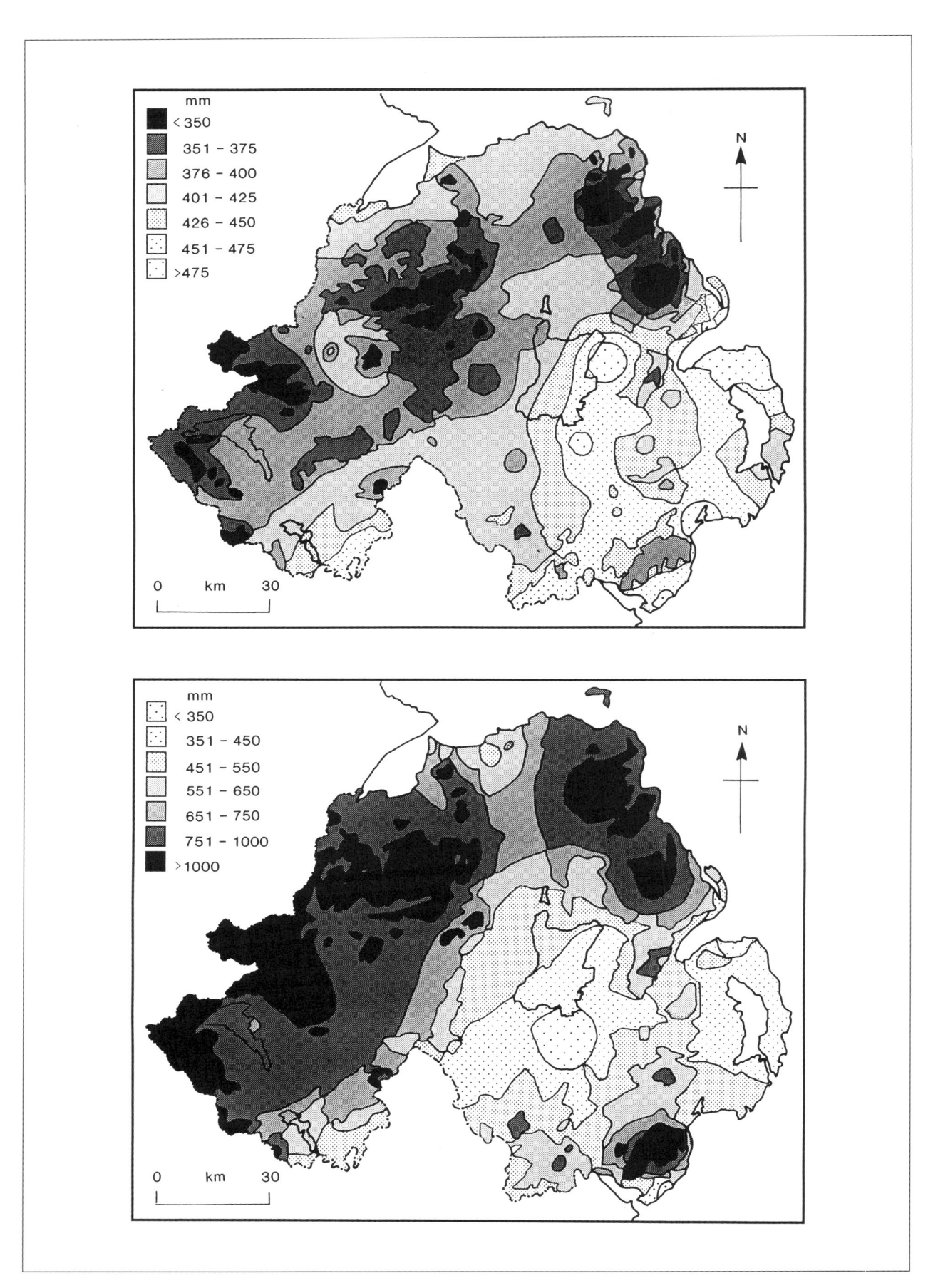

Figure 4.13(a) GIS derived map of mean annual potential evapotranspiration (PT) in mm, 1969–94.
Figure 4.13(b) GIS derived map of mean annual moisture excess (P-PT) in mm, 1969–94.

Moisture balance equations in applied terms provide valuable indications of atmosphere-soil-crop relationships. If the PT requirement of a crop is not fully supplied by precipitation, a crop will begin to suffer moisture stress as the available water capacity (AWC) of a soil becomes depleted and growth will be retarded if irrigation is not employed. Such conditions occasionally occur in Northern Ireland, particularly in the Newtownards, Lisburn and North Armagh areas (Wilcock, 1982). The demand for moisture by a vigorously growing green crop is considerable, for the volume of water required to produce a season's grass, hay, root crops or cereals is 450 mm (Collins et al, 1986).

In contrast, under conditions of moisture excess, if the soil texture and structure encourages good drainage, then leaching and soil acidification result. Soils with poor internal drainage, may have a drainage requirement which necessitates providing an artificial land drainage system. Where land drainage is required, it should be carried out when soil moisture is at its lowest level, generally in August and September in most parts of the country. In practice, both irrigation and drainage requirements may occur on the same site in one year.

Acknowledgement

The author wishes to thank The Met. Office, Belfast Climate Office, for permission to consult climatological records.

References in Appendices

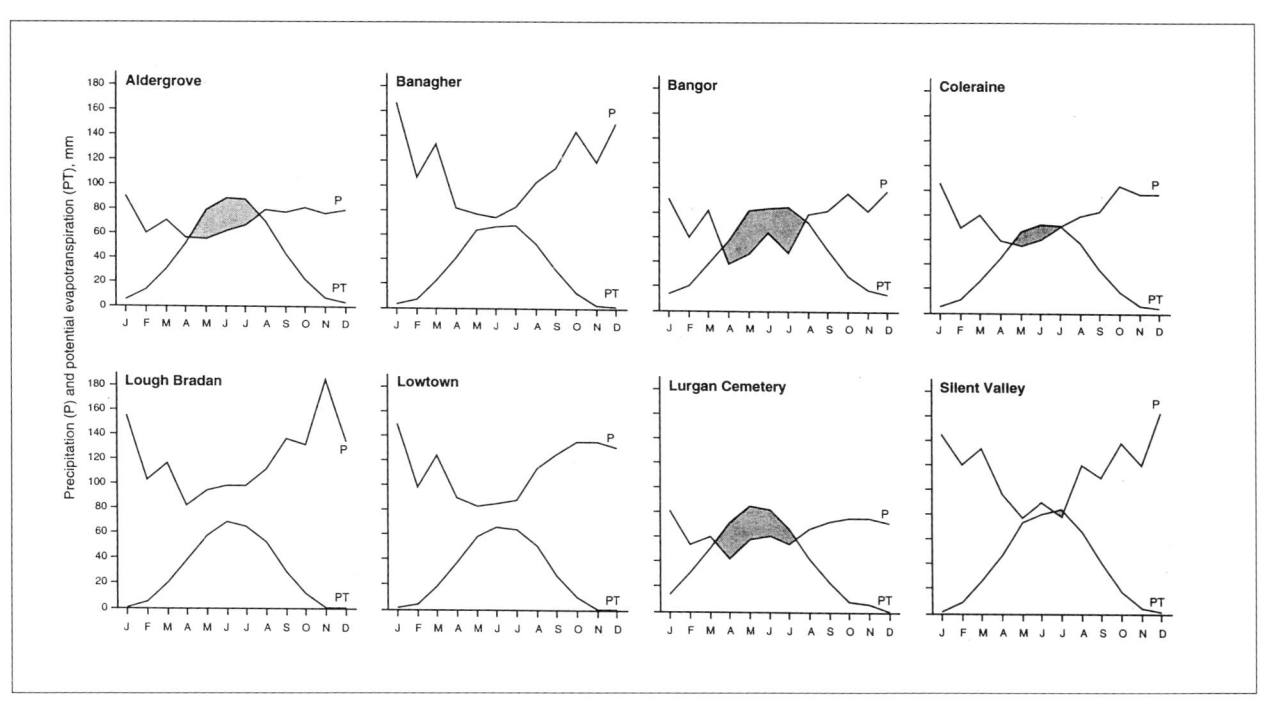

Figure 4.14   Mean monthly values for precipitation (P) and potential evapotranspiration (PT) at eight climatological stations, 1969–94.

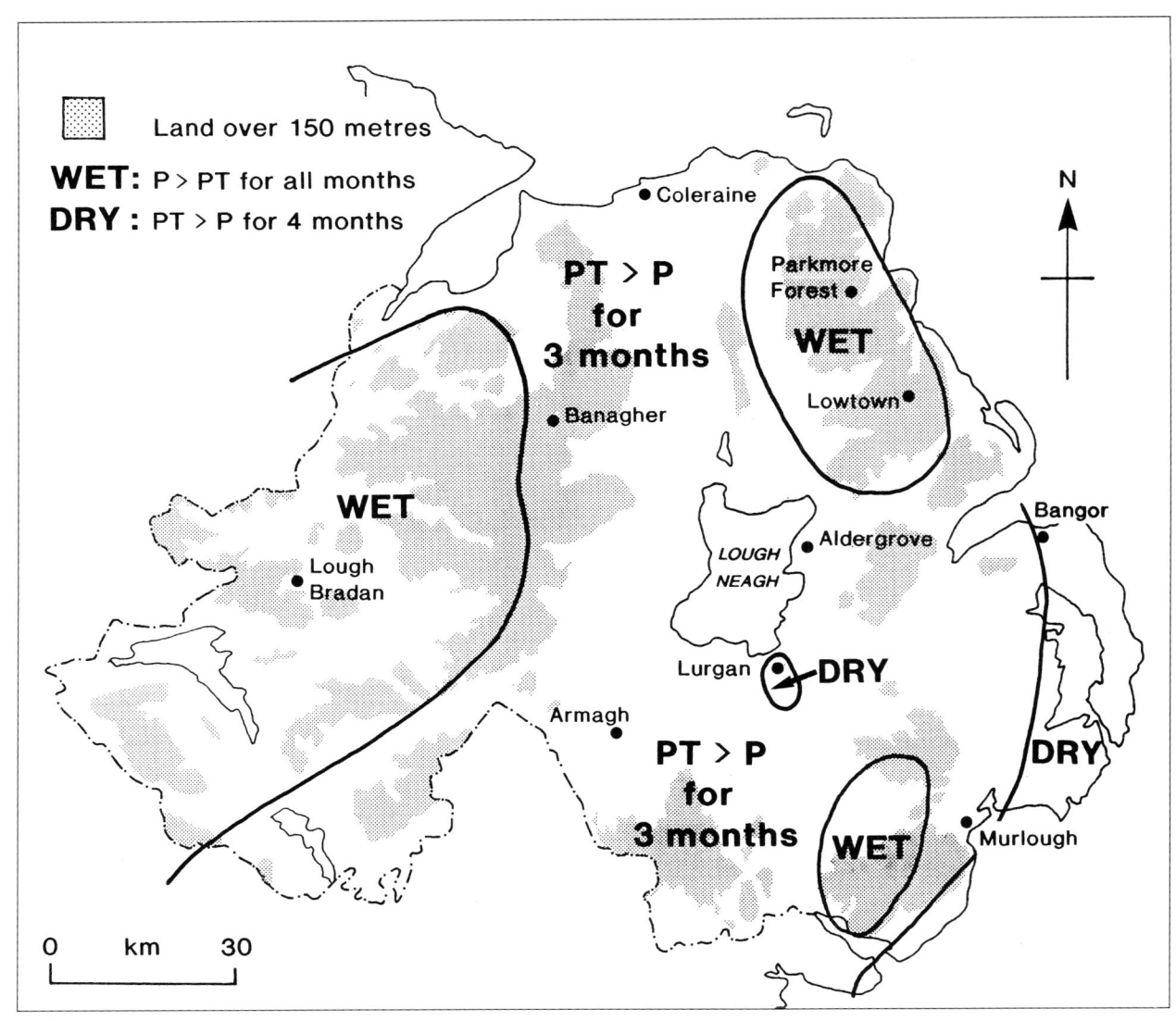

Figure 4.15 Generalised distribution of the mean number of months in the year when soil moisture deficit prevails (PT>P).

TABLE 4.1

Extreme wind speeds in m s^{-1} recorded by anemograph stations in Northern Ireland
(Data from The Met. Office, Belfast Climate Office)

Station	Height of Anemograph m O.D.	Recording Period	Highest Gust Speed	Month/year occurrence	Highest Hourly Mean Speed	Month/year occurrence
Aldergrove	80	1927-93	39.1	9/1961	25.2	9/1961
Ballykelly	12	1958-71	47.3	9/1961	27.3	2/1963
Ballypatrick Forest	165	1968-93	42.2	1/1968	27.3	1/1968
Belfast Harbour	21	1967-93	41.2	1/1976	24.2	12/1966 2/1991
Carrigans	129	1964-93	41.7	11/1965	23.2	1/1965
Castle Archdale	79	1964-93	44.8	1/1974	24.7	1/1974
Kilkeel	36	1964-93	55.6	1/1974	27.8	1/1974
Orlock Head	51	1968-93	41.2	1/1974	28.8	1/1978

TABLE 4.2

Estimated extreme wind speeds in m s^{-1} at 10 m above the ground at Aldergrove (after Hardman *et al.* 1973)

Maximum gust speeds					Maximum hourly mean speeds				
Av. annual max.	Speeds likely to be exceeded only once in stated no. years				Av. annual max.	Speeds likely to be exceeded only once in stated no. years			
	10	20	50	100 years		10	20	50	100 years
31.7	72	77	82	87	18.1	42	45	49	51

TABLE 4.3

Average daily bright sunshine in hours, 1961-90. (Data from The Met. Office, Belfast Climate Office)

Station	J	F	M	A	M	J	J	A	S	O	N	D	Year
Aldergrove	1.46	2.37	3.15	5.13	5.93	5.67	4.76	4.50	3.63	2.75	1.90	1.25	3.54
Armagh	1.43	2.27	3.09	4.56	5.35	4.96	4.28	4.14	3.48	2.65	1.92	1.16	3.27
Ballywatticock	1.75	2.56	3.48	5.20	6.21	6.07	5.23	4.92	3.92	3.00	2.13	1.45	3.83
Carmoney	1.22	2.28	3.03	4.95	5.63	5.14	4.11	4.19	3.40	2.52	1.70	0.93	3.26
Castle Archdale	1.45	2.26	3.10	4.62	5.01	4.80	4.00	3.93	3.43	2.55	1.81	1.08	3.17
Coleraine	1.24	2.17	3.00	4.86	5.55	5.17	4.09	4.05	3.26	2.42	1.64	0.93	3.20

TABLE 4.4

Average amount of cloud in oktas at Aldergrove 1961-90. (Data from The Met. Office, Belfast Climate Office)

J	F	M	A	M	J	J	A	S	O	N	D	Year
6.0	5.7	5.5	5.2	5.5	5.7	6.3	6.2	5.8	5.8	6.0	6.1	5.8

TABLE 4.6

Extreme maximum and minimum temperatures (^{0}C) at Armagh 1901-93. (Data from Monthly Weather Report, The Met. Office)

	J	F	M	A	M	J	J	A	S	O	N	D
Extreme max. temp.	13.9	14.4	21.7	22.7	26.2	30.0	30.6	29.0	27.8	22.8	16.7	15.0
Year of most recent occurrence	1957	1953	1965	1975	1989	1989	1934	1975	1906	1908	1948	1948
Extreme min. temp.	-11.1	-10.6	-12.2	-7.2	-2.4	-0.6	3.9	2.2	-4.4	-4.4	-8.3	-10.0
Year of most recent occurence	1963	1969	1947	1917	1982	1930	1918	1944	1919	1926	1919	1909

Table 4.5

Monthly and annual values of mean daily maximum, mean daily minimum and mean daily air temperature for selected stations, 1961-90. (Data from The Met. Office, Belfast Climate Office)

Station	Altitude m O.D.	Mean Daily Values °C	J	F	M	A	M	J	J	A	S	O	N	D	Year
Aldergrove	68	Max Min Mean	6.5 1.1 3.8	6.8 1.1 4.0	8.8 2.0 5.4	11.5 3.5 7.5	14.4 6.1 10.3	17.3 9.1 13.2	18.5 10.9 14.7	18.2 10.7 14.5	15.9 9.0 12.5	13.0 6.9 10.0	8.9 3.1 6.0	7.3 2.0 4.7	12.3 5.5 8.9
Armagh	62	Max Min Mean	6.7 1.3 4.0	7.1 1.2 4.2	9.5 2.3 5.9	12.3 3.6 8.0	15.2 6.1 10.7	18.0 9.1 13.6	19.4 10.9 15.2	18.9 10.6 14.8	16.5 8.9 12.7	13.2 6.8 10.0	9.0 3.0 6.0	7.4 2.0 4.7	12.8 5.5 9.2
Carrigans	113	Max Min Mean	6.4 0.8 3.6	6.8 0.7 3.8	8.9 1.8 5.4	11.7 3.2 7.5	14.5 5.6 10.1	17.2 8.5 12.9	18.3 10.3 14.3	18.1 10.0 14.1	15.8 8.3 12.1	12.9 6.3 9.6	8.8 2.5 5.7	7.2 1.7 4.5	12.2 5.0 8.6
Castle Archdale	66	Max Min Mean	6.7 0.8 3.8	7.0 0.8 3.9	8.9 1.9 5.4	11.5 3.1 7.3	14.2 5.4 9.8	16.8 8.5 12.7	17.9 10.4 14.2	17.8 10.0 13.9	15.6 8.4 12.0	12.9 6.5 9.7	9.0 2.7 5.9	7.5 1.8 4.7	12.2 5.0 8.6
Coleraine	23	Max Min Mean	7.0 1.3 4.2	7.1 1.2 4.2	8.8 2.4 5.6	11.0 3.7 7.4	13.7 5.9 9.8	16.2 8.6 12.4	17.4 10.6 14.0	17.4 10.3 13.9	15.6 8.8 12.2	13.1 6.9 10.6	9.4 3.4 6.4	7.8 2.2 5.0	12.0 5.4 8.7
Helen's Bay	43	Max Min Mean	6.8 2.5 4.7	7.0 2.4 4.7	9.0 3.3 6.1	11.5 4.7 8.1	14.4 6.9 10.7	17.1 9.5 13.3	18.6 11.3 14.9	18.3 11.3 14.8	16.3 9.9 13.1	13.2 7.9 10.5	9.2 4.7 6.9	7.7 3.5 5.6	12.4 6.4 9.4
Hillsborough	116	Max Min Mean	6.5 1.4 4.0	6.4 1.3 3.9	8.4 2.1 5.3	10.7 3.4 7.1	13.6 5.7 9.7	16.5 8.6 12.6	17.9 10.4 14.2	17.6 10.3 14.0	15.5 8.8 12.2	12.5 6.8 9.7	8.7 3.3 6.0	7.3 2.3 4.8	11.8 5.4 8.6
Parkmore Forest	235	Max Min Mean	5.3 0.1 2.7	5.2 0.0 2.6	7.0 0.8 3.9	9.4 2.1 5.8	12.2 4.3 8.3	14.9 7.2 11.1	16.2 9.0 12.6	16.0 8.9 12.5	14.0 7.3 10.7	11.3 5.5 8.4	7.6 2.2 4.9	6.2 1.1 3.7	10.4 4.0 7.2
Silent Valley	129	Max Min Mean	6.7 1.7 4.2	6.4 1.5 4.0	8.1 2.4 5.3	10.2 3.8 7.0	12.9 6.2 9.6	15.8 9.0 12.4	17.4 10.7 14.1	17.2 10.6 13.9	15.3 9.1 12.2	12.5 7.1 9.8	8.9<>3.8 6.4	7.5 2.6 5.1	11.6 5.7 8.7

TABLE 4.7

The average duration of the growing season for periods 1951-60,1961-70,1971-80,1981-90 at selected stations. (Derived by the author from Bailie, 1980; Woods, 1995, and data from The Monthly Weather Report, The Met. Office)

Meteorological Station	Average growing season in days			
	1951-60	1961-70	1971-80	1981-90
Aldergrove	266	244	259	272
Armagh	276	252	259	278
Castle Archdale	262	242	258	254
Hillsborough	257	237	251	271
Moneydig	256	244	257	259

TABLE 4.8

Monthly and annual averages of rainfall (mm) for selected stations 1961-90. (Data from The Met. Office, Belfast Climate Office)

Station	Altitude m O.D.	J	F	M	A	M	J	J	A	S	O	N	D	Y
Aldergrove	68	86	58	68	53	60	63	64	80	85	89	78	78	862
Annalong Valley	130	130	92	104	81	78	79	75	103	108	126	118	123	1217
Armagh	62	80	58	65	55	60	59	52	77	71	87	73	76	813
Ballymena	38	98	70	82	58	65	66	66	86	93	103	94	92	973
Ballywalter	12	85	58	64	51	55	61	52	75	79	88	81	76	825
Banagher	216	156	106	119	80	88	77	76	100	111	140	128	141	1322
Belfast P. Stn.	5	95	65	75	55	57	61	48	78	84	93	89	85	885
Castle Archdale	66	112	82	94	63	74	79	68	95	104	118	110	111	1110
Coleraine	23	99	70	73	56	65	60	69	80	90	108	100	94	964
Crom Castle	58	107	77	82	58	72	75	62	86	90	108	94	102	1013
Glenderg Forest	180	220	137	175	108	112	117	121	160	183	200	208	209	1950
Hillsborough	116	87	60	70	57	62	64	57	83	85	94	82	84	885
Labbyheige	347	182	134	146	98	106	97	101	127	145	169	153	172	1630
Parkmore Forest	235	180	123	149	109	101	102	103	133	154	177	185	172	1688
Woodburn North	217	118	84	96	73	78	85	81	109	120	130	118	113	1205

CHAPTER 5

RIVERS, DRAINAGE BASINS AND SOILS

DR. DAVID N. WILCOCK
SCHOOL OF ENVIRONMENTAL STUDIES, UNIVERSITY OF ULSTER

A. SOME THEORETICAL CONSIDERATIONS; RIVERS IN THEIR CATCHMENTS

(i) Forcing Functions

Rivers, in Northern Ireland as in any modern society, are managed to improve agriculture, to provide water supply, energy and recreation, to control pollution, and to enhance fisheries and environmental quality generally. Central to managing any river as a multi-purpose resource is the idea that what happens in the river is controlled by what happens in its drainage basin, i.e. the area inside the watershed contributing water and sediment to the river channel (Figure 5.1). The terms 'drainage basin' and 'catchment', incidentally, may be used synonomously. Since drainage basins are covered by soils of varying thickness, chemistry, and physical composition, there are close links between drainage basin soils, river flows and river water quality.

A river channel's natural cross-sectional shape, flow velocities and nutrient load are responses to the gravitational downstream flow of water and sediment delivered from its catchment. In turn, the characteristics of water and sediment flows are determined by the climate, geology, soils and topography of the tributary drainage basin. Other energy flows, related to photosynthesis and respiration, also take place in rivers and lakes. These flows are important for the life in our natural water bodies and are closely linked to the water, sediment and nutrient flows. Under natural conditions channel morphology becomes adjusted to the flows of water and sediment through the river network. In turn plant productivity, plant communities and river ecosystems become adjusted to streamflow characteristics, sediment transport, nutrient supplies and channel morphologies in a complex series of inter-dependent linkages which maintain the essential equilibrium between a river's ecology and its drainage basin. In this conceptual model of how rivers relate to their drainage basins, river morphology and ecology are viewed as dependent on the water and sediment loads delivered from the catchment slopes. The stream flow and sediment loads are the *forcing functions*. Changes to these, brought about for example by changing climate, afforestation, deforestation, land drainage, urbanization, acidification, or cultural eutrophication, bring about adjustments in the rivers themselves.

A drainage basin's soil mantle is seen as having a vital role in this model, modifying the gravitational delivery of water and sediment to the river in different ways, at different times, and at different places within a catchment. It sometimes acts as a buffer on extreme rainfall, storing heavy rainfall and releasing it later. It may act as a filter on suspended or dissolved particles in soil water moving through the soil. This is one way in which soil protects rivers from pollution, though its capacity to act as a filter in this way is not infinite. Soils can become saturated with adsorbed chemicals, lose their filtering capacity, and become a direct source of pollution.

(ii) The Runoff Cycle:

Some of the *precipitation* on a drainage basin (mainly rainfall in Northern Ireland) goes into *surface storage* on vegetation canopies, roofs and depressions on the ground before being *evaporated* back to the atmosphere. This water is effectively lost to the local river system (Figure 5.2).

A second fraction may be transmitted across surface slopes as *overland flow*. If it takes place at all in Northern Ireland, overland flow occurs during the most intensive or prolonged rainfall, ideally when soils are already saturated. Overland flow is rapid and usually produces flood flows in the river channel, usually lasting only hours or days at most. Frequent floods, those of a size which occur several times a year on average, would not fill a naturally-formed channel. Floods of a size which occur every year fill such a channel to bankfull capacity, and less frequent floods progressively overtop the channel, spilling on to the flood plain. These floods, of course, cause a lot of physical damage but they also recharge shallow groundwater supplies.

A third fraction may enter the soil either to be stored in the *unsaturated* zone as soil water or rapidly transmitted to the river through the soil profile as

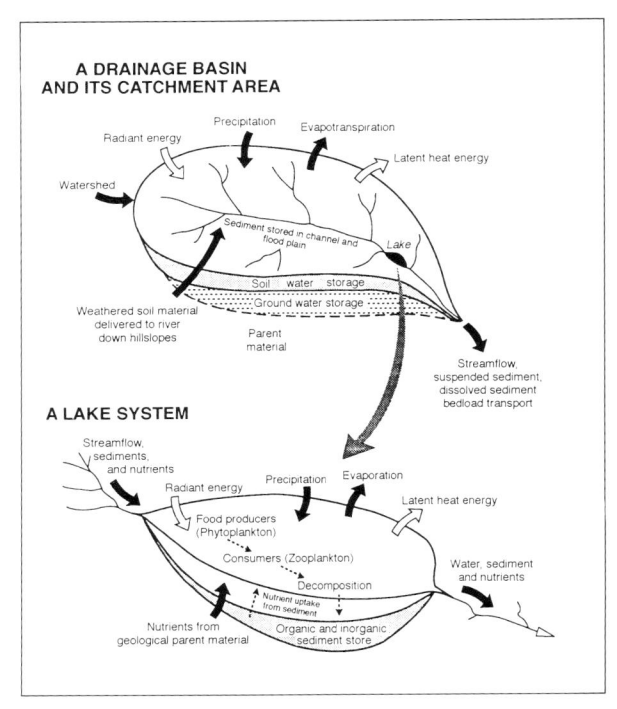

Figure 5.1. Some of the principal material and energy fluxes in a drainage basin which affect river and lake processes.

throughflow. In an individual rainfall event, subsurface throughflow may reach the river shortly after overland flow. But it still contributes to flood flows. Overland flow and throughflow combined, together with precipitation falling directly onto a river channel, are sometimes called *quickflow*. *Soil water*, held against the force of gravity between soil particles, may be taken up by plants and eventually removed into the atmosphere through their leaves as *transpiration*. Since transpiration and evaporation are difficult processes to identify and measure separately, they are collectively called *evapotranspiration*. Soil water is important in that it represents the main source of water for plants. Too little, as, perhaps, in a coarse, sandy, soil may leave plants short of water and nutrients in dry weather. Too much, as perhaps in a clay soil, can drive out air and atmospheric oxygen, leaving plants unable to take up nutrients. The main soil characteristics determining soil water content are soil structure, texture, mineral composition and organic content.

The fourth route in this simplified model of the runoff cycle is the deep percolation of water into that part of the soil profile where all pore spaces between soil particles are filled with water. This is the *groundwater* zone or *zone of saturation*. Water here moves slowly, perhaps eventually becoming streamflow only days, weeks, even months after it first fell as precipitation. Groundwater is thus the source of dry weather flows in a river, most obviously those which occur in summer.

(iii) Infiltration Capacity

One of the most important soil properties affecting the hydrology of any catchment is infiltration capacity (often measured in mm hour^{-1}), which measures the rate at which water enters a soil from the surface. Affected by soil structure, texture, vegetation, organic matter, compaction and antecedent rainfall, the infiltration of any dry soil is initially quite high, but decreases with time as the soil surface layers become saturated. The infiltration rate ultimately equilibrates at a relatively low, constant, value for any particular soil and this is the infiltration capacity. High infiltration capacities may be up to 40 + mm hour^{-1} while low infiltration capacities may be 1 mm hour^{-1} (Mulqueen, 1978).

Infiltration capacity determines how much of any rainfall event infiltrates the soil and how much, if any, travels across the surface or is stored on it. If, for

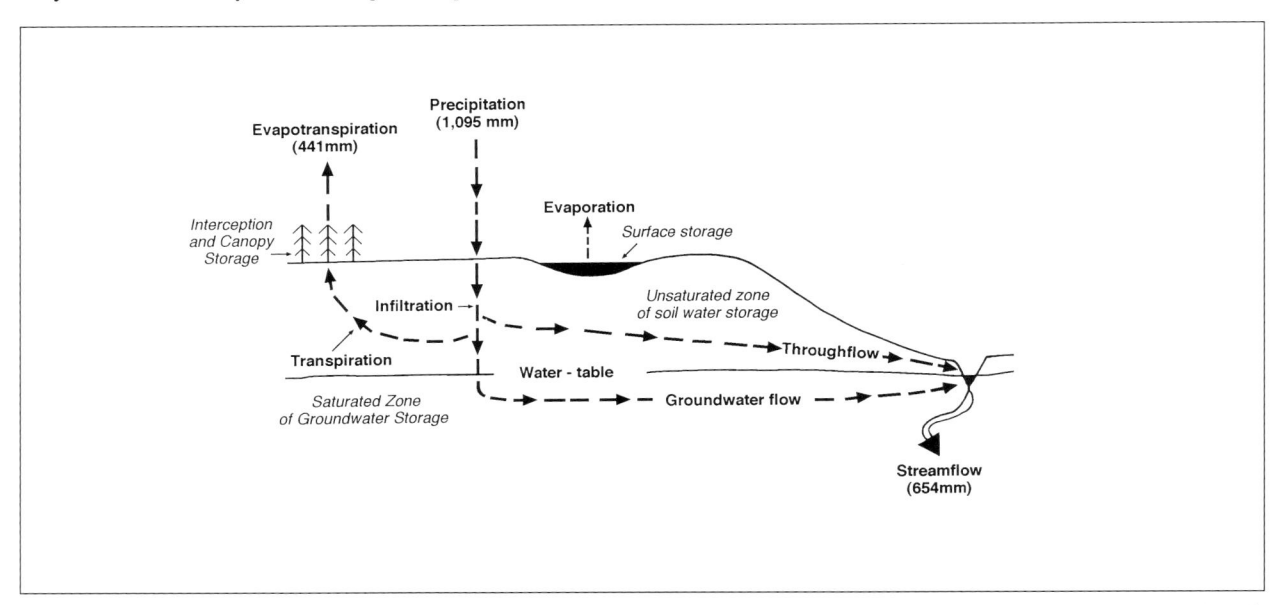

Figure 5.2 Schematic diagram of the runoff cycle in Northern Ireland. The main inputs and outputs of water in the runoff cycle are represented in millimetres.

example, precipitation rate (in mm hour^{-1}) exceeds infiltration capacity during an individual rainfall event, some water must remain on the surface or run off as overland flow, because it cannot be absorbed into the soil. This type of overland flow is termed *Horton overland flow*, after R.E.Horton an American hydrologist who first wrote extensively on this topic. Horton overland flow may be distinguished from *saturated overland flow* which occurs when water-tables are at or close to the ground surface. High water tables and saturated overland flow are often produced by long periods of low intensity rainfall. Because rainfall in Britain and Ireland is generally associated with low intensity, frontal, rainfall rather than high intensity convective rainfall, it is usually argued that Horton overland flow in these areas is rare. In that overland flow, moreover, is rarely observed, even saturated overland flow is not regarded as the normal flood producing mechanism. Instead, conventional wisdom argues that subsurface throughflow is the main mechanism producing flood events in humid areas like Northern Ireland (Hewlett and Hibbert, 1967: Ward, 1984).

Few measurements of infiltration capacity have been made in Northern Ireland. This is unfortunate because infiltration rates help determine *rates of recharge* of groundwater aquifers and the *generation of flood flows*. An annual cycle of infiltration measurements has been undertaken on a 15 km^2 catchment on the outskirts of Coleraine. Land use is mainly agricultural grassland and soils are clay loams and sandy loams. Infiltration rates throughout the year are low, mean monthly values at eleven sites ranging from 9.01 mm hour^{-1} in June to 0.64mm hour^{-1} in January (Figure 5.3). The highest individual reading was 19.7 mm hour^{-1} in June. Frosts in January and the associated cracking of surface soils also produced increases in infiltration not normally associated with wet winter conditions. An interesting observation was that winter rainfall intensities often exceed infiltration capacities. In January, 1982, for example, nearly 70 per cent of the rainfall was of an intensity greater than the catchment's measured mean infiltration capacity. Some of this rainfall must have been removed as Horton overland flow or stored on the surface prior to subsequent evaporation and/or infiltration (Wilcock and Essery, 1984).

(iv) Permeability

Soil permeability is, perhaps, the most important soil characteristic affecting the movement of water in Irish soils. Permeability is related to but distinct from porosity. Porosity refers to the volume of pores in a soil as a percentage of the soil's total volume. Permeability refers to the interconnectedness of those pores and determines how freely soil water and groundwater are able to move vertically and laterally towards the river. Hydrologists use the term hydraulic conductivity (k) to quantify this property, soils with low k values having restricted soil water movement, high soil water retention, and little water movement within soil peds. Cracking may occur in such soils due to high clay content, and when this occurs soil water often moves at accelerated rates between the soil peds. Catchments containing extensive areas of such soils tend to have large quickflow components, large areas of badly-drained land and relatively low dry weather river flows. In contrast, catchments containing extensive areas of soils with high k values tend to have small quickflow components, well-drained soils and high dry weather flows.

Peat hydrology represents a special case of low

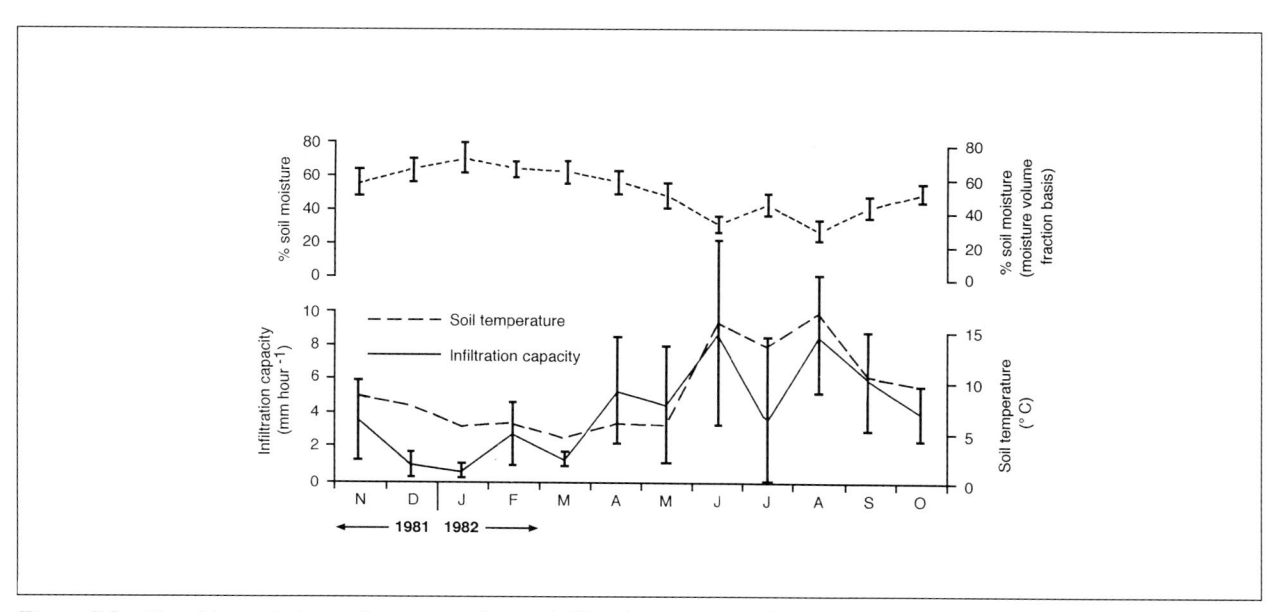

Figure 5.3. Monthly variations of mean catchment infiltration capacity, % antecedent soil moisture and soil temperature between November 1981 and October, 1982 in the University of Ulster's study catchment at Coleraine. Standard deviation for monthly values of infiltration capacity and soil moisture are shown to illustrate variations of both variables across the catchment.

permeability soils. A conventional view is that peat acts as a sponge, absorbing rainfall in winter and releasing it in summer. This view is too simplistic, for many peat deposits have low hydraulic conductivities and high water retention which preclude large volumes of water release by gravity flow. An important property affecting the hydraulic conductivity of peat is its degree of decomposition. This usually increases with depth so that in the lower, 'catotelm', layers of raised bogs, below about 70cm, hydraulic conductivities may be 10^{-6} cm sec^{-1}. Such k values would be classified as 'very slow' (Bailey, 1979). Even when decomposition does not increase with depth, saturated hydraulic conductivities are often below 70 cm due to compaction. In the top 50 cm or so of raised bogs, the 'acrotelm', peat is not well decomposed, compaction is limited, and hydraulic conductivities may rise to 10^{-3} cm sec^{-1} (Higgins, 1988). These would be classified as 'moderate' (Bailey, 1979). On the blanket bogs of Long Mountain and the Antrim Plateau in the upper River Main valley, hydraulic conductivities measured by Higgins, at between 10^{-7} cm sec^{-1} and 10^{-8} cm sec^{-1}, were often lower than in raised bogs. Similarly low catotelm hydraulic conductivity values have more recently been recorded by Hammond *et al*, 1990.

Given the relatively small depth of acrotelm layers relative to total peat thickness and the small variation in water-tables in surface peats it is doubtful if the regulating effect of peats on streamflow is very significant. Peat catchments tend to have steep flow duration curves and low groundwater flow (Wilcock, 1977). Twenty years' hydrological research in this field has been summarized as follows: 'The hydraulic conductivity of the acrotelm is much higher than than that of the catotelm, but since the storage capacity of the acrotelm is limited, once the water-table drops below the most permeable layers (usually the active root zone), throughflow discharge falls to a very low level. Thus the very thinness of the acrotelm and the impermeability of the catotelm determine that peat-covered catchments are poor suppliers of baseflow. Unlike the acrotelm, the catotelm remains permanently saturated, and rates of water movement through it are very low indeed' (Burt, 1995). It may be that the regulating effect claimed for peat is short-term rather than seasonal and confused with surface water storage in flood plain wetlands where many raised bogs in Northern Ireland typically occur (Essery and Wilcock, 1990a).

The influence of soil on catchment hydrology is clearly complex. Impermeable iron pans act as barriers to downward-moving water, deflecting it horizontally. Weathering can reduce grain size and increase compaction, both of which might in turn reduce permeability. Despite this complexity, it is possible to generalize the relationships in Ireland between geomorphology, soils and drainage: 'Heavy soils with impervious layers are associated with shales, iron pans with granites, schists and some sandstones ... glacial features such as kames and ridges induce seepage into the lower ground ... (a process) ... greatly facilitated by the occurrence of stratified sands and gravels. When the geology and topography are known, the drainage problems are known too in a general way' (Mulqueen and Gleeson, 1981, p.30).

Groundwater vulnerability maps:
Infiltration, leaching, and the permeability characteristics of soil and rock aquifers in a catchment can be combined to provide a measure of the vulnerability of groundwater supplies to pollution from such sources as farms, factories and landfill sites. A Groundwater Vulnerability Map for Northern Ireland (British Geological Survey, DANI Soil Survey and DoE Environment Service, 1994) combines three broad geological classes of permeability with three broad leaching (or infiltration) soil classes to produce seven classes of groundwater vulnerability. The most vulnerable areas occur where soils with a high leaching potential overlie highly permeable formations such as chalk, Triassic Sandstone and Carboniferous Limestones. Northern Ireland's most vulnerable groundwaters, therefore, are in the Lagan valley between Lisburn and Belfast, in a narrow and discontinuous north-south running belt to the west of Lough Neagh between Tobermore and Armagh, in the lowlands adjoining Upper and Lower Lough Erne, and in the lower reaches of the Roe valley north of Limavady. Much of County Down, the south of County Armagh, all of the Sperrins above the glacial sands and gravels in the lower valleys, and the north-east of County Antrim have little or no vulnerability to groundwater pollution either because the rocks have limited permeability or are protected by thick soils with a low leaching potential.

Clay soils, generally, provide good protection for they have low permeability and a chemical composition which filters out many contaminants and binds them to the clay particles. Care has to be taken, however, even with clay soils. In very dry weather clay soils may crack, and surface water, perhaps containing pesticides, might move down to groundwater. Saturation of a soil's ability to adsorb pollutants can also be reached and 'soil may be converted from a sink to a source of pollutants. The term "chemical time bomb" was first used in the Netherlands to describe the accumulation and subsequent release of phosphate and heavy metals from heavily manured agricultural soils. The concept has now been extended to cover potential long-term and delayed effects of all types of chemical pollution in soils and sediments' (Royal Commission on Environmental Pollution, 1996, p. 27).

(v) Sediment loads in Northern Ireland rivers.
Little is known of the sediment loads in Northern Ireland rivers. Because vegetation cover is extensive

and almost permanent, the agricultural economy is pastoral rather than arable, and rainfall rarely intensive, sediment loads are relatively small. Measured suspended loads on the River Clogh (95km^2) and River Main (205km^2) suggest that loads in excess of 100mg l^{-1} at bankfull stages are rare. (Essery and Wilcock, 1990b). Dissolved load is usually at its most concentrated during low flows (Figure 5.4) since it is principally derived from groundwater flow which has the greatest opportunity to remove materials in solution from catchment soils and bedrock.

A significant effect on river channel processes in Northern Ireland's upland catchments results from rapid mass movements of peat – so-called bog bursts, bog flows and bog slides – which transport large volumes of organic soils and slurries down steep slopes into upland river channels. A very large bog burst occurred in 1993 in the upper headwaters of the Clady River, on the eastern slopes of Carntogher. A heavy daily rainfall of 60mm following a period of sustained wet weather caused highly humified peat to liquefy and flow for a distance of 2.5km downslope and along the flood plain, thereafter being confined to the river channel. Fish kills extended 8km downstream (Wilson, Griffiths and Carter, 1996). The possible environmental impacts of such events include the siltation of spawning gravels and damage to invertebrate and fish populations (Plate 5.1).

B. SOME CHARACTERISTICS OF NORTHERN IRELAND RIVERS

(i) Hydrology

Best practice in river management requires accurate information on river flows and water quality, and Northern Ireland is fortunate in having an excellent network of streamflow gauging stations installed and maintained by the Rivers Agency in the Department of Agriculture for Northern Ireland. The current network is shown in Figure 5.5. Most types of large catchment are well represented in the network, but small upland catchments, in which many damaging flood flows are generated by intense rainfalls (Betts, 1992), are not particularly well represented, nor are catchments in urban areas.

Most river management problems are concerned with extreme flows. Floods cause physical damage to the channel, to agricultural land and to flood plain property. They also have important natural functions, removing debris and pollutants and attracting game fish into the river to spawn. Low flows limit the amount of water available for abstraction and the amount of effluent that can be diluted. In allocating consents to discharge wastes into rivers, therefore, the Environmental Protection Division has to ensure 'that the discharge (of effluent) can be sustained by the receiving waterway without damaging the quality of the aquatic environment or breaching national or EC standards' (Department of the Environment and Department of Agriculture for Northern Ireland, 1993, p.7). Low flows also limit the amount of water available for power generation, an increasingly important consideration if run-of-the-river power generation schemes are to play a significant role in the Government's Non Fossil Fuel Obligation target of 45 MW by 2005 (Energy Technology Support Unit, 1993).

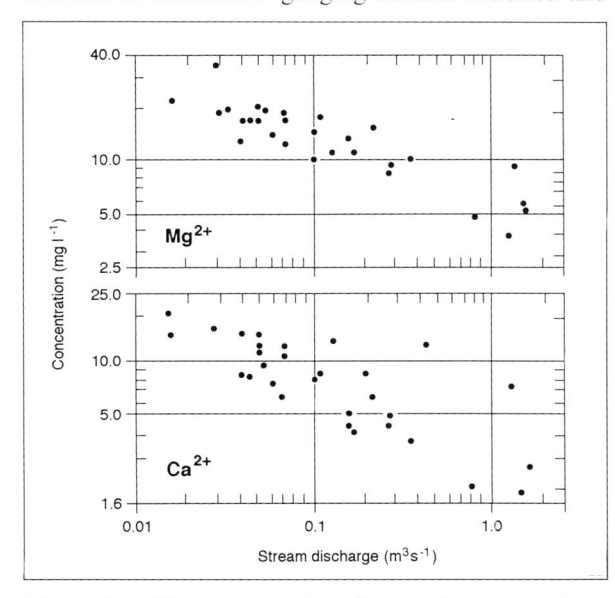

Figure 5.4 The concentration of magnesium and calcium ions in relation to stream discharge on the upper Agivey River, drainage area 15 km^2.

Plate 5.1 A view of the bog flow on Carntogher, September 1993. Note the dense herring bone pattern of drains on the hillside and the downstream extent of peat slurry in the river valley (Courtesy, Nigel McDowell and Peter Wilson, University of Ulster).

Streamflow characteristics are important and under-emphasized ecological factors in river ecosystems. A river basin characterized by permeable soils and geology permits water to infiltrate easily into the soil moisture and groundwater stores. Passage of water through these stores delays runoff, buffers the river and its animal life against the physical effects of high flood flows, and probably sustains a high dry-weather flow relatively rich in dissolved solids. On the negative side, high infiltration rates may make groundwater vulnerable to pollution from agricultural fertilizers. Impermeable soils and/or steep-sided catchments, by contrast, induce rapid runoff, high flood peaks AND low dry weather flows. Such catchments may be vulnerable to bed and bank erosion, to relatively high sediment loads, and to pollution in summer when low flows may be unable to dilute effluent loads from industry and/or agriculture.

Mean daily streamflows in a catchment are graphically summarized by flow duration curves, six of which are illustrated in Figure 5.6. Interesting features of a catchment's hydrology can be deduced from even a cursory examination of such curves. For example, low flows are not always ranked directly in relation to catchment size, the Lagan, with a catchment area at New Forge of 490 km² having lower Q_{90} and Q_{95} flows than the Bush which has a catchment area some 40 per cent smaller. Clearly, other factors than area determine low flow characteristics. Equally clearly, the Lagan has a small dry weather flow with which to dilute urban effluent from the 600,000 people in its catchment, and it is little surprise that the Lagan has a water quality problem.

Extreme flows in different catchments can be standardized with regard to drainage area (Figure 5.7) and Q5/Q95 ratios calculated. These may be used to indicate how much seasonal groundwater transfer takes place in a catchment, low ratios indicating high transfer rates, high ratios indicating limited transfer. In rivers like the Itchen in southern England, where 83 per cent of total annual flow is from groundwater, the Q_5/Q_{95} ratio is 3.5 whereas in the River Ter in Essex, which has an annual groundwater contribution of only 39 per cent, the Q_5/Q_{95} ratio is 25 (Ineson and Downing, 1965). None of the ten catchments illustrated in Figure 5.7 suggest significant seasonal groundwater transfers and it is noticeable that the Main, Six Mile Water and Bush, three of the four

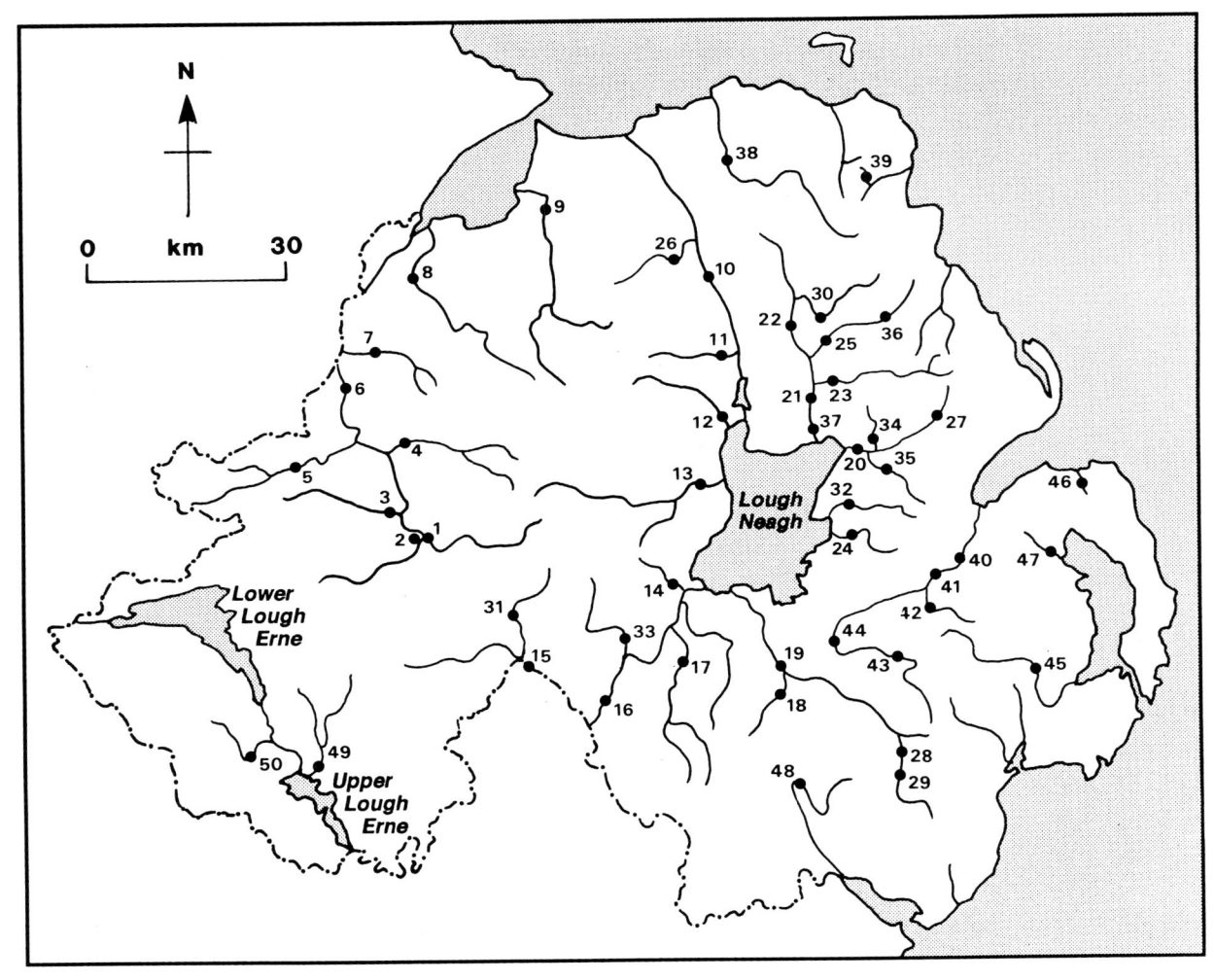

Figure 5.5 The hydrometric network in Northern Ireland. (Source: Rivers Agency, Department of Agriculture for Northern Ireland).

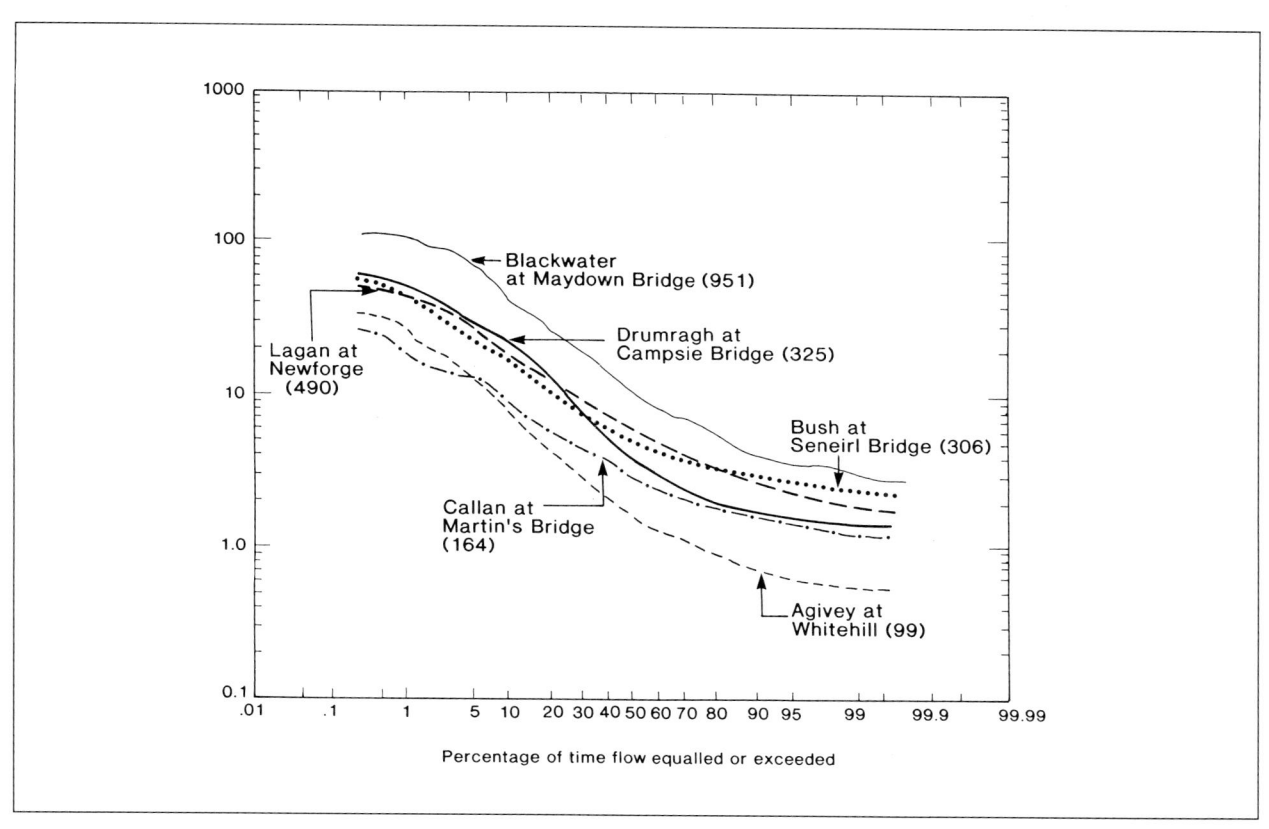

*Figure 5.6 Flow duration curves (January – December, 1993) for six Northern Ireland rivers. Catchment areas in km²
(Source: Rivers Agency, Department of Agriculture for Northern Ireland).*

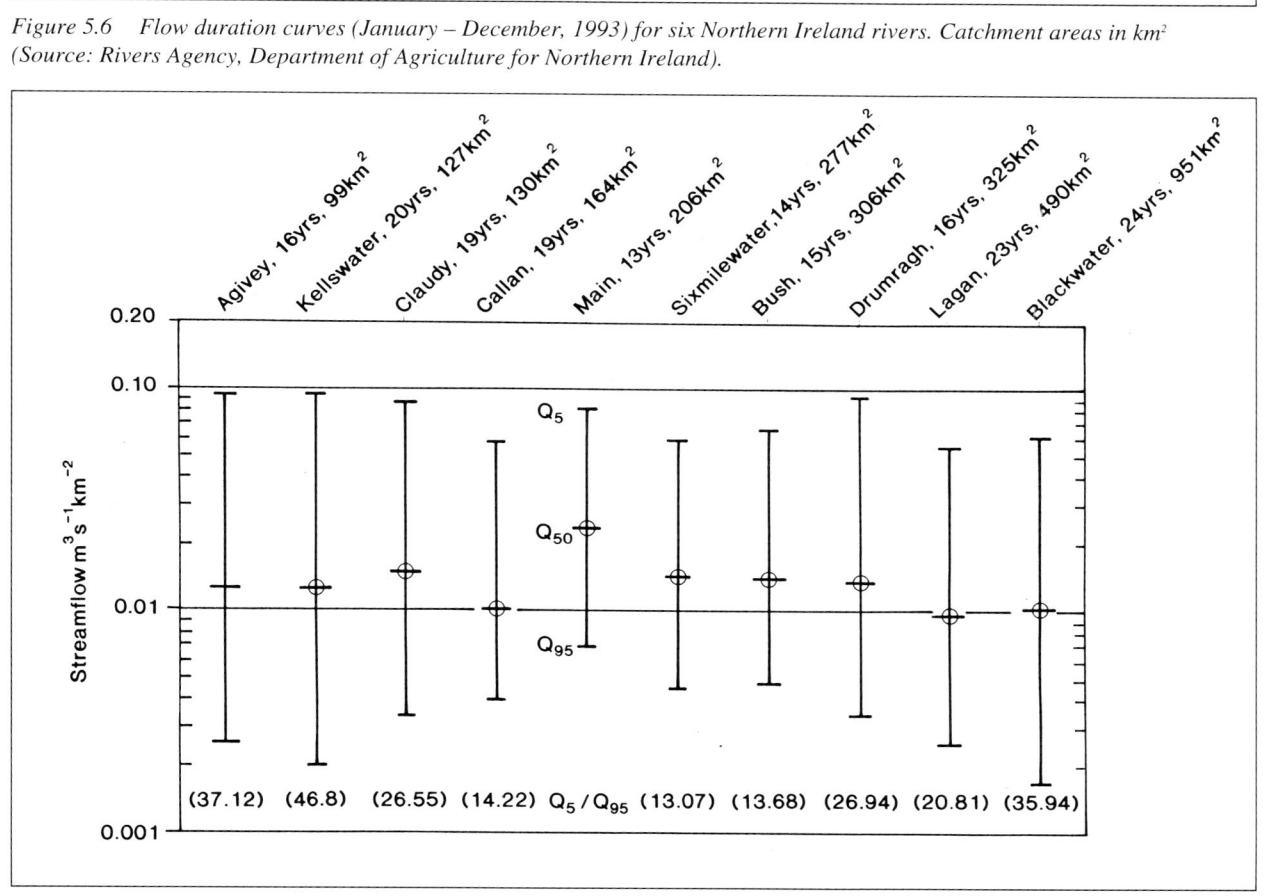

Figure 5.7. Mean annual Q_5, Q_{50} and Q_{95} flows for ten gauging stations in Northern Ireland. The Q_5 flow is the flow equalled or exceeded 5% of the time. This is a high flow but would not normally fill the channel. The Q_{95} flow is the flow equalled or exceeded 95% of the time. This is a low flow and provides little dilution of any effluents. A polluted river experiences its worst water quality at these low flows, and below. Also shown are Q_5/Q_{95} ratios. Note that the percentile flows calculated for each river are not calculated for the same years and that only general comparisons between drainage basins can be based on these data. (Source: Rivers Agency, Department of Agriculture for Northern Ireland.)

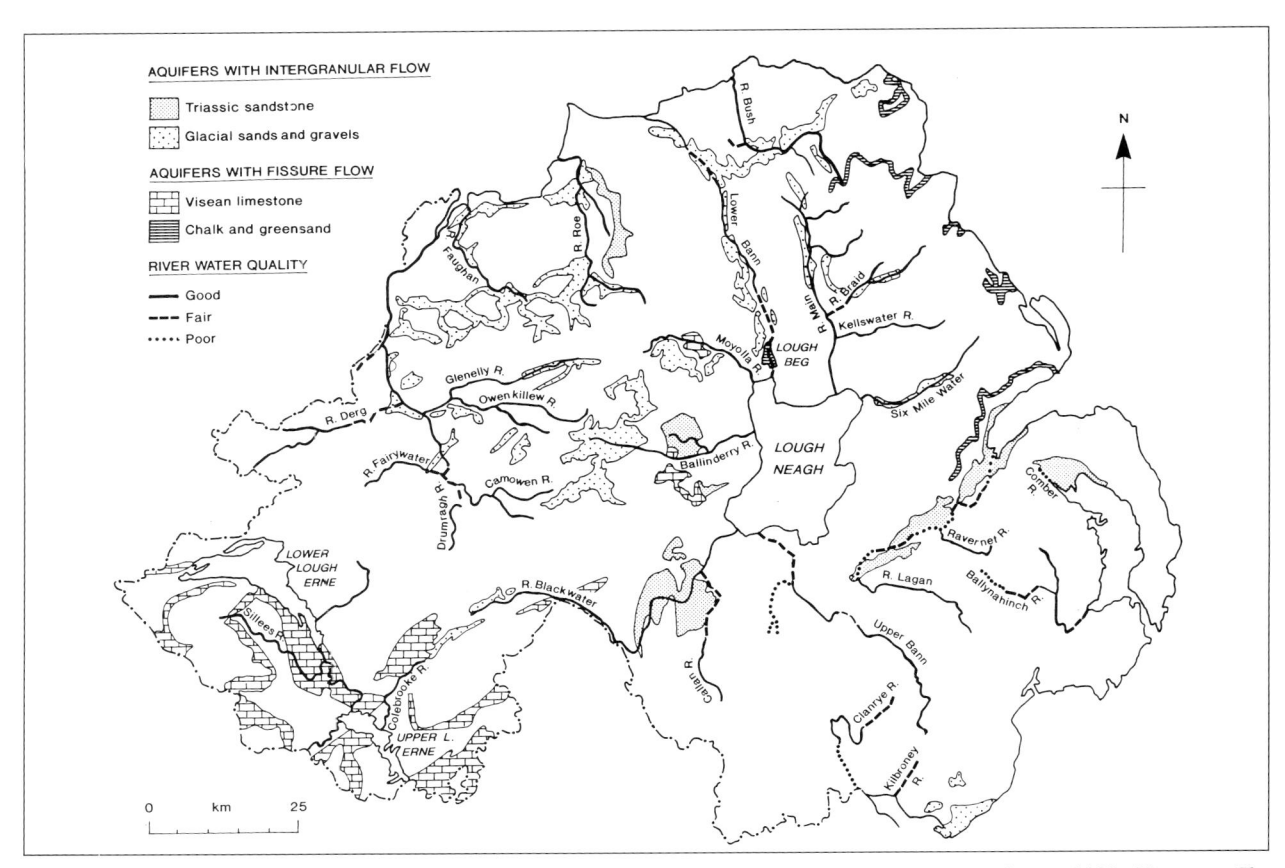

Figure 5.8. A simplified map of the principal aquifers in Northern Ireland and of river water quality in 1985. [Sources: The Hydrogeological Map of Northern Ireland, British Geological Survey and Environment Service, (DoE(NI)), 1994 and River Water Quality in Northern Ireland, 1985, Environmental Protection Division, DoE(NI), HMSO, 1987].

catchments to have Q_5/Q_{95} ratios below 15, contain extensive areas of very permeable glacial sands and gravels (Figure 5.8) which probably account for the high dry weather flows. These sands and gravels are vulnerable to surface pollution, and nitrate concentrations can reach 10mg NO_3–N l^{-1} (British Geological Survey and Environment Service, 1994).

Flow duration curves are important water management tools and attempts have been made throughout the British Isles to categorize and synthesize them for use in ungauged catchments (Institute of Hydrology, 1980). A similar study of 14 catchments in Northern Ireland appeared to show that Q95 is more closely related to topography than to soil, which is further evidence of the general impermeability of Northern Ireland catchments (Wilcock and Hanna, 1987)

Attempts have also been made to predict flood peaks in Northern Ireland from catchment characteristics (Cochrane and Wright, 1983; Hanna and Wilcock, 1984). The standard flood, from which floods of different magnitudes and return periods are usually calculated (Natural Environment Research Council, 1975), is the mean annual flood which has a return period of 2.33 years ($Q_{2.33}$). The mean annual flood is the average of all annual floods in a gauging station record, and an annual flood is the largest individual flood in any particular year. The mean annual flood would overtop the banks of most natural river channels. In their study, Hanna and Wilcock found that drainage area (AREA), channel slope (S1085), annual rainfall (SAAR) and a term measuring soil winter rainfall acceptance potential (SOIL), most successfully predicted the mean annual flood in Northern Ireland. This four variable model takes the form:

$$Q_{2.33} = 2.0999 \times 10^{-4} \, AREA^{\,0.99} \, S1085^{\,0.37}$$
$$SAAR^{\,1.11} \, SOIL^{\,0.96}.$$

It was, incidentally, the imperfections of the SOIL term in the above studies, pioneered by the United Kingdom Institute of Hydrology, which led to the HOST (Hydrology Of Soil Types project) described in Chapter 9C.

(ii) Water Quality

Water quality in Northern Ireland's rivers has been measured by physico-chemical criteria since 1980 (Environmental Protection Division, 1987) and by both physico-chemical and biological criteria since 1990 (Water Quality Unit, 1993). Physico-chemical criteria include dissolved oxygen (DO), biochemical oxygen demand (BOD) and ammonia (as NH4), and are determined from river water samples collected (in 1985) from 132 sampling sites (Table 5.1). Samples in 1985 were collected at fortnightly intervals from 49

Table 5.1 Selected characteristics of the National Water Council's 1978 classification of river water quality used in the 1980 and 1985 classifications of Northern Ireland rivers

Class	Description	DO	BOD	NH_4	Drinking Water	Fish	Amenity Value
					Uses		
1A	Good, unpolluted	>80%	< 3	< 0.4	Yes	Game	High
1B	Good. Cannot be placed in 1A because of high quality effluent, canalization, low gradient, or eutrophication	>60%	< 5	< 0.9	Yes	Game	High
2	Fair. No signs of pollution other than humic colouration & a little foaming below weirs	>40	< 9	n/a	After treatment	Coarse	Moderate
3	Poor	>10	< 17	n/a	No	No	Low grade industrial abstraction
4	Bad	Likely to anaerobic at times		n/a	No	No	Nuisance

DO – Dissolved Oxygen (mg l^{-1}) as a percentage of saturation. BOD – Biochemical Oxygen Demand (mg l^{-1}). NH4 – Ammonia. Rivers are allocated to a given class if 95% of all water samples analysed fall within the relevant thresholds. [Note that these 95 percentile standards are not flow-related].

sites, at monthly intervals from 55 and at quarterly intervals from 28. Collected in this way, samples are not necessarily flow-related and classifications between individual years may not be strictly comparable in that one year's data may accidentally reflect higher flows (and, therefore, greater dilution) than data from another year. But physico-chemical methods are important in providing information on water quality in potential sources of drinking water and potential receiving waters for effluent discharges.

A weakness of physico-chemical methods is that sampling is too infrequent to catch individual pollution events, water quality evidence for which might only last hours or days. It was partly to overcome this latter weakness that biological monitoring has now been introduced to complement physico-chemical methods. Biological monitoring has been used in Ireland for several years (Richardson, 1987) and the Northern Ireland Water Quality Unit now uses RIVPACS, the River Invertebrates Prediction and Classification System (Royal Commission on Environmental Pollution, 1992) to classify river reaches into four classes: good, moderate, poor, and bad. Good rivers are those on which surveyed invertebrate populations are unstressed and close to what would be expected. Moderate quality exists on rivers which show slightly stressed fauna in relation to what would be expected, while poor and bad quality indicate highly and extremely stressed invertebrate fauna.

Results of recent river water quality surveys in Northern Ireland are presented in Table 5.2. Care has to be taken in drawing conclusions about trends from this Table because physico-chemical data presented for 1980 refer to a shorter total surveyed river length then for 1985 and 1991, while rivers in the 1991 biological survey included some not included in the physico-chemical surveys. The early 1980s, however, appear to have witnessed a fall in the percentage of rivers in the highest water quality class (much of it due to individual rivers oscillating between 1A and 1B grades) and an increase in the percentage of rivers in the poor category. Individual rivers giving cause for concern in 1985 were: the lower Fairywater, 'dropped to Class 3 due to depressed dissolved oxygen levels the reason for which is unknown' (Environmental Protection Division, 1987, p.7); the Glenelly and Derg, dropped to Class 2, again for unknown reasons;

Table 5.2. Physico-chemical and biological water quality in Northern Ireland rivers, 1980–1992

Physico-chemical Class	Biological Class	1980 km	%	1985 km	%	1991 km	%
1A	Good	291	23.3	204	14.7	386	27.8
1B	Good	767	61.4	954	68.6	764	54.9
2	Fair	158	12.6	157	11.3	222	16.0
3	Poor	34	2.7	75	5.4	18	1.3
4	Bad	0	0	0	0	0	0
	Good					1227	54.4
	Moderate					774	34.4
	Poor					216	9.6
	Bad					36	1.6

the Ballybay river (Class 3); middle reaches of the Ravernet (Class 2 / Class 3) due to sluggish flow, prolific weed growth and deoxygenation; the Ballynahinch river, partly polluted by Ballynahinch sewage treatment works; the Kilbroney River, due to industrial discharges; the Comber River; and, of course, the lower reaches of the Lagan, for reasons already mentioned. Perhaps the main reason for the apparent deterioration between 1980 and 1985 was the 77 per cent increase (from 388 to 689) in the number of agricultural pollution incidents over the same period, due principally to silage effluent and the winter spreading of slurry on saturated, impermeable soils. Slurries sometimes have BOD values above 4,000 mg l^{-1} while silage BOD values can exceed 50,000 mg l^{-1}. These BOD values illustrate the importance of sustaining dry weather streamflow in order to provide an acceptable level of dilution in rivers and lakes.

Between 1985 and 1991, corresponding improvements in water quality appear to have taken place, perhaps due to improved farming practices following pollution awareness campaigns by Environment Service, DANI and officers of the Fisheries Conservancy Board and Foyle Fisheries Commission. Between 1987 and 1992 the number of agricultural pollution incidents in Northern Ireland decreased from 568 $year^{-1}$ to 247 $year^{-1}$. A noteworthy feature of Table 5.2 is the 11.2 per cent of rivers allocated to the 'poor' and 'bad' categories by the biological monitoring system as opposed to the 1.3 per cent allocated on the basis of physico-chemical criteria.

Lakes represent one of Northern Ireland's principal amenity assets. Altogether, there are 1668 lakes in the Province covering 4.4% of the land area. Three-quarters of these lakes are less than 2ha in area (Smith et al, 1991). Concern for water quality in Northern Ireland's lakes has principally focussed on Lough Neagh and upper and lower Lough Erne. Lough Neagh, the largest freshwater lake in Britain and Ireland, has been studied intensively for the past twenty-five years. During this century it has become very eutrophic. Early studies (Wood and Gibson, 1973) described the blue-green algae crops which made the Lough unsightly, blocked water treatment filters, and lowered oxygen levels in the late 1960s. The importance of Lough Neagh as a water supply, as well as its role as an eel fishery and recreational amenity stimulated intensive research which soon identified urban point sources of phosphorus as the major problem. Tertiary sewage treatment, to remove P, was implemented at nine major population centres in the Lough Neagh basin in 1981 and appeared to improve the phosphorus loadings on the Lough, the mean concentration of which fell from .063 mg l^{-1} in 1981 to .036 mg l^{-1} in 1987. Oxygen levels as low as 3mg l^{-1} have been recorded in Lough Erne (Jordan and Smith, 1987).

The management of water quality in lakes and rivers is becoming increasingly important because of the Nitrates Directive (91/676/EEC) and the Urban Waste Water Directive (91/271/EEC). These Directives respectively require member states to identify areas 'vulnerable' to nitrate pollution and water bodies 'sensitive' to urban waste water pollution, principally from phosphates. Methodologies to define such areas in Northern Ireland have recently been developed (Environment Service, 1993). Studies of similar problems in Pennsylvania emphasize the importance of soil hydrology, particularly high surface runoff patterns within a catchment, as the critical controls on phosphorus exports. Factors affecting such patterns are 'soils, topography, groundwater and moisture status over the watershed' (Zollweg et al, 1996, p. 31). The study concludes that the attainment of water quality objectives in a catchment is not achieved most efficiently by chemical and/or land use management applied over the entire watershed area but by directing control towards the more localized source areas of surface runoff: '... Control directed toward these specific landscape positions offers the greatest and most efficient opportunity for controlling P export from upland agricultural watersheds' (Zollweg et al, 1996, p.38).

The importance of local hydrology and soil conditions is also important in the siting of septic tanks, effluents from which 'must not be discharged direct to a river or aquifer without further treatment, such as passage through an unsaturated soil zone' (Wood and Sheldon, 1980, p.199). Fulfilling strict hydrological criteria like these ideally requires soil texture analyses, and both groundwater monitoring and piezometric studies in the area of proposed septic tanks, before licenses should be approved.

C EVOLVING PRIORITIES IN THE MANAGEMENT OF NORTHERN IRELAND'S RIVERS

The early priority : agricultural improvement

Given Northern Ireland's large areas of impermeable soils, high annual rainfall ($c.1,200$mm year^{-1}), low evapotranspiration ($c.400$ mm year^{-1}), and the gentle gradients of many of its major rivers (the Bann, for example, falls only 15m in the 64km between Lough Neagh and its estuary at Coleraine) it is unsurprising that agricultural output was limited until the mid twentieth century and the arrival of systematic arterial and field drainage.

Throughout the nineteenth century (Lyn, 1980; Dooge, 1987) and for most of the twentieth century, the need for agricultural improvement has been the main driving force in river management. Poor natural drainage and flooding are related but separate aspects of surplus water in the soils of Northern Ireland and strategies to improve agricultural output therefore involve (i) field drainage to evacuate water from soils (ii) ditching and/or ditch clearance to provide the connectors between field drains and rivers and (iii) arterial drainage, or channelization, schemes to prevent inundation of low-lying land and flood plains. Other economic factors, of course, have been important in river management : the need for river and canal transport, and for water and power supplies in rural areas. Hydro-power required the building of dams and weirs on many of our major rivers (Gribbon, 1969), many of which only compounded problems of drainage caused by low channel gradients.

The first major drainage scheme of the twentieth century in Northern Ireland was on the lower Bann between 1930 and 1942. Designed to give drainage benefit to 57,000 ha of agricultural land by lowering the level of water in Lough Neagh, the scheme lowered outfalls on all major rivers flowing into Lough Neagh and increased bankfull flows in the lower Bann from 330 m^3 sec^{-1} to 425 m^3 sec^{-1}. Altogether, some 3.8 million m^3 of bed and bank material were excavated from the river (Wilcock, 1979). The second world war further emphasized the need for increased agricultural production, and the origin of an integrated government strategy linking arterial drainage to field drainage can be traced back to 1944 : 'If farmers cleared their own land and removed the danger of flooding in their own area, the quicker flow of water would often inundate the land of farmers downstream and result in acrimony and even litigation ... It (is) obvious that local schemes must be related to one another and that the only efficient method of restoring the river system would be to begin works at an outfall near the sea and carry them backwards to the source' (Government of Northern Ireland Parliamentary Paper, 1944, p.189).

To implement this strategy the 1947 Drainage Act established a Drainage Council to determine the 'main' rivers to be improved as the Bann had been. Engineering work started at the mouth of each determined 'main' river and moved progressively upstream, thus providing efficient outlets for any increased streamflows produced by subsequent channel improvements on tributary rivers and/or field drainage. By 1977, 1,250km of main river had been improved and the 1964 Drainage Act initiated the second phase of arterial drainage on tributary streams, the so-called 'minor' rivers. To date, more than 6,000 km of arterial drainage works have been completed, of varying intensities in different catchments (Department of Agriculture for Northern Ireland, 1991). In contrast to arterial drainage, all of which has been implemented by the Department of Agriculture, field drainage has always been the responsibility of individual farmers, aided by a succession of grant-aid schemes between 1929 and the early 1980s. Under these schemes, about half of the total agricultural land in Northern Ireland has received some form of field drainage since 1947 (Green, 1979).

Three types of field drainage problem are recognised by Galvin, 1979: (i) artesian seepage or springs, (ii) high water-table and (iii) impermeable soils. Artesian seepage occurs where water is forced to the ground surface under pressure. In these highly localized situations, deep drains are often required. High water-table problems can occur, even in permeable soils, when ground is low-lying relative to river outfalls and water table gradients are low. Drainage problems usually set in if water-tables rise to within 0.5 m of the ground surface. Empirical formulae for the design of pipe drainage systems in such situations require information on hydraulic conductivities, the thickness of the permeable layer relative to the impermeable layer, the minimum depth at which water-tables are to be controlled, and the design drainage coefficient – the amount of water to be removed day^{-1}. This is usually up to about 25mm. Impermeable soils may require pipe drains no deeper than 75cm and 10m apart. Drains in permeable material may be 1m deep and up to 40m apart (Robinson, 1990).

Impermeable soils represent the most extreme drainage problems since their drainable pore space may be only 3–10 per cent in comparison with 20 per cent–30 per cent in more permeable soils. Impermeable soils thus retain large amounts of water and soil disruption is often necessary to increase pore space. Methods include *subsoiling* and/or *ripping, mole drain installation and deep ploughing.* Subsoiling and ripping involve mechanically rupturing the soil to create fissures and the loosening of compacted areas, especially iron pans. Mole drains are pipeless channels about 75mm in diameter drawn through the soil at about 50cm depth (Figure 5.9). Impermeable clay soils are ideal for this form of drainage since their high clay content (often > 35%) gives them considerable structural strength which prevents collapse. Where collapsing might occur, the mole can be filled with gravel to extend its working life, possibly up to 15 years or more.

The polders at Myroe, Ballykelly, Longfield and Black Brae on the south side of Lough Foyle also represent part of the attempt to improve agricultural output. Originally constructed in the 1850s at a time of great social and economic stress, these sea embankments have a combined total length of 16 km, an average height of about 3m, and protect from the sea some 1,700 ha of high quality agricultural arable land as well as the railway line from Belfast to Londonderry. They are exposed to a wind fetch of 11 km and after 140 years show signs of deterioration. The 4.4 km Myroe stretch of the embankment was refurbished in 1995 at a cost of £2.5 million (Private communication, Rivers Agency).

The land drainage policy implemented after 1947 produced drier soils, better trafficability and a longer growing season. But drainage alone did not improve soil structure and fertility. In the 1940s, the then Ministry of Agriculture estimated that 931,000 ha in the Province were lime-deficient and an Agricultural Lime Scheme was introduced, grant-aiding farmers for up to 60 per cent of the costs of liming their soils. Lime usage rose from 100,000 tons year^{-1} in 1947 to a peak of 800,000 tons year^{-1} in 1962, by which time 57 per cent of the original lime-deficiency had been made good. The final component of the agricultural improvement strategy involved application of fertilizers to raise soil fertility level. At its peak in 1973 the average application of fertilizers in Northern Ireland was 294kg ha^{-1} year^{-1} of which the nutrients nitrogen (N), phosphorus (P_2O_5) and potassium (K_2O) accounted for 35 per cent (Wilcock,1979). These figures are very similar to those for the same period in England, Wales and Scotland (Green, 1974).

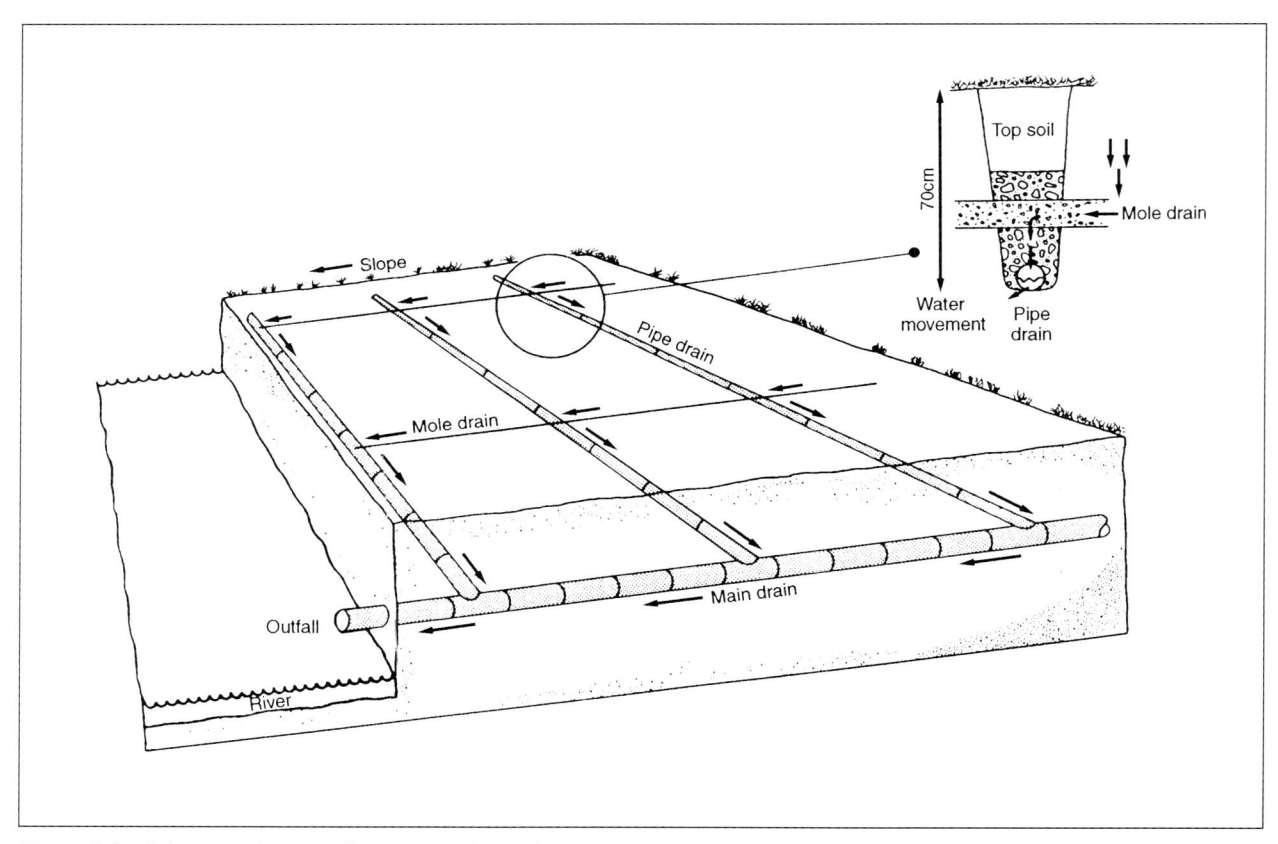

Figure 5.9 Schematic diagram illustrating relationships between pipe and mole drainage in relation to outfalls into a river channel.

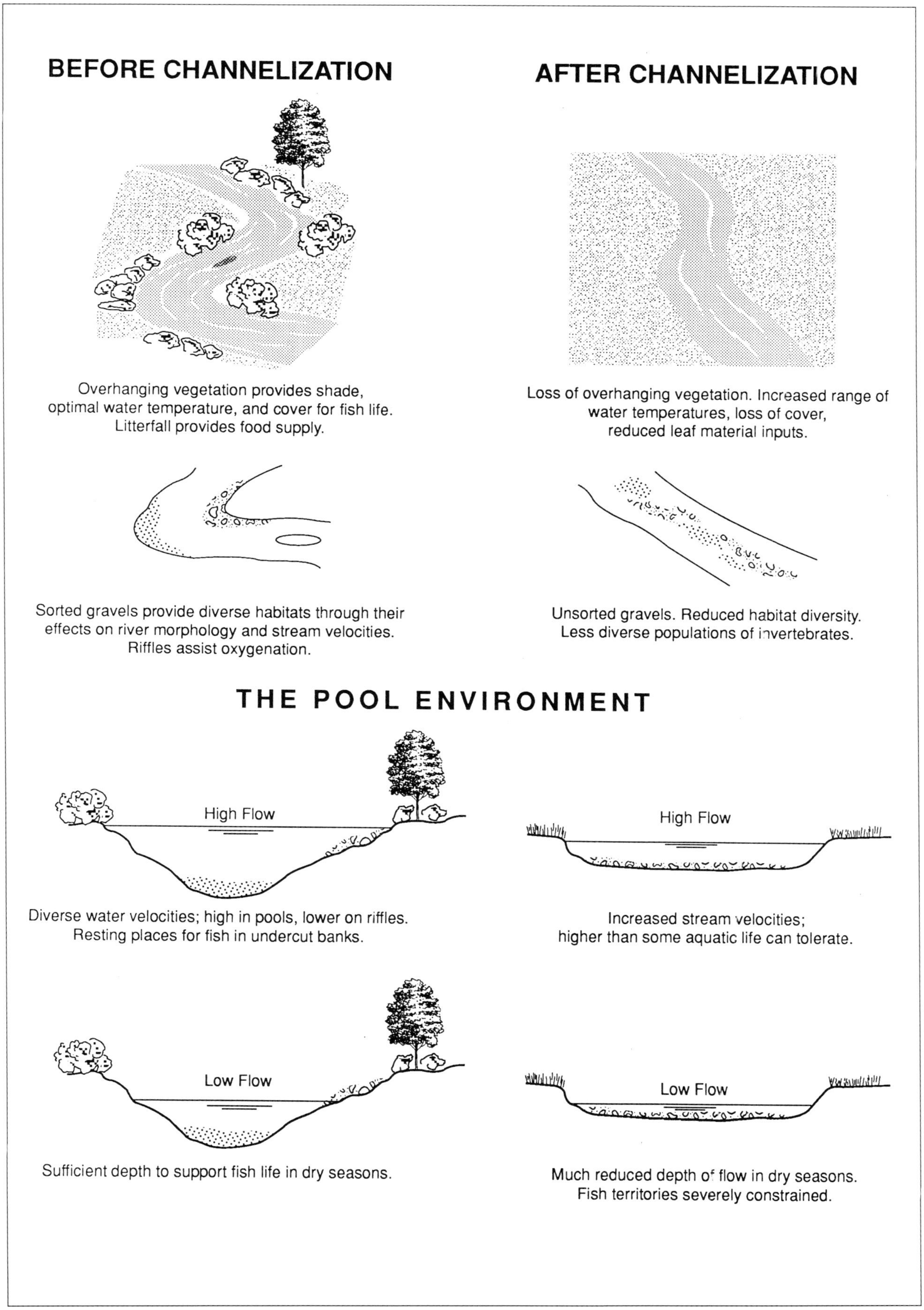

BEFORE CHANNELIZATION

Overhanging vegetation provides shade, optimal water temperature, and cover for fish life. Litterfall provides food supply.

Sorted gravels provide diverse habitats through their effects on river morphology and stream velocities. Riffles assist oxygenation.

AFTER CHANNELIZATION

Loss of overhanging vegetation. Increased range of water temperatures, loss of cover, reduced leaf material inputs.

Unsorted gravels. Reduced habitat diversity. Less diverse populations of invertebrates.

THE POOL ENVIRONMENT

High Flow

Diverse water velocities; high in pools, lower on riffles. Resting places for fish in undercut banks.

High Flow

Increased stream velocities; higher than some aquatic life can tolerate.

Low Flow

Sufficient depth to support fish life in dry seasons.

Low Flow

Much reduced depth of flow in dry seasons. Fish territories severely constrained.

Figure 5.10 Schematic diagram illustrating flood plain, bedload, and channel morphology in relation to stream discharge before and after channelization (Source: Brookes, 1988).

A new priority; environmental quality

The intensification of Northern Ireland's agriculture, underpinned by the drainage strategy outlined above, had many environmental consequences, as elsewhere in Britain and Ireland. The general hydrological, geomorphological and ecological impacts of channelization on rivers (Figure 5.10) are outlined by Brookes (1988). The effect of arterial drainage on streamflow is to increase the size of flood peaks (Bailey and Bree, 1981), though following arterial drainage larger floods than formerly can usually be accomodated within the enlarged channel. The effects on low flow are more difficult to establish. A recent attempt uses modelling techniques to simulate streamflow and suggests that hydrograph recession limbs are steeper following arterial drainage than before (Wilcock and Wilcock, 1995).

The effects of field drainage may be to delay surface and rapid subsurface runoff in impermeable soils and to accelerate it in more permeable soils. Field drainage may thus modify or amplify the effects of arterial drainage, much depending on soil type and location of field drainage within the catchment as well as catchment size (Robinson, Eeles and Ward, 1990). Arterial drainage also increases suspended sediment loads dramatically in the short term (Wilcock and Essery, 1991) and reduces the area of flood plain wetland. A delayed environmental impact of the agricultural intensification in Northern Ireland since 1947 may be the increasing amounts of soluble reactive phosphorus (SRP) recently detected in rivers of the Lough Neagh basin. These may be due to small but persistent leakages of SRP from steadily – accumulating levels of phosphorus in the soils derived from fertilizer applications over many years (Foy *et al.* 1995).

By the mid 1980s, policy towards land drainage began to change due to a combination of factors: 'concern about the environmental impact of drainage; the Government's desire to reduce public expenditure; doubt about the economic viability of agricultural drainage projects; and concern that drainage schemes were contributing to chronic European Community food surpluses' (Williams and Browne, 1987, p.8). A new emphasis on the management of rivers from a broader perspective than drainage alone resulted in the use of river ecological surveys to guide engineering works (Department of Agriculture, 1991) and recent attempts to integrate water quality and ecologically-enhancing design elements into river engineering works on the Blackwater and other rivers (Higginson and Johnston, 1989: Royal Society for the Protection of Birds *et al*, 1994).

Elsewhere in the United Kingdom, attempts are now being made to manage streamflow, water quality and river ecology within the framework of integrated catchment plans (Gardiner, 1991). In Northern Ireland, three non-statutory water quality management plans are currently being prepared for the Lagan, Foyle and Erne catchments, and will 'relate identified catchment uses to water quality standards and objectives' (Water Quality Unit, 1993, p. 28). Soil characteristics and processes clearly affect the amount and quality of water in our rivers, and, hopefully, the soil data now available for all of Northern Ireland through the HOST project will be integrated into catchment water management strategies at the earliest opportunity.

References in Appendices

CHAPTER 6

LAND COVER – BASED ON CORINE LAND COVER PROGRAMME

DR. R. W. TOMLINSON, SCHOOL OF GEOSCIENCES, THE QUEEN'S UNIVERSITY OF BELFAST

Introduction: land cover as seen from space

In the late 1980s and early 1990s, the European Union (EU) began to assemble data on the environment of Europe under the CORINE project (Co-ordination of Information on the Environment). Mapping land cover of the EU and adjacent countries to a common system formed part of this project and was based on interpretation of LANDSAT TM images at a scale of 1:100,000. Ireland joined this mapping programme in 1992 (O'Sullivan *et al*, 1994). Such satellite images can provide a continuous record of the land cover of the world, although if only reflected energy is sensed, areas with frequent cloud cover, as in Ireland, have few cloud-free, useful images. The most recent, mainly cloud-free images of Ireland suitable for land cover mapping, were from May 1990; only small areas required patching with 1989 data.

The classification is designed to include the diversity of land cover in the EU and adjacent countries. This large area has several climatic regions and differing topography so that the content (plant species etc) of an individual class will vary, but in all countries the structure of that land cover class must be sufficiently similar for an area to be allocated to it rather than to other classes. The minimum area delimited (the polygon) is 25 ha and boundaries must be placed within 1 mm of where they are seen on the image. Polygons from the completed image interpretation are then digitized into a GIS and labelled. This allows maps of land cover to be produced and data to be manipulated for specific demands.

The CORINE classification is hierarchical, with 5 classes at Level 1 (Table 6.1). These are divided into 44 classes by Level 3, which is the level of interpretation required by the EU. Only 33 of the Level 3 classes were used in Northern Ireland; the redundant classes in part were caused by the environmental differences between the Mediterranean region and the Atlantic northwest of Europe. Inadequacies in the classification for particular countries can be overcome by subdividing classes to Level 4. This facility was adopted in Ireland for pasture (2.3.1) and for peat bogs (4.1.2), both of which are extensive and variable in Ireland.

Agricultural land

Arable land 2.1.1 is recognized on the May images by the blue colour of fields of bare soil (Colour Plate 20) and the brownish-red of winter sown cereals. However, the apparently bare fields include not only spring sown cereals and land for root and horticultural crops, but also some which are resown to grass; hence there is some inflation of the area classed as arable. In any arable landscape, not all fields will be for crops, some will be in pasture. There is a need for a threshold beyond which an area is no longer classed as arable. In the CORINE programme in Ireland, if more than 25 per cent of an area was pasture then it was classed as 2.4.2 *complex cultivation patterns*. With the exception of the Foyle – Mourne valley and the polders north of Limavady (for locations mentioned in the text see Figure 6.1), parcels of arable land greater than 25 ha are rare in the west of the Province (Figure 6.2). In the Lower Bann and Main valleys there is a scattered distribution which continues down the west side of Lough Neagh as far as Cookstown. The remainder of the arable land is in Co. Down where it forms a ring around the outer edge of the county. From the horticulture near Comber, arable land continues westward along the Lagan valley, then south to Newry and east to Lecale. Finally, in the Ards peninsula it is associated often with large estates, such as at Ballywalter.

Class 2.4.2. *complex cultivation patterns*, which in Northern Ireland is a mix of arable and pasture fields, occupies a greater area than the arable (2.1.1), but follows a similar distribution (Figure 6.3). It is extensive in Co. Down, following CORINE it is almost 37 per cent of the area, but is absent from the centre of the county where rankers and shallow brown earths occur on Silurian shales, together with surface water gleys on silty-clay, drumlin tills. The Mourne plain is mainly within this class.

Class 2.4.1. *annual crops associated with permanent crops*, was used for the apple orchards of

Table 6.1 CORINE classification of land cover (classes not used in Ireland in italics)

Level 1	Level 2	Level 3
1. Artificial surfaces	1.1. Urban fabric	1.1.1. Continuous urban fabric
		1.1.2. Discontinuous urban fabric
	1.2. Industrial, commercial and transport units	1.2.1. Industrial or commercial units
		1.2.2. Road and rail networks and associated land
		1.2.3. Sea ports
		1.2.4. Airport
	1.3. Mines, dumps and construction sites	1.3.1 Mineral extraction site
		1.3.2 Dump
		1.3.3 Construction site
	1.4. Artificial non-agricultural vegetated areas	1.4.1 Green urban areas
		1.4.2 Sport and leisure facilities
2. Agricultural areas	2.1. Arable land	2.1.1. Non irrigated arable land
		2.1.2. Permanently irrigated land
		2.1.3. Rice fields
	2.2. Permanent crops	*2.2.1. Vineyards*
		2.2.2. Fruit trees and berries plantations
		2.2.3. Olive groves
	2.3. Pastures	2.3.1. Pastures
	2.4. Heterogeneous agricultural areas	2.4.1. Annual crops associated with permanent crops
		2.4.2 Complex cultivation patterns
		2.4.3 Land principally occupied by agriculture with significant areas natural vegetation
		2.4.4 Agro-forestry areas
3. Forest and semi-natural areas	3.1. Forests	3.1.1. Broad-leaved forest
		3.1.2. Coniferous forest
		3.1.3. Mixed forest
	3.2. Scrub and/or herbaceous vegetation associations	3.2.1. Natural grassland
		3.2.2. Moors and heathlands
		3.2.3. Sclerophyllous vegetation
		3.2.4. Transitional woodland-scrub
	3.3. Open spaces with little or no vegetation	3.3.1. Beaches, dunes, sand
		3.3.2. Bare rocks
		3.3.3. Sparsely vegetated areas
		3.3.4. Burnt areas
		3.3.5. Glaciers and permanent snowfields
4. Wetlands	4.1. Inland wetlands	4.1.1. Inland marshes
		4.1.2. Peat bogs
	4.2. Coastal wetlands	4.2.1. Salt marshes
		4.2.2. Salines
		4.2.3. Intertidal flats
5. Water bodies	5.1. Continental waters	5.1.1. Stream courses
		5.1.2. Water bodies
	5.2. Marine waters	5.2.1. Coastal lagoons
		5.2.2. Estuaries
		5.2.3. Sea and ocean

Level 4 subdivision in Ireland:-
2.3.1. Pasture: 2.3.1.1. High productivity. 2.3.1.2. Low productivity. 2.3.1.3. Mixed productivity
4.1.2. Peat bogs:- 4.2.1.1. Not exploited (image). 4.2.1.2. Exploited, image evidence

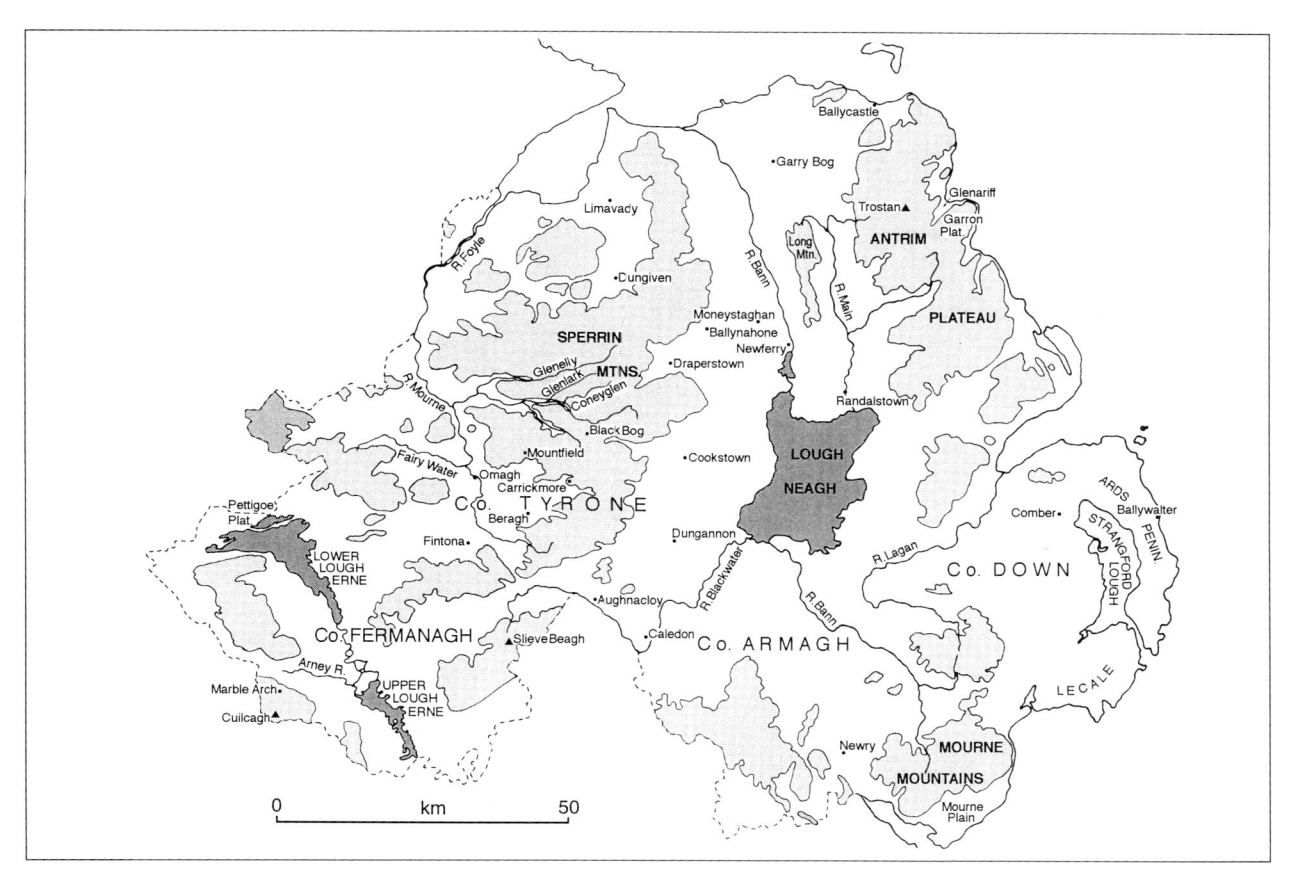

Figure 6.1 Location of places mentioned in Chapter 6.

Co. Armagh, where orchards with an understorey of grass are frequent amongst mainly pasture fields. This area is distinct, but within the broad band of 2.4.2 which extends from the Lagan valley westward to a line running from Dungannon south to Caledon.

Pasture (2.3.1) is the only Level 2 class in the CORINE classification that is not divided at Level 3. There is a considerable extent of pasture land in Northern Ireland – according to CORINE 53 per cent of the total area and 76 per cent of the agricultural area – and it is known to vary in quality. The interpretation team used image evidence to divide the pasture at Level 4 into three types. *High productivity grasslands (2.3.1.1)* are sown or well managed and have a uniform sward. They are shown on the image in a strong red colour, indicating high levels of near infra-red reflectance. *Low productivity grasslands (2.3.1.2)* have a predominance of sown grass, but there can be significant rushes, thistles, nettles and ragwort which might reduce near infra-red reflectance. This can be lowered also by dampness, which additionally diminishes middle infra-red reflectance, or by shortness of the sward – perhaps due to heavy grazing or recent cutting (though less likely in early May) – as well as to physical factors such as thin, dry soils. These pastures show on the image as pale greeny-yellow or, in the case of dampness, by a greyness to the red colour. Areas (<25 ha) with an intimate mix of fields of the two main pasture types are classed as 2.3.1.3.

High productivity pastures (2.3.1.1), as defined in CORINE, may be found throughout the lowland areas of Northern Ireland (Figure 6.4a). However, it is noteworthy that south Antrim, the valleys of the Main, Lower Bann and Strule, an area centred on Cookstown and another around Fintona – Beragh, are predominantly in this class of pasture. Pastures of low productivity (Figure 6.4b) have a scattered distribution as a result both of local physical and management factors. Concentrations may be noted on the shallow, rocky soils of central Down, in the extreme west and south of Fermanagh, to the north of Lower Lough Erne and at the foothills of the Sperrins. Those areas with a patchwork of high and low productivity pastures (2.3.1.3) are distributed widely, as might be expected from the definition.

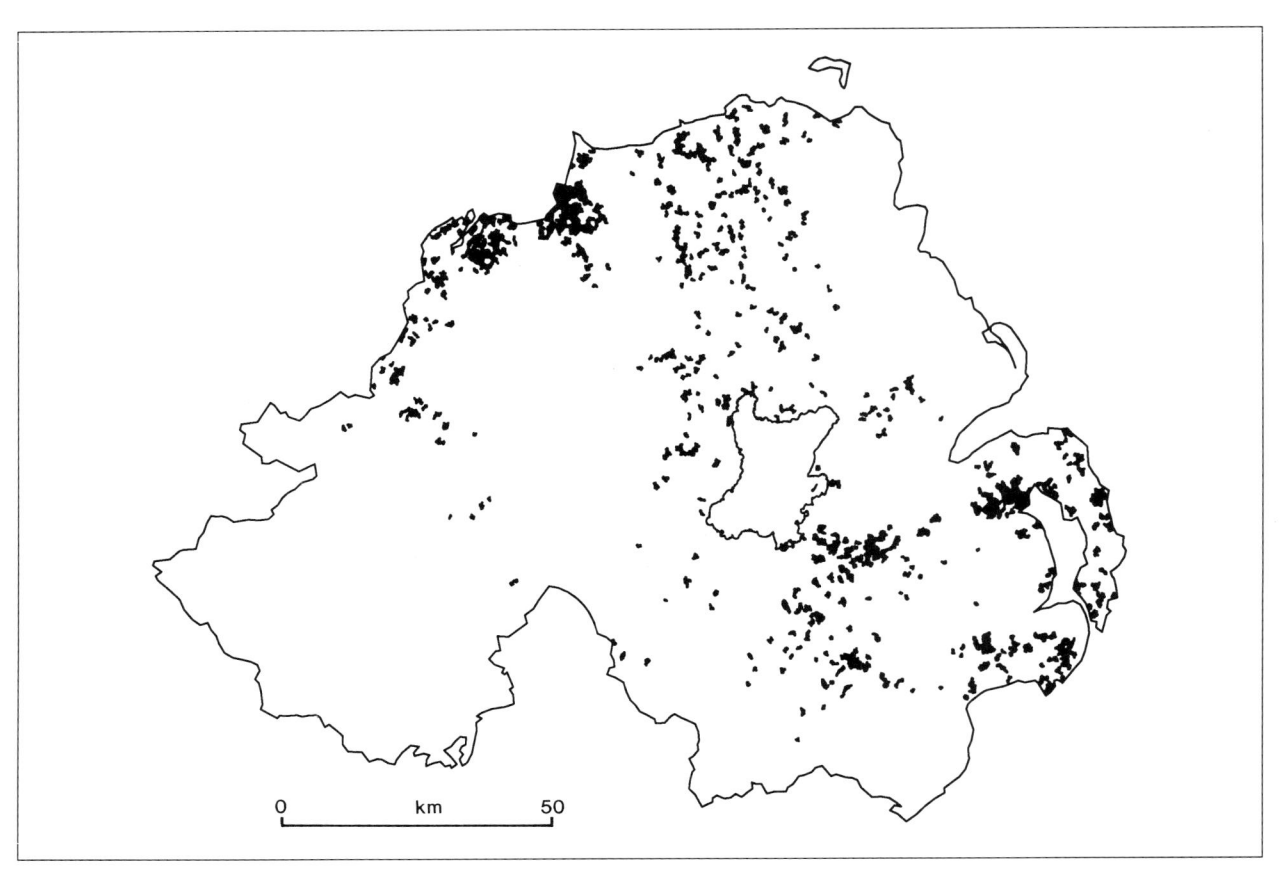

Figure 6.2 Arable land, CORINE 2.1.1

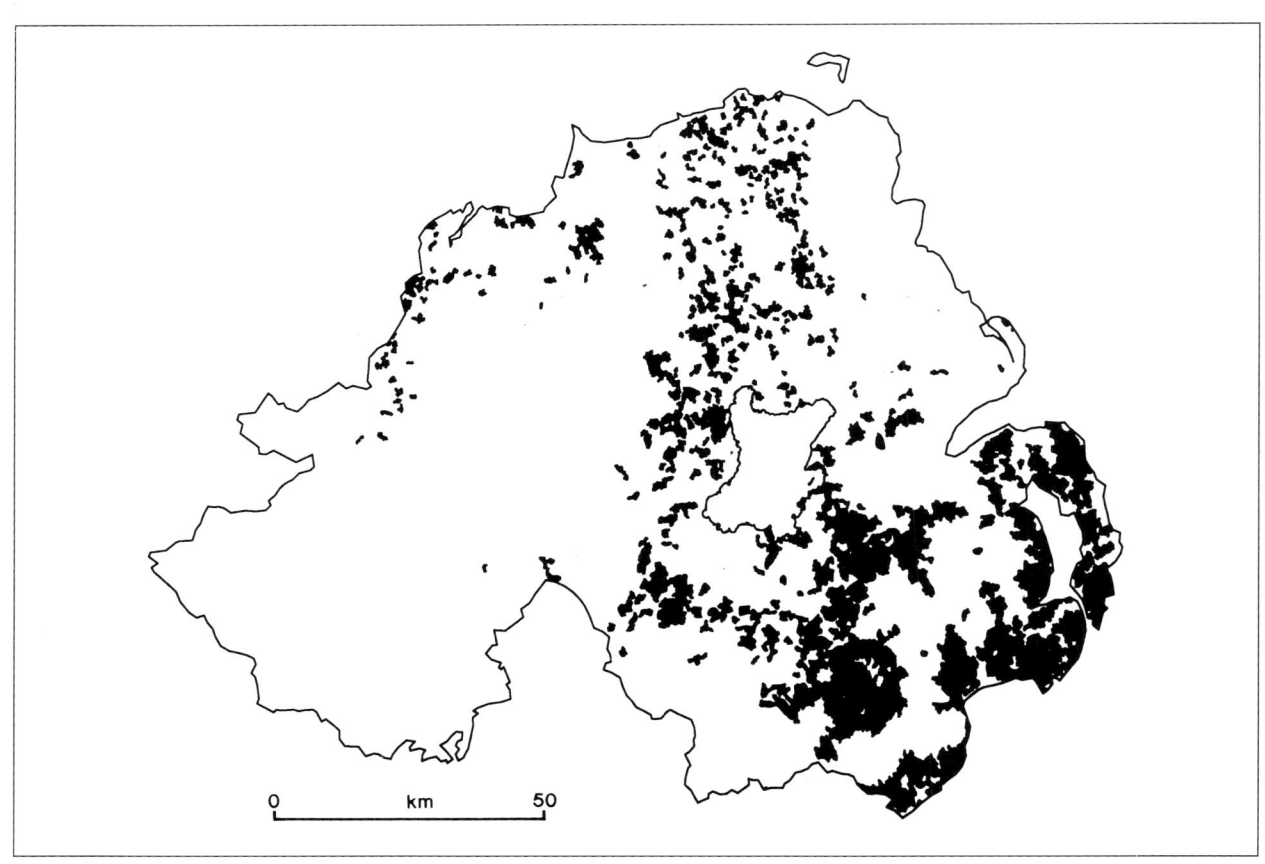

Figure 6.3 Complex cultivation patterns, CORINE 2.4.2

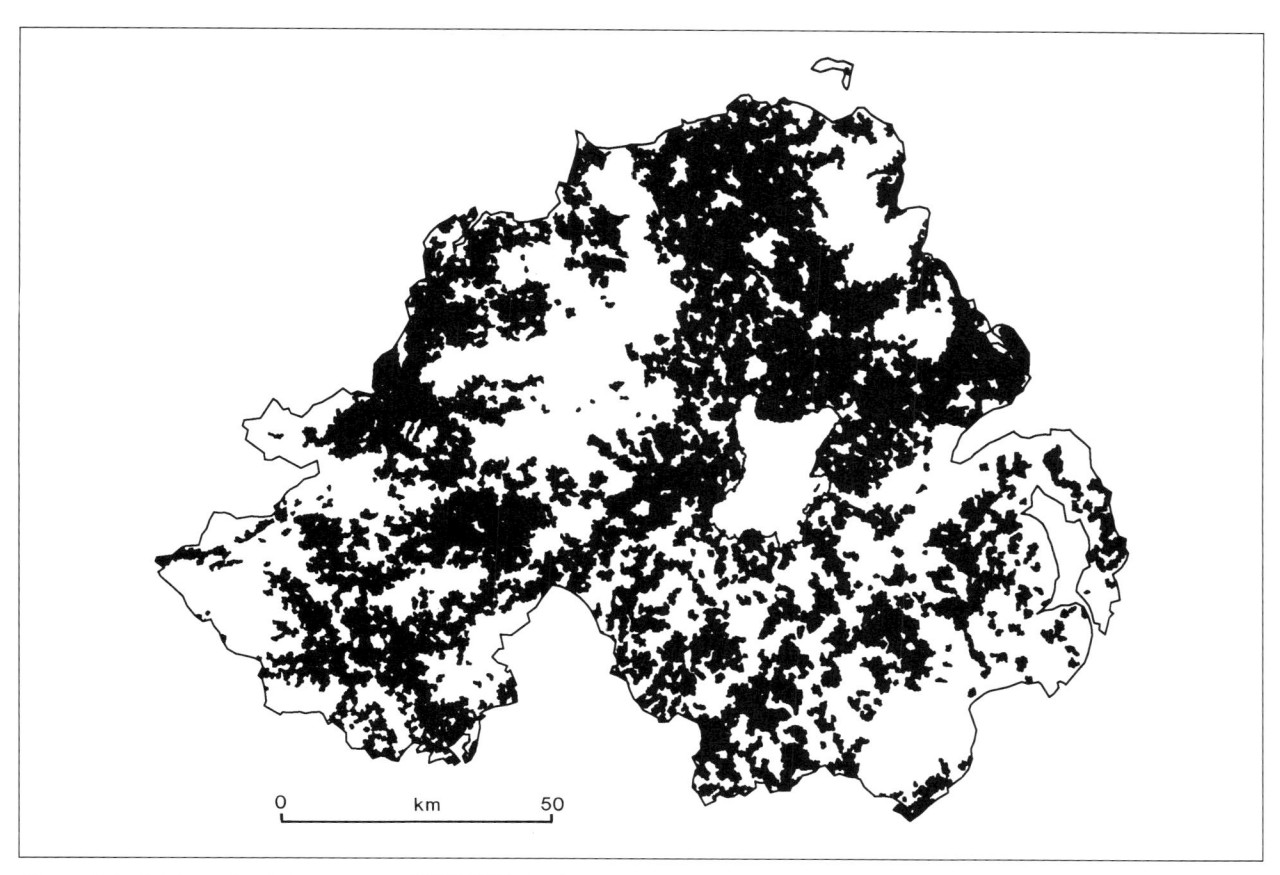

Figure 6.4a High productivity pastures, CORINE 2.3.1.1

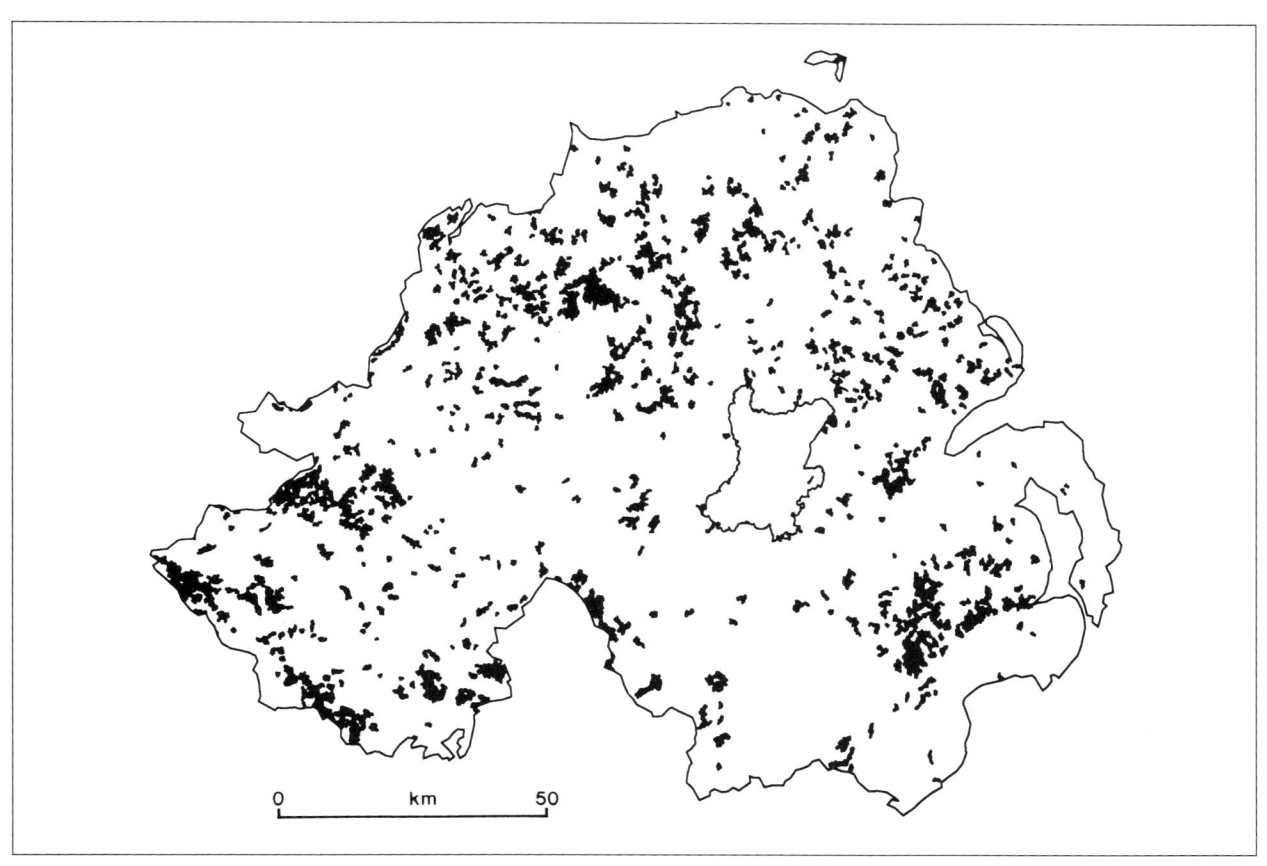

Figure 6.4b Low productivity pastures, CORINE 2.3.1.2

Comparison of the CORINE land cover with a simplified Soil Map 21 (Figures 6.5a, 6.5b) shows some relationship between soil types and the land cover in east Down.

The horticultural area east of Comber occurs on some of the best soils in Northern Ireland (brown earths [BE] and gleyed soils [G1, SWG1] on sands and Triassic Sandstone till). Other arable areas (2.1.1) and mixed arable and pasture (2.4.2) also occur on similar soils; for example, the large area of 2.4.2 in the Comber-Lisbane-Killinchy area on surface water gleys [SWG1] and BE developed over shale and red Triassic Sandstone mixed tills. Another extensive area of 2.4.2 is found on the predominantly shallow brown earths [SBE] and SWG1 on shale till which extends along a coastal belt south of Killinchy. In contrast, poor quality pastures (2.3.1.2) are often on the brown ranker complex on shale (SH). Along the east coast north of Portavogie, SWG1 on shale and Red Triassic Sandstone mixed till dominates, but in fact the Triassic component is clay from a marl or mudstone off the coast. Understandably, the majority of these SWG1 clay-rich soils are associated with high quality pasture for dairy cows. However, the relationships between agricultural land cover and soils should not be taken too far because management practices are also important.

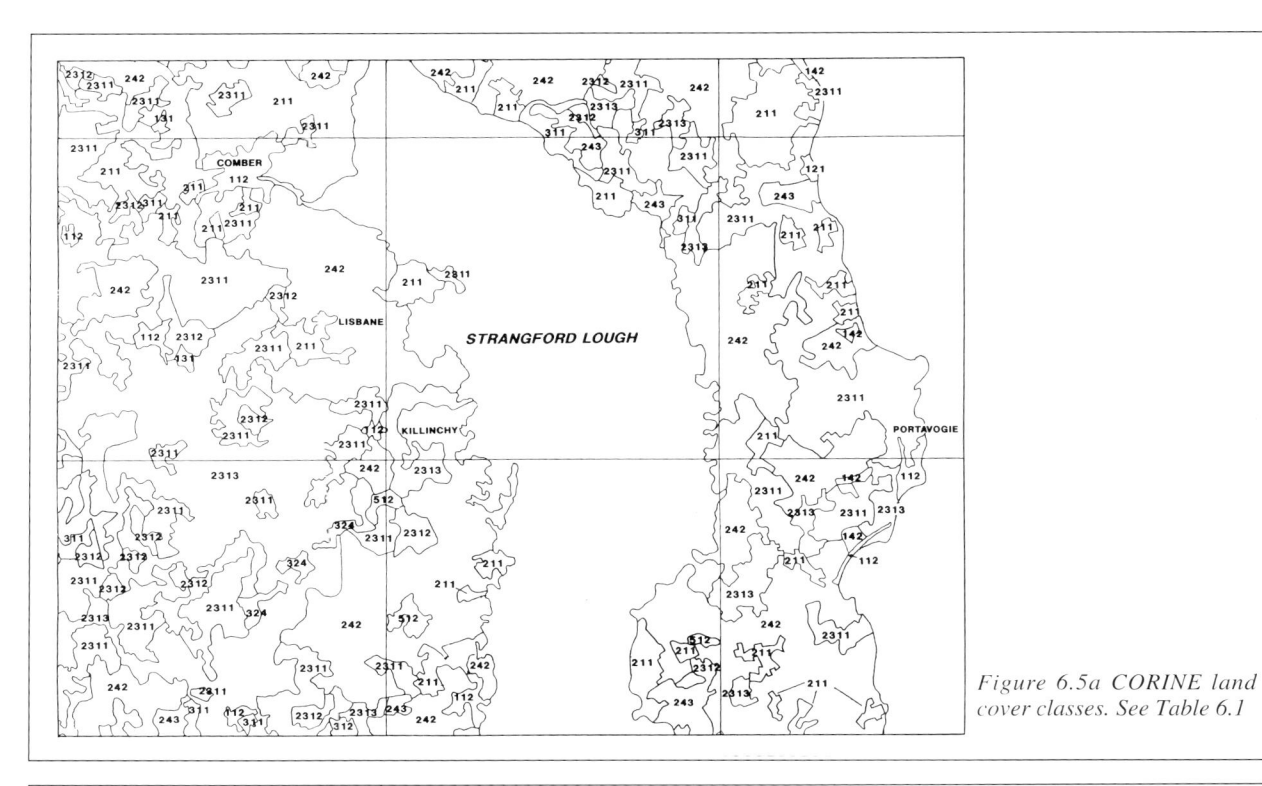

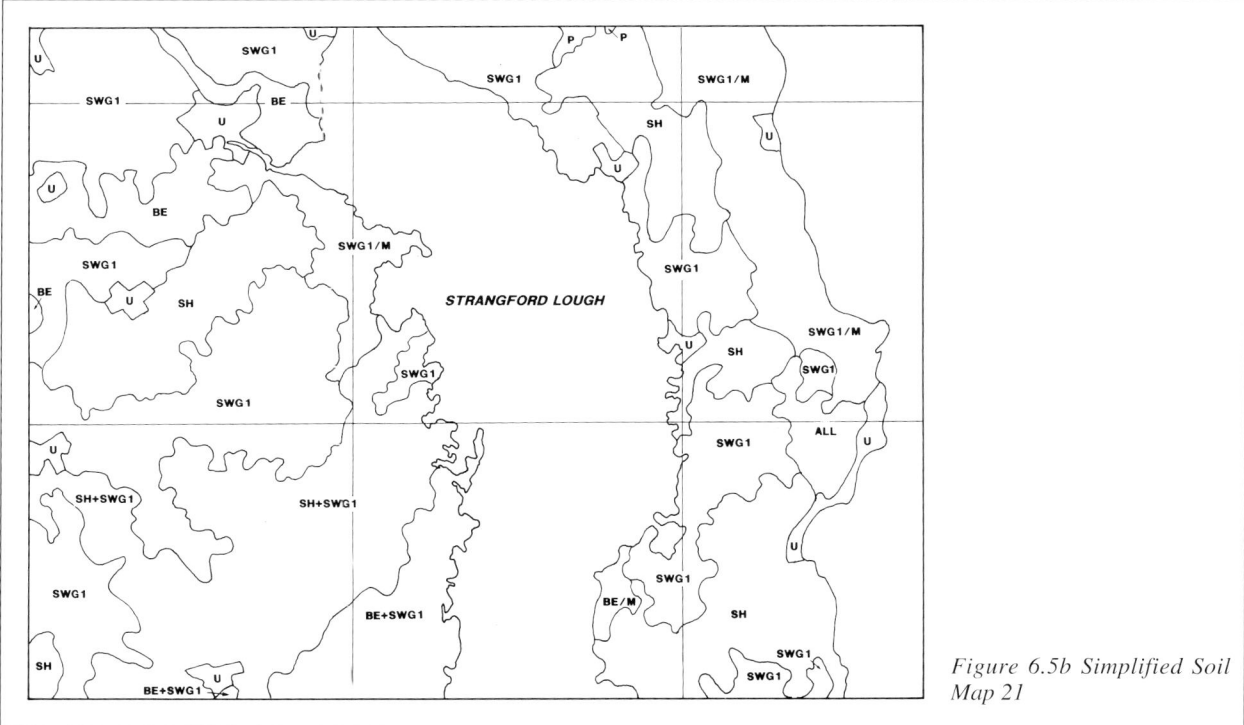

Figure 6.5a CORINE land cover classes. See Table 6.1

Figure 6.5b Simplified Soil Map 21

Peatbogs

Figure 6.6 shows all peatbogs over 25 ha – as mapped in CORINE. The blanket peatland is enclosed within a boundary to separate it from lowland, predominantly raised peat bogs. Blanket peatland is primarily in uplands, but westward the lower limit descends, so that on the Pettigoe plateau it can be found below 100 m; this is the only example in Northern Ireland of Atlantic bog. In the mid-west, many peatland areas are transitional between lowland raised bogs and northwestern blanket bog; Black Bog at Creggan, between Omagh and Cookstown, is typical of this transitional region.

The LANDSAT TM images at a scale of 1:100,000 do not allow detailed division of the peatland, other than between non-exploited (4.1.2.1) and exploited (4.1.2.2), nor was this required for the CORINE land cover programme. However, for many purposes, such as wildlife conservation, more detailed mapping is essential. With limitations, this can be done from satellite data (Cruickshank and Tomlinson, 1990a), but to facilitate conservation programmes a peatland survey was conducted from aerial photographs in the late 1980s (Cruickshank and Tomlinson, 1990b). Blanket peatland was divided into eight major classes using tone, texture and pattern. The delimited types were transferred to 1:20,000 scale maps, their areas measured and summed to give subtotals for regions as well as totals for Northern Ireland (Table 6.2).

The western and northern bias in the distribution of blanket peat apparent on the satellite image and the map (Figure 6.6) can be seen in the measured totals; 73 per cent of blanket peatland is in the west where upland is more extensive and climatic conditions result in a higher input of moisture than is lost by evapotranspiration, see Chapter 4.

In the east, the two main areas of blanket peat are on the Antrim Plateau and the Mourne Mountains. In Antrim, almost horizontal basalt flows form extensive plateaux which have allowed water to accumulate and peat to develop. However, sharp breaks of slope resulting from the several flows of basalt, some steep hillsides and high summits encourage extensive peat erosion. This takes two main forms; on the slopes there is gullying, where subparallel, steep sided channels cut the peatland into large rafts, whereas on flatter surfaces, generally above 360 m and where the peat is deep, channels wind and branch seemingly with no direction. Between these freely branching channels there is a myriad of tiny islands of peat. In the most severe examples of this erosion, where the channels have widened, these tiny islands (haggs) are surrounded by bare peat and even bedrock as a result of wind and frost action. Peatland has been reduced in Antrim not only through erosion, cutting for fuel (approximately 25 per cent of present peatland has been cut at some time in the past) and reclamation for agriculture, but also by forest planting, Figure 6.7.

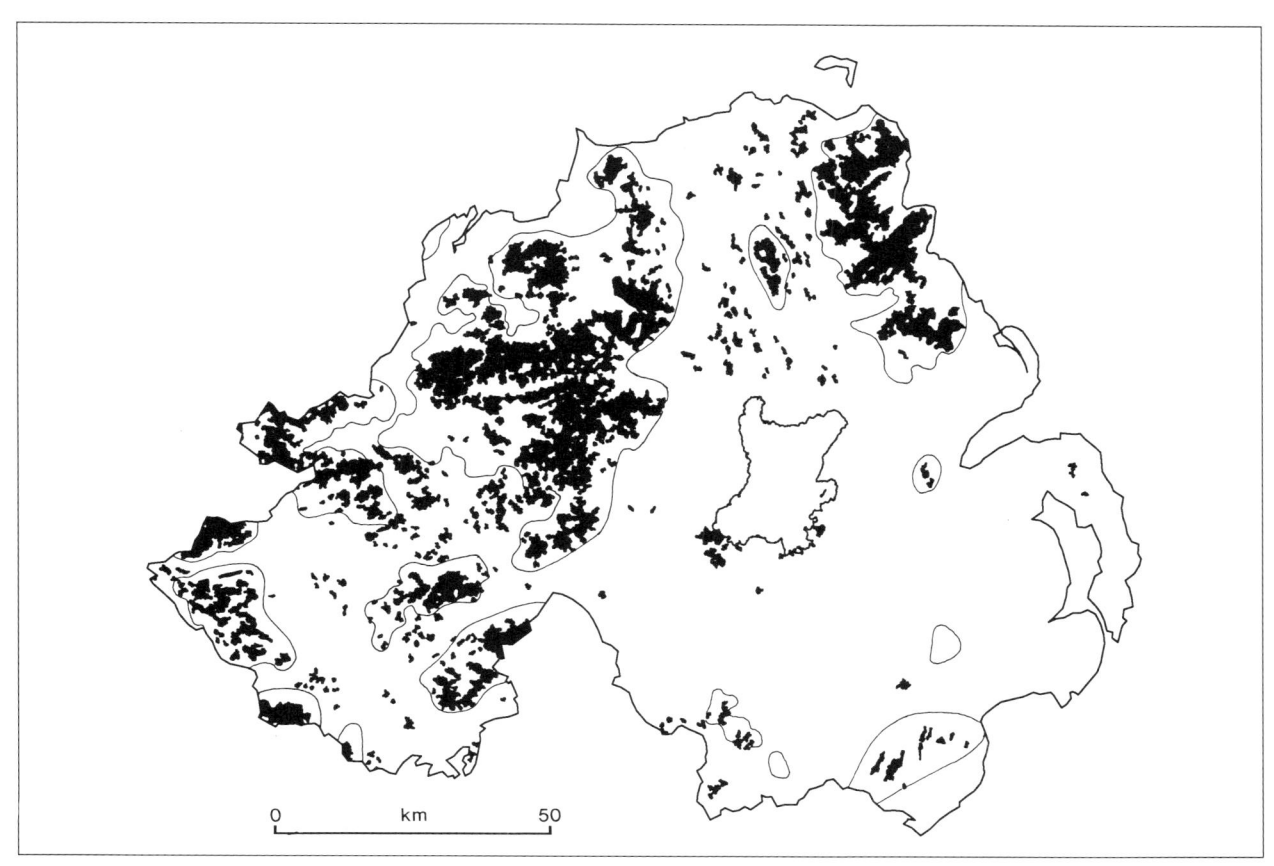

Figure 6.6 Peatbogs (Blanket peat areas outlined) CORINE 4.1.2

Table 6.2 Areas of peatland types in Northern Ireland

	East total	% East	West total	% West	NI total	% NI
			LOWLAND	PEATLAND		
Intact Mixed	1,131	8		11	2,270	9
Cut	11,568	78	8,474	81	20,042	80
Other	2,026	14	858	8	2,884	11
Total ha	**14,725**	**58**	**10,471**	**42**	**26,196**	**100**
			BLANKET	PEATLAND		
Intact Mixed	4,457	12	10,132	10		10
Intact diverse	326	0.8	3,990	4	4,316	3
Molinia + flush	272	0.8	807	0.7	1,079	0.8
Calluna	950	2	1,241	1	2,191	1
Other	9,129	24	12,708	12	21,837	15
Drained	4,706	12	7,672	7	12,378	9
Cut	10,052	26	56,683	54	66,735	47
Eroded	8,359	22	10,900	10	19,259	14
Total ha	**38,251**	**27**	**104,133**	**73**	**142,384**	**100**
			ALL	PEATLAND		
Total ha	**52,976**	**32**	**114,604**	**68**	**167,580**	**100**

Conifer forestry in blanket peat is extensive in some parts, as for example between Glenarrif and Ballycastle, Figure 6.7. As a consequence of these reductions, intact peatland is quite rare; the Garron plateau has the largest expanse and is the area of most interest to conservation. This broad plateau is bounded by steep slopes so that inaccessibility has protected all but the edges from cutting. Erosion is not widespread because of the gently undulating surface and moderate altitude; it is confined to short, steep slopes on higher residual summits on the northern side. In the Mournes, the small area of blanket peatland is restricted to high, flat interfluves and almost none has escaped cutting and erosion.

The west not only has 73 per cent of blanket peatland, but also 70 per cent of intact blanket peatland. In the Sperrins, all the peripheral hills and the lower slopes with blanket peat are affected by hand (spade) cutting. The highest ridges have intensive erosion of freely branching channels and hagging, whereas extensive areas of the upper slopes have gully erosion. Losses of peatland through past cutting and erosion are compounded by afforestation, both State Forests and private forestry (e.g. Glenlark and Coney Glen). Machine cutting of fuel peat is widespread in the Sperrins, in contrast to Antrim where it is significant only in the northern part of the plateau and on Long Mountain. The machines used (compact harvesters), commonly rather like a large chain-saw attached to the power take-off of a farm tractor, extrude peat from slits cut to about a metre in depth across the surface of the bog. Although the majority of these cutting sites are small, less than 1 ha and many around 0.1 ha, their density, especially near to roads, gives them landscape importance. During the summer the surface of the site may be covered by drying 'sausages' of peat creating harsh black squares on the hillside. This contrasts with spade-cutting which has slow areal expansion over many years as the peat is cut from a vertical face. The living top layer of the face is often thrown back on to the earlier cut area and this aids recolonization in all except the most heavily utilized sites. Lines of recycled white plastic fertilizer bags filled with the dry turf from machine cutting also can present a jarring appearance in the landscape. Many of the small machine cutting sites will be re-colonized by bog plants, in particular by cotton sedge (*Eriophorum angustifolium*), but larger sites, which tend to have more intensive extraction, may remain bare of plant cover throughout the year. High density of machine cutting sites may have significant negative effects on landscape quality. For example, in the Sperrins, the Area of Outstanding Natural Beauty also has a plentiful supply of accessible sites for peat extraction. Additionally, the impact on habitats, for example those of wading birds, on hydrology and on the chemistry and sediment load of rivers, especially in regard to salmonid fish, must be considered.

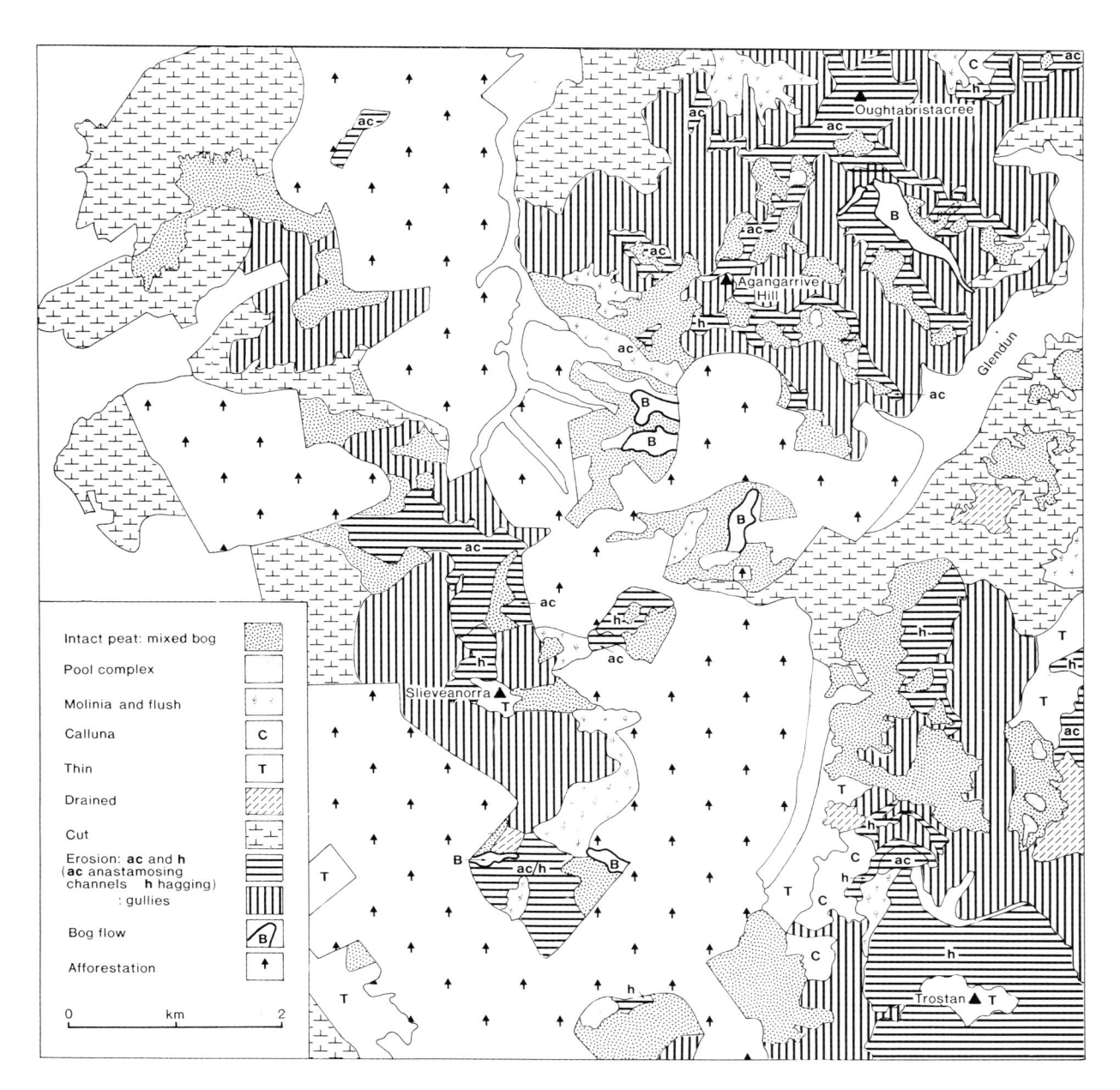

Figure 6.7 This small section from the Northern Ireland Peatland Survey maps of an area around Slieve an Orra, Co. Antrim demonstrates the complex spatial pattern of the various types of blanket peatland. Intact peatland occupies flatter surfaces although some of the pool complexes probably are situated over shallow basins in the underlying topography. Anastomosing (freely branching) channels are found on the higher, flatter summit areas whereas gullying is extensive on steeper slopes. Some of the bog bursts which cluster in this region also may be seen. Past hand cutting is in the more accessible parts and present machine extraction of fuel peat often takes place on formerly hand cut peatland. State Forests are extensive in this area, but new policies initiated in 1993, prevent new planting on intact peat and there is a 'general presumption against afforestation of oligotrophic blanket bog' (DANI, 1993)

On the gentler slopes of the lower hills in mid-Tyrone, blanket peat is widespread and extends to low altitudes. Erosion is less developed than in the High Sperrins and losses are mainly from cutting and forestry. Intact peatland occupies a larger proportion of the blanket peatland (15 per cent) than in areas described previously. Fermanagh is varied in its blanket peatland; to the north of Lower Lough Erne, is the eastward continuation over the Irish border of the Pettigoe plateau. Although cut around the edge, this unique area of Atlantic bog has extensive intact bog at its centre.

Cuilcagh also is affected by cutting only at the edge and has much intact, relatively high level blanket bog. Slieve Beagh completes the quartet of areas of prime interest to conservation (the others are the Garron and Pettigoe plateaux and Cuilcagh). On Slieve Beagh over 20 per cent of peatland is intact and large pool complexes may be found, although some individual circular pools are a legacy of World War II bombing practice. Slieve Beagh, like Slieve an Orra, has a concentration of bog bursts rarely found elsewhere in Northern Ireland.

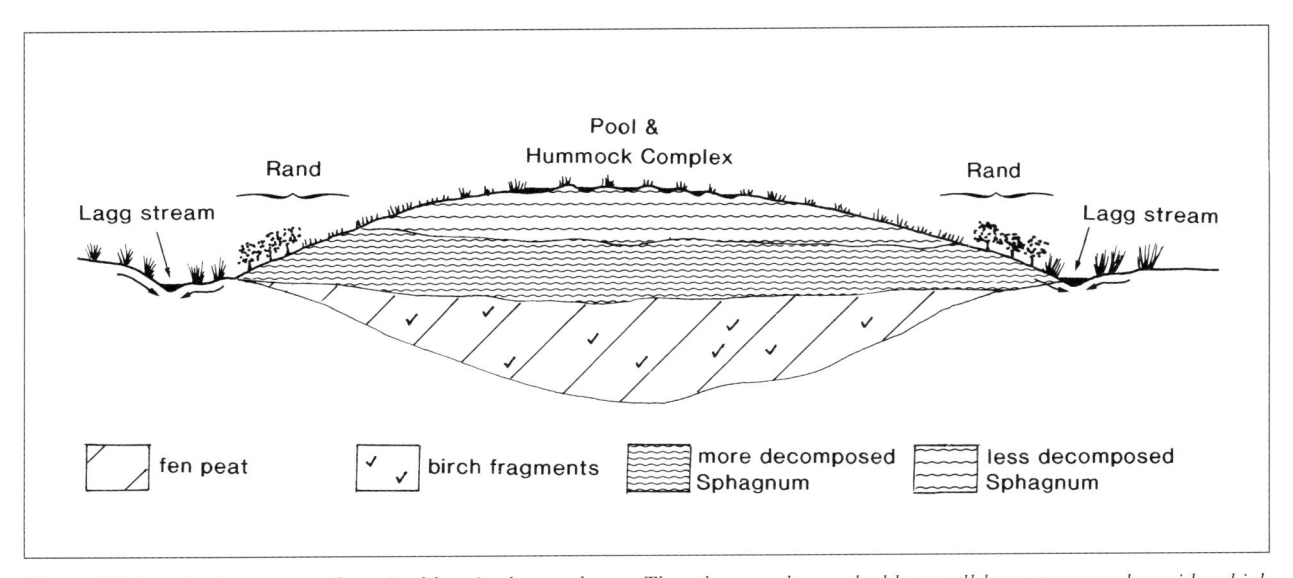

Figure 6.8 A typical structure of a raised bog is shown above. The edge may be marked by small lagg streams alongside which the ground has lateral and vertical water movement so that there is constant replenishment of plant nutrients and a supply of oxygen. Large tussocks of purple moor grass (Molinia caerulea) may be dominant. The lagg leads into steeply sloping edges (rand) of the raised bog which give it the classic convex profile. In the rand, the water table is further from the surface than at the centre of the bog and there is some water movement; sweet gale or bog myrtle (Myrica gale) is often found on this rand. The centre of the bog is typically flat or very gently sloping and has a water table at or just below the surface. In some instances, the centre may be marked by a micro-topography of pools, either of open water or with aquatic bog mosses, and intervening hummocks with species requiring drier conditions may be found, for example common heather (Calluna vulgaris), deer sedge (Trichophorum cespitosum), single-headed cotton sedge (Eriophorum vaginatum), woolly hair moss (Racomitrium lanuginosum) and hummock forming bog mosses (e.g. Sphagnum rubellum).

Lowland peatland comprises 15 per cent of the total, of which 58 per cent is in the drier east and 42 per cent in the west. However, this breakdown is partly a product of the classification since many lower blanket peatlands and transitional areas could be allocated to lowland. In the east, the larger surviving lowland bogs are found in the Main and Lower Bann valleys and to the south of Lough Neagh. These broad, flat lowlands have allowed bog development even though annual rainfall is low (c. 850 mm). Outside these locations, bogs are often confined to inter-drumlin hollows, especially in counties Armagh and Down where, after generations of peat cutting, only isolated examples remain. A smaller percentage of the lowland bogs (9 per cent) remains intact than of blanket bog (15 per cent), and even though some large bogs have survived, for example Ballynahone, Garry and Moneystaghan, intact crowns are disturbed by discontinuous forest patches and by drainage. In addition, because of past cutting which has eaten into the bogs, few retain the classic features of rand and lagg.

Machine cutting for fuel is found on lowland bogs as well as in the blanket peatland, but more concern has been expressed about the role of extraction for horticulture. This comprises two main types, machine cutting of large sods which are subsequently milled and vacuum extraction of a previously harrowed, therefore dry and fluffy peat surface. The first method has been used extensively on bogs around the southern shore of Lough Neagh, but the second has become more common and is associated especially with some of the largest lowland bogs, as for example at Sluggan near Randalstown and at Newferry in the Lower Bann valley. The rarity of large lowland bogs gives them conservation importance and there is a clear conflict of interest with the desire of extracting companies to maximize their returns on investment in expensive plant and machinery; this may be achieved best on larger, deep bogs with potential for a long period of extraction.

In the west, several bogs over 25 ha are concentrated around Omagh. To the northwest of the town, the Fairy Water valley has 4 of the top 10 Northern Ireland lowland bogs listed for conservation. Individually these are not large (Figure 6.9.), but their position in an ecological gradient, they are near together and frequent in an extensive cut-over peatland, gives them added conservation value. Southeast of Omagh, the valleys broaden and larger bogs have developed, but because of the higher population density and ease of access, little has survived intact. Presently, many of these bogs are sites of machine extraction of fuel peat.

In Fermanagh, small bogs are widely scattered and normally cut-over, but the Arney valley has a concentration of bogs which occupy most of the inter-drumlin hollows. Few of these have intact peatland. In the Erne valley there is relatively little peatland, possibly because of reclamation and arterial drainage. (DANI Soil Survey mapped peat soils, as organic alluvium, over a more extensive area.) The Strule valley with its well drained glacial sands and gravels is generally peat free, as are lowlands around Draperstown and Dungiven.

The total area of peatland in Northern Ireland can be obtained from several different sources. Even if the estimate made by Hammond (1979) as part of an all-Ireland survey is ignored, this leaves three recent attempts – the Northern Ireland Peatland Survey, CORINE land cover mapping, and the Soil Survey – which give three totals, 167,580; 134,317; 193,061 ha respectively. Initially this appears confusing, but it is understandable because each estimate had its own terms of reference and methodology. The Peatland Survey recorded peatland as seen on the most recent readily available air photographs. Most of the photographs were of the 1970s, but some areas had not been photographed since the mid 1960s. Clearly, soils under forest cannot be seen and therefore forest areas were not included in the peatland totals. In addition, no account could be taken of agricultural fields with peat beneath the pasture or other land cover because, as with forest cover, the peat cannot be seen from the air. This applies also to CORINE mapping – a satellite-borne sensor records reflectance only from surface materials so that peat occupied by forest or agricultural crops goes undetected. Also, CORINE land cover mapping followed a classification that applies to the EU and this can mean that peatland is over- or under-represented. For example, some poor, upland 'grasslands' with a high sedge content have the image appearance of peatland and thus 'inflate' the total. The Soil Survey, in contrast, is not concerned with the cover, but with the material beneath; any 'soil' composed principally of plant remains, and is greater than 50 cm depth, is classified as peat.

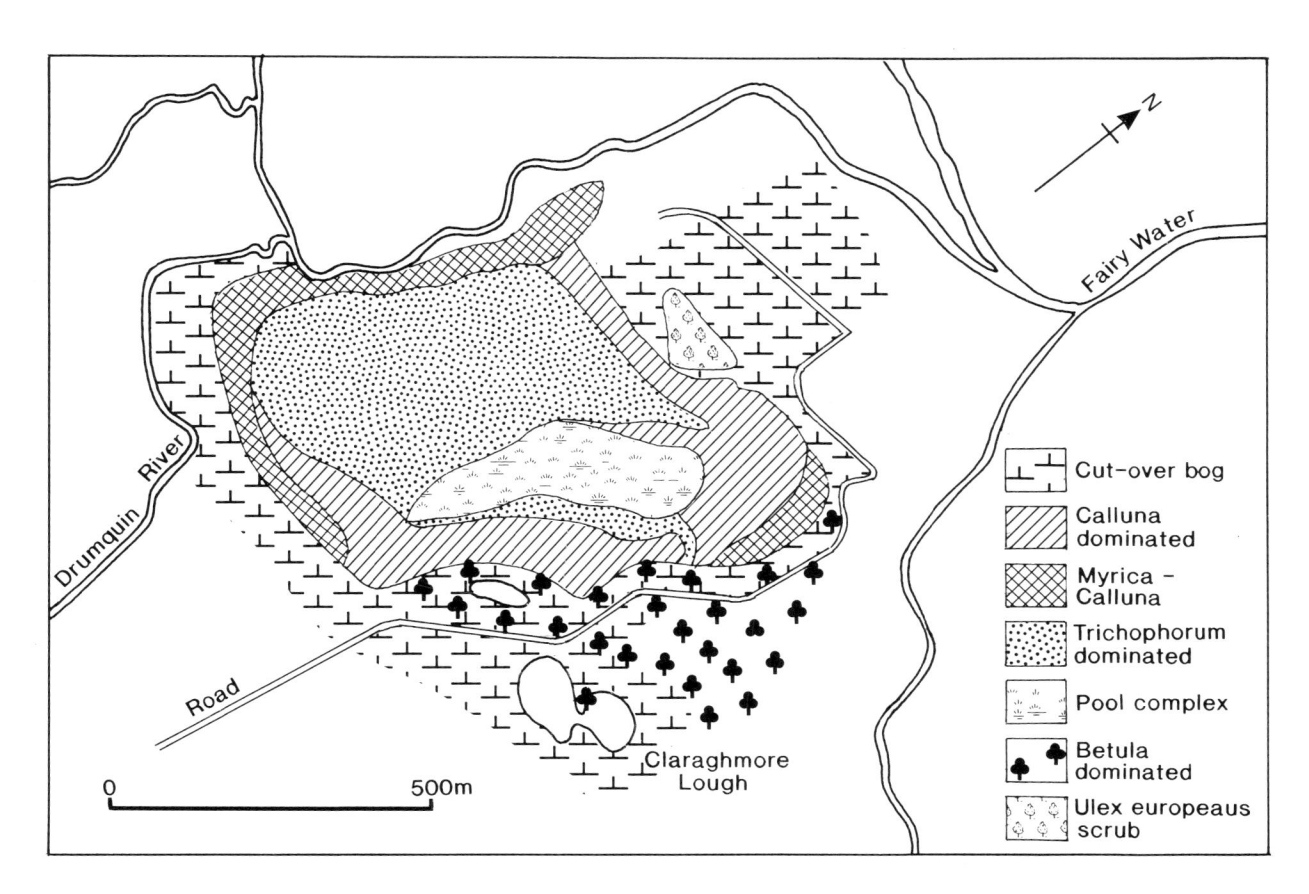

Figure 6.9 Vegetation of Claragh Bog, Fairy Water, Co. Tyrone

Claragh bog is one of a cluster around the confluence of the Fairy Water and Drumquin rivers (Morrison, 1954) which have formed within depressions in the limestone till (Soil Map 12). Many of these bogs have been cut over, but Claragh has suffered comparatively little and shows several of the main elements of a raised bog. Although it has lost its lagg, along the western and southern edge there are remnants of a rand with Myrica gale and Calluna vulgaris. Most of the bog surface, inside this Myrica-Calluna fringe, is occupied by a Trichophorum cespitosum dominated community, but towards the north-centre is a good example of a pool and hummock complex. Claragh is one of 4 bogs in this area designated as ASSIs; although this may limit disturbance, the isolation of the bogs in a surround of unprotected land, can mean that the bogs are in danger. For example, over Easter 1996, burning of surrounding vegetation spread onto the bog, damaging delicate plant assemblages.

The growth of Claragh bog is representative of many raised bogs formed over a basin. On a surface of clay a dark, dense amorphous peat forms all but the top metre of peat. These dense layers are mainly of fen peat with remains of common reed (Phragmites communis); abundant birch fragments may be found in the upper limits. A marked change then takes place to a lighter, more reddish peat with a high content of bog mosses (Sphagnum species), heather and sedges. These plants indicate a wetter environment and more acid conditions.

Reclaimed agricultural land and forest areas would be included if they meet the definition. On the other hand, much of the marginal hill-land classified as peat in CORINE or in the Peatland Survey is excluded by the Soil Survey because it does not in general meet the 50 cm rule – even though there might be evidence of past cutting and a plant species composition indicative of peatland. A considerable proportion of the marginal hill land therefore is grouped into surface water humic gleys and humic rankers.

Throughout the 1980s, and indeed to the present, there has been a growing concern about the loss of peatland – both national and provincial. This began with concern over extraction for horticulture (mainly from lowland bogs), but soon extended to losses resulting from fuel extraction in the blanket peatland, especially by compact harvesters. However, because much of this extraction takes place near roads, its perceived influence has been exaggerated; in 1991 less than 3 per cent of the blanket peatland area was affected by such fuel extraction and approximately 80 per cent occurred on peatland that had been hand cut at some time in the past (Cruickshank *et al.* 1995). The role of forestry also has tended to be over-emphasised; it occupies only 20 per cent of the blanket peatland. Nevertheless, partly in response to pressures from conservation interests and partly because of the economic and physical difficulties of growing conifers in upland peat areas, the policy of the Forest Service is not to plant new areas on oligotrophic blanket bog (DANI, 1993). More recently, in response to concern about global warming and the role of 'greenhouse gases', attention has been directed to peatlands as stores of carbon sequestered from the atmosphere. Current research indicates that peatlands in Northern Ireland account for 42 per cent of the carbon stored in the soils. An as yet unkown amount may be released by drainage for forestry or other purposes, but it may be significant in comparison with the amount of carbon sequestered by forest trees. Although reduction of fossil fuel use would make the largest contribution to lessening the dangers of global warming, the role of peatlands in sequestering and storing carbon is another argument for their conservation – alongside those of ecology and landscape (see Chapter 9A).

Transitional vegetation

Examination of Figures 6.10a and 6.10b shows that CORINE classes 2.4.3 *land principally occupied by agriculture with significant amounts of semi-natural vegetation* and 3.2.4 *transitional woodland scrub* frequently are located around the mountains, as for example in the western Sperrins, or in the more gentle uplands, as in central Tyrone or south Armagh, where improved land merges with semi-natural grasslands, heaths and peatlands. Surface water gleys and surface water humic gleys often predominate in these areas. Patches of these cover classes are found in the lowlands, especially where there is high soil moisture and a mixture of scraps of bog and fen with pastures. For example, 2.4.3 is common in east Fermanagh and Tyrone where small remnant bogs and damp ground are scattered between low hills. There is a large area also between Dungannon, Aughnacloy and Caledon where gentle uplands and inter-drumlin lakes provide for a mix of peat, small woodland patches and damp ground. Transitional scrub woodland is particularly noteworthy on the slopes around Lower Lough Erne and near the shore of Upper Lough Erne, as well as on cut-over bogs in the Main and Lower Bann valleys.

In the past the accent in agriculture was on production and these 'poorer' lands with difficult soil were grant-aided for improvement, for example through drainage schemes. However, with increased environmental awareness through the 1970s and 1980s, the continuation of 'mountains' of food in EU storage and the perceived over-expenditure on the Common Agricultural Policy (CAP), the purposes of the CAP and member countries began to change. Intensive agriculture results in loss of ecological diversity; Tubbs and Blackwood (1971) in their ecological evaluation for land use planning, argue that value varies inversely with intensity of use and Mather (1986), using Nature Conservancy Council data (1977), shows that the numbers of species of animals, birds and insects are lower on 'modernised' farms than on 'unmodernised' ones. The importance of less-intensively farmed areas to European wildlife has become a focus for research (Curtis *et al.* 1991; McCracken, 1993) and since the late 1980s, some are being conserved through designation as Environmentally Sensitive Areas (ESAs). For example, the West Fermanagh and Erne Lakeland ESA has been selected because the Erne basin is regarded as one of the best wetland sites within the UK, containing a range of habitats particularly important for breeding waders. It is important for over-wintering birds and to the west of the county, there are impressive hay meadows. Bordering the fields are woodland and scrub, blanket bogs, mountains and moorlands. All of these are important for wildlife and landscape and there is clearly a close intermixture of agricultural cover types with semi-natural ones.

Comparison of the land cover types inside the ESA with those in the rest of Fermanagh (and considering only rural land, i.e. land under artificial surfaces and under lakes was excluded) shows that within the ESA agricultural land occupies 60 per cent of the rural land whereas in the rest of Fermanagh it accounts for over 80 per cent – the difference is accounted for by the higher percentage of semi-natural land cover. Further, the ESA has almost twice the percentage of land classed as 2.4.3 or 3.2.4 than does the rest of the county. The ESA seems to have been delimited wisely. It has selected areas once thought of as agriculturally unproductive and in which

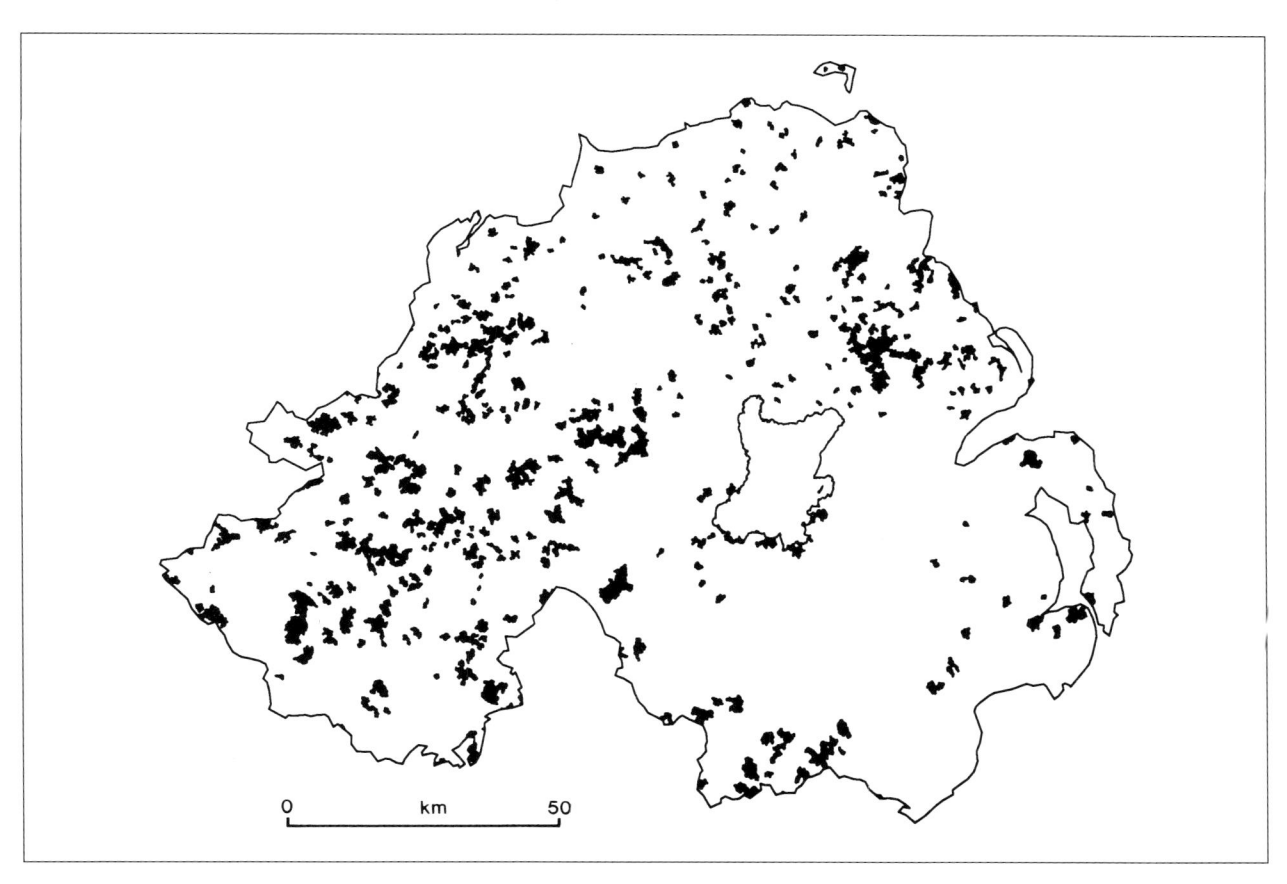

Figure 6.10a Agriculture with significant natural vegetation, CORINE 2.4.3

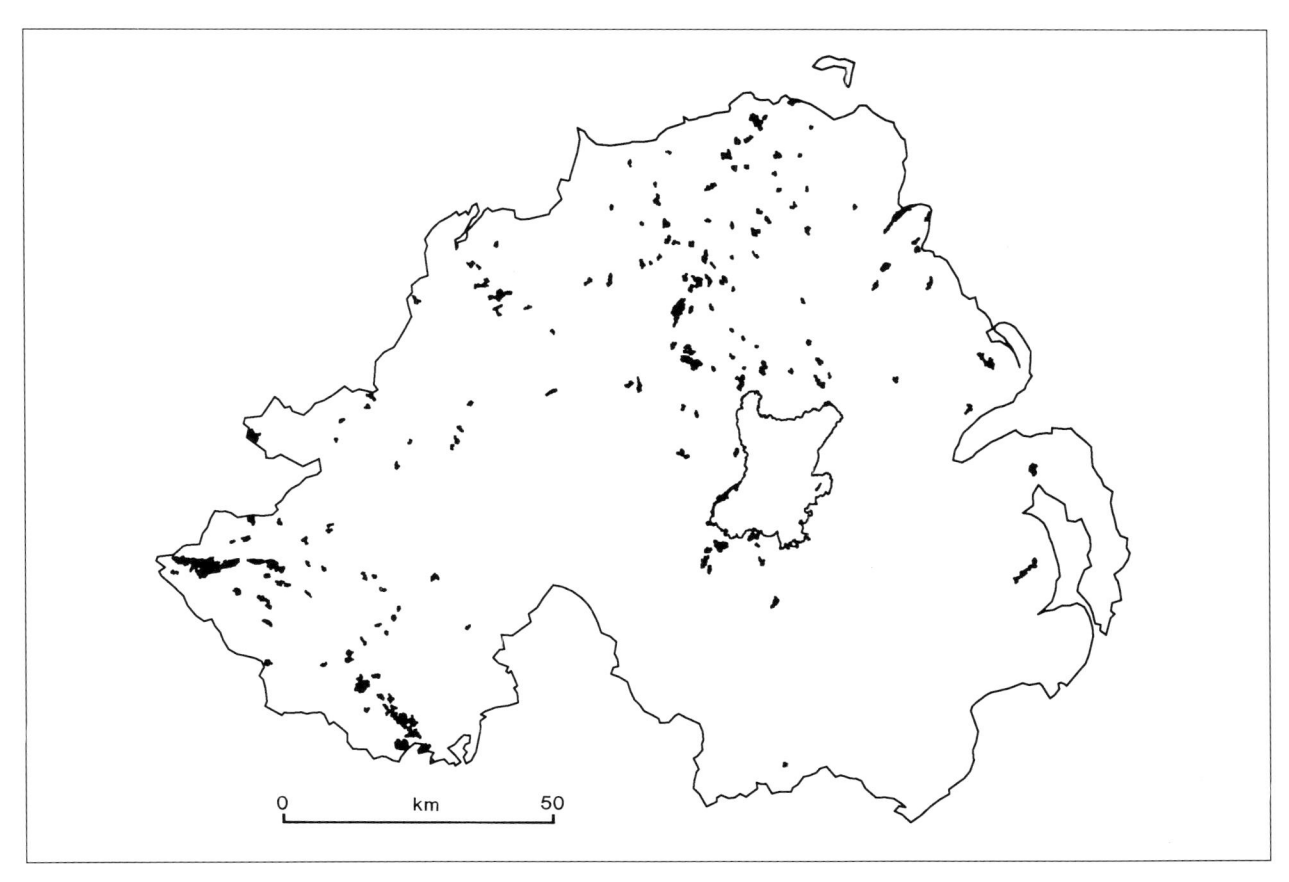

Figure 6.10b Transitional woodland scrub CORINE 3.2.4

the soils needed considerable investment if they were ever to be regarded as of much agricultural value. Now, these areas have value precisely because the soils (which include poorly and very poorly drained SWG and SWHG, peat, organic alluvium and humic and other rankers) can sustain only poor grasslands, scrubland, and other semi-natural vegetation. Additionally, these soils and their vegetation are some of those which can store higher amounts of carbon and have a greater capacity than agricultural crops to sequester carbon dioxide from the atmosphere and thereby offer some amelioration of the 'greenhouse gas' problem (see Chapter 9A).

Forests and Woodlands

In the CORINE land cover classification, three forest/woodland types are recognized – broad-leaved (3.1.1), coniferous (3.1.2) and mixed (3.1.3). In Ireland, many individual broad-leaved and mixed woodlands are less than the 25 ha minimum unit to be identified and are unrecorded in CORINE. Additionally, at a scale of 1:100,000 and in early May, when canopies are not fully developed, small broad-leaved and mixed woodlands can be difficult to distinguish on satellite images from some coniferous woodlands. Generally, coniferous State Forests may be delimited readily, partly because of their large size and geometric outlines, but also because they contrast with their surroundings. The majority (64 per cent) is planted on peat and peaty soils (i.e. humic rankers, peaty podzols, surface water humic gleys, humic gleys and organic alluvium) so that the purplish-red of conifer trees contrasts with the green and blue-green of peatland vegetation. The distribution of conifer forests (Figure 6.11a) shows a clear association with blanket peatlands (compare with Figure 6.6); large forests occur on the north Antrim plateau, on the mountains between the Glenshane Pass and Binevenagh, in the east and west Sperrins, in west Tyrone, on the mountains south of Lower Lough Erne and on Slieve Beagh. The forests are not continuous, but either concentrated into large blocks or in scattered patches; the ability of the Forest Service to acquire land is a major factor in this pattern. The distribution on Slieve Beagh, with a ring of separate plantings around the edge of the hill mass, is notable.

Coniferous forests contrast with their surroundings in the real landscape as well as on satellite images. The species planted, in particular Sitka spruce, Norway spruce and Lodgepole pine are alien to Ireland both in form and colour. The conical outline of the trees and abrupt linear edges to the plantations contrast sharply with the rounded, almost featureless slopes of the peatland. The predominant dark blue-green foliage is very different from the dull brownish-greens of the heather, sedge and bog moss communities. In some areas of Northern Ireland, State Forests have been criticised strongly because of this impact on the landscape. Their effect on wildlife and the environment also has been considered detrimental. Streams can have enhanced acidity and associated increases in aluminium concentrations are detrimental to salmonid fish. Drainage for the plantations may change river regimes, leading to more flashy discharge patterns. There is a loss of habitat for wildlife, not only plant populations but fauna – including a loss of territory for birds of prey and for summer visiting waders. The addition of fertilizers can lead to the loss of plant species adapted to the poor nutrient conditions of the peat.

In 1970 a White Paper (Cmd Paper 550) proposed that 120,000 ha of forest should be planted (90,000 in State Forests) by 2,000 AD. This target seldom looked like being met; the planting rate has been insufficient because small, family farms have been unwilling to sell land, hence the patchwork on Slieve Beagh, and the more available hill-land is in blanket peat where a poor supply of plant nutrients limits growth. Together with the environmental concerns, these problems prompted changes in policy; from 1987, 5 per cent of trees planted by the Forest Service were to be broad-leaves, to soften plantation edges and increase the wildlife value of forests. As discussed above, since 1993, there has been a general presumption against afforestation of heather moorland, blanket bog and raised bog.

The annual planting rate for private woodland implied by the 1970 White Paper has not been achieved; between 1970 and 1986 it was about 112 ha per annum (Guyer and Edwards, 1989). In contrast to the recent limitation placed on afforestation of peatland, planting of better quality lowland is being encouraged; for example, the Farm Woodland Scheme was introduced in 1988 to boost development of farm woodlands and to assist in taking land out of production. Incentives to farmers included annual payments over the life of the woodland in addition to planting grants. Higher payments became available to those farmers planting better land and for longer, especially if beech and oak were planted. It will take many years before landscape impacts of any planting under this scheme become evident and even then they are unlikely to appear on any future CORINE type map if the same minimum polygon size is used. Today, broad-leaved and mixed woodlands are seen to be scattered (Figure 6.11b) and largely confined to demesnes and their surroundings. Some isolated patches of broad-leaved woodland may be successors of original woodland or may have colonized slopes too difficult for agriculture. The species content of these will be affected by the soils on which they occur; for example in the limestone country around Marble Arch in western Fermanagh ash often forms the dominant tree. In the Glens of Antrim, the soils of high base status below the basalt cliffs and on the drift covered chalk, support hazel woodland with a very rich ground cover – especially of spring flowering plants.

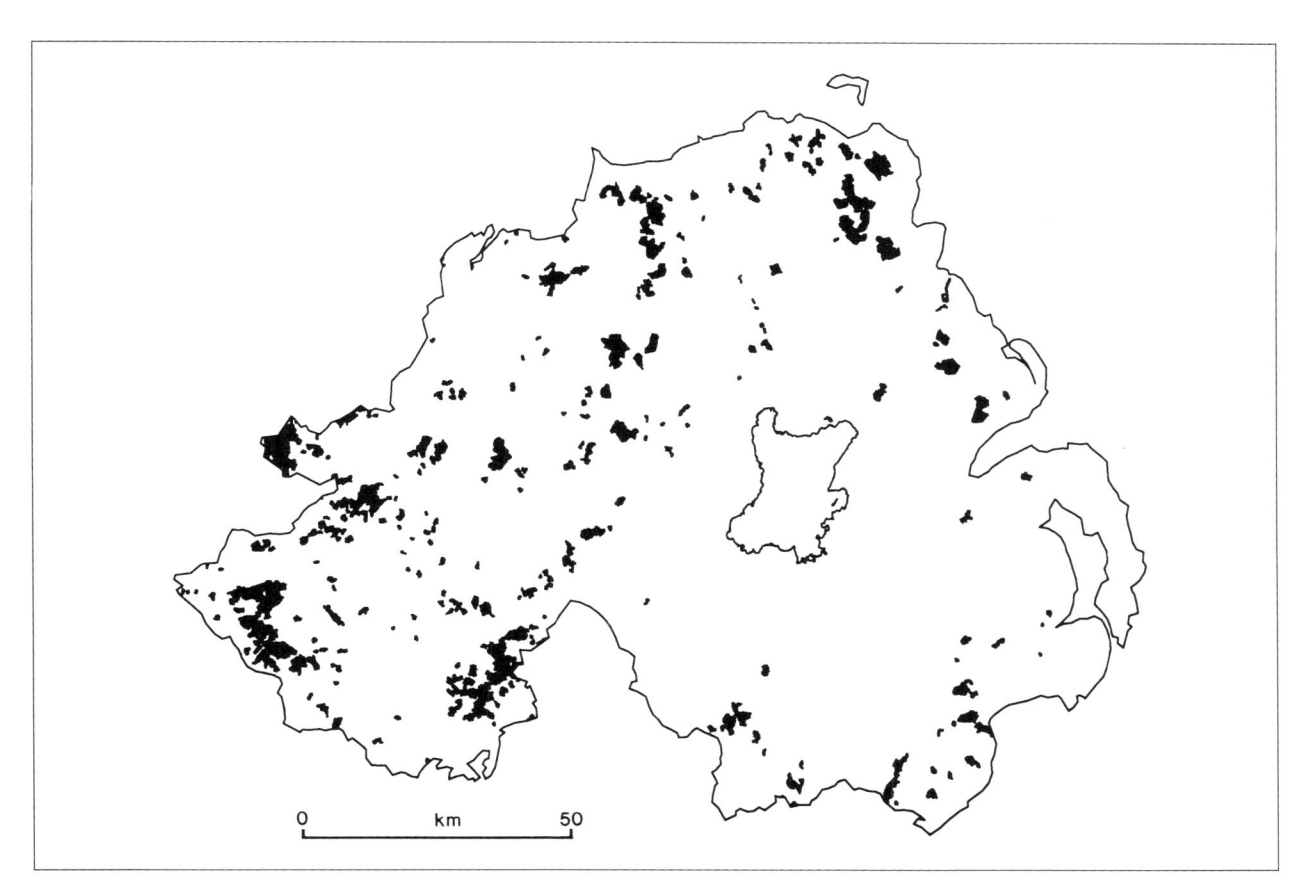

Figure 6.11a Conifer forest, CORINE 3.1.2

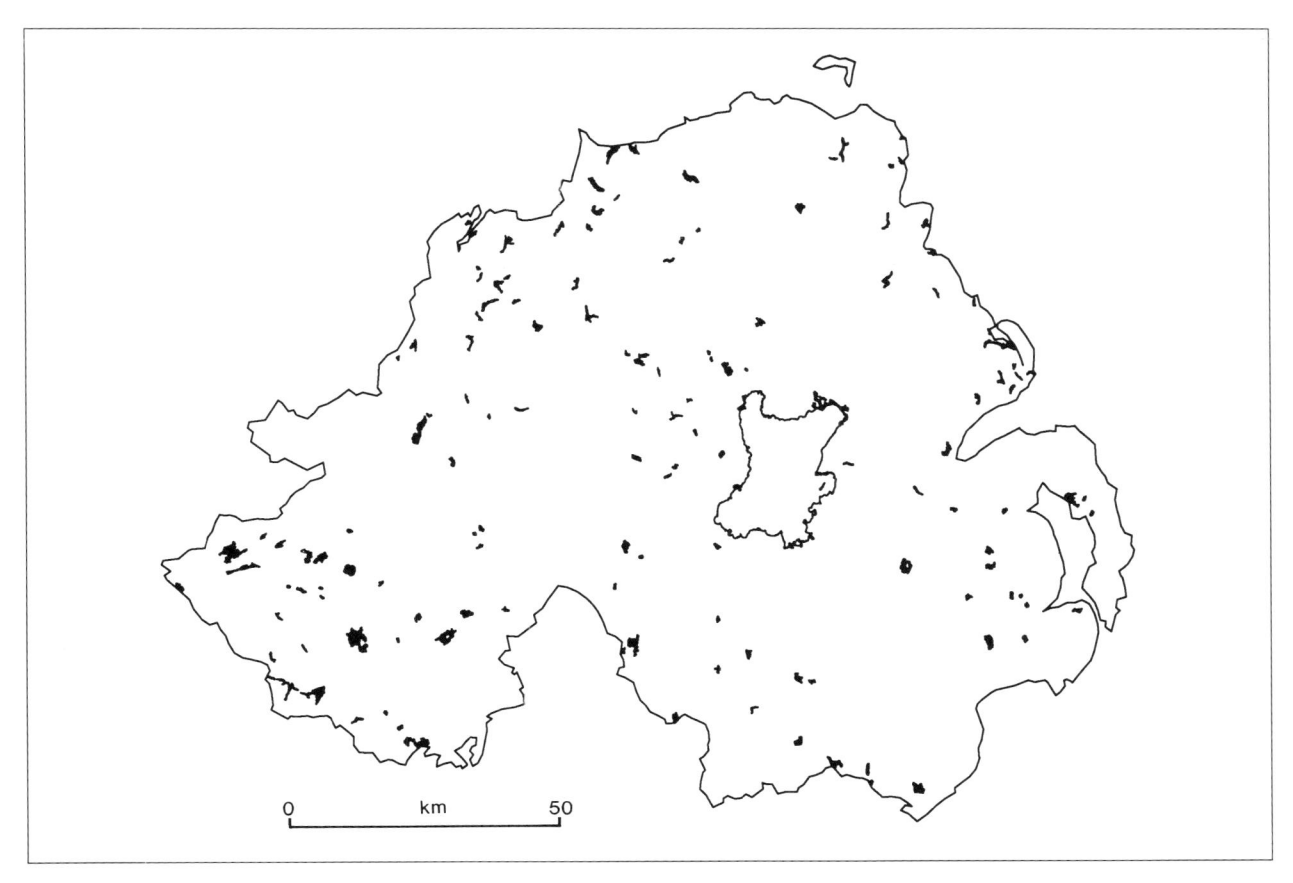

Figure 6.11b Broadleaf and mixed forests, CORINE 3.1.1 and 3.1.3

113

Upland grass and heath

CORINE class 3.2.1 *natural grasslands*, is found predominantly on the margins of the uplands (Figure 6.12a). Here the peat thins, either naturally because of the steeper slopes and improved drainage, or because of past peat cutting. The peat depth is generally less than 50 cm and the Soil Survey classifies these areas into humic gleys and humic rankers. However, because CORINE is based on land-cover reflectance data recorded by satellite borne sensors, some areas may have been classed as peat rather than as natural grassland because of their similar colours and tones on the images. For example, on the south facing slopes of the Glenelly valley in the Sperrins, the Soil Survey is more restrictive of peat than is CORINE. Whereas most upland grassland is composed of coarse grasses that are still senescent in early May and are a blue-ish colour on the image, in the chalk (e.g. Antrim coast) and limestone areas (e.g. Marble Arch) they may appear in red tones, indicating continued growth through the winter.

Moors and heathlands (3.2.2) in CORINE have a restricted occurrence in Northern Ireland (Figure 6.12b) with concentrations on steep slopes with thin peat in the Mournes and on the glacial sands and gravels between Mountfield and Carrickmore (which include the Murrins NNR). However, heather is more widespread than indicated by CORINE mapping. This is because many areas are classed as peatbog (following the definitions of CORINE) even though heather is a high percentage of the vegetation cover.

Some areas of 'natural' grassland may develop from heather as a result of different grazing policies. Examples occur on the Antrim plateau and on Slieve Croob where land on either side of a boundary has different cover; one with heather and the other with coarse grasses (e.g. *Molinia caerulea* or purple moor grass and *Nardus stricta* or mat grass) which became dominant as heather was grazed out.

In Northern Ireland there are general attempts to reduce stocking levels through overgrazing rules under HLCA/SAPS (Hill Livestock Compensatory Allowance/Sheep Annual Premium payments), but these affect a small minority of farmers. The Moorlands Scheme, introduced in 1995, goes further than this general requirement to farm responsibly. It offers farmers incentives to reduce ewe numbers even lower than levels necessary to protect the environment; these lower stocking levels aim to encourage heather regrowth. The heather moorlands are defined for this scheme as those in which heather (common, bell or cross-leaved) as well as commonly associated crowberry, bilberry and dwarf whin, comprise more than 25 per cent of the plant cover. The Moorland Scheme is part of the Northern Ireland Agri-Environment Programme and is employed outside the ESAs in areas where stocking levels have resulted in overgrazing and where the farmer has more than 5 ha of eligible moorland. Participation is voluntary, but the farmer receives payment for each eligible ewe. The farmer agrees to reduce stocking levels for 5 years to no more than 0.3 Livestock Units/ha (2 ewes) between 1 March and 31 October. In addition no stock are permitted to graze the moorland from 1 November to 28 February.

Within ESAs, farmers joining the voluntary ESA scheme must agree to take part in Tiers 1 and 2, which for moorlands freezes sheep numbers at those of 1 January 1993 or 1994 (depending on the particular ESA) and places further controls; for example winter grazing is prohibited, just as it is under the Moorland Scheme. During the remainder of the year the heather must be grazed, otherwise it may become 'leggy' and open allowing other plants to enter the vegetation cover – even including birch. Stocking levels are controlled at 10 ewes/5 ha on dry heath and 8 ewes/5 ha on wet heath. In recognition of their contribution to maintaining heather, farmers receive annual payments, declining as total hectarage increases. Under the optional 'Enhancement Plan', heather regeneration may be grant aided. The aim is to establish areas of different age structures, either by burning or by flailing. This ensures that there are always some areas of young nutritious heather for grazing and a variety of cover and heights of heather for bird, animal and invertebrate populations.

For those farmers not part of the ESA or Moorland schemes, a heather management code of practice is being drawn up currently.

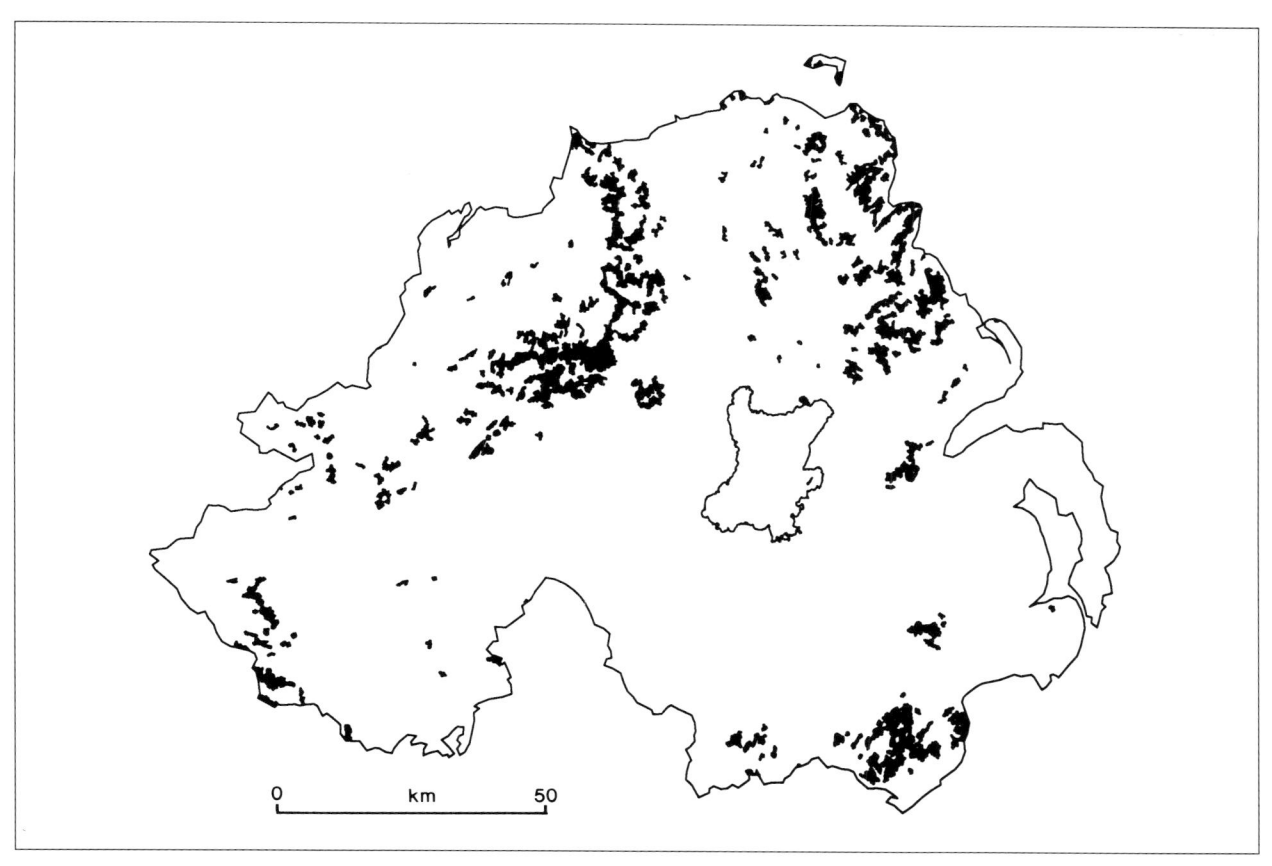

Figure 6.12a Natural grassland, CORINE 3.2.1

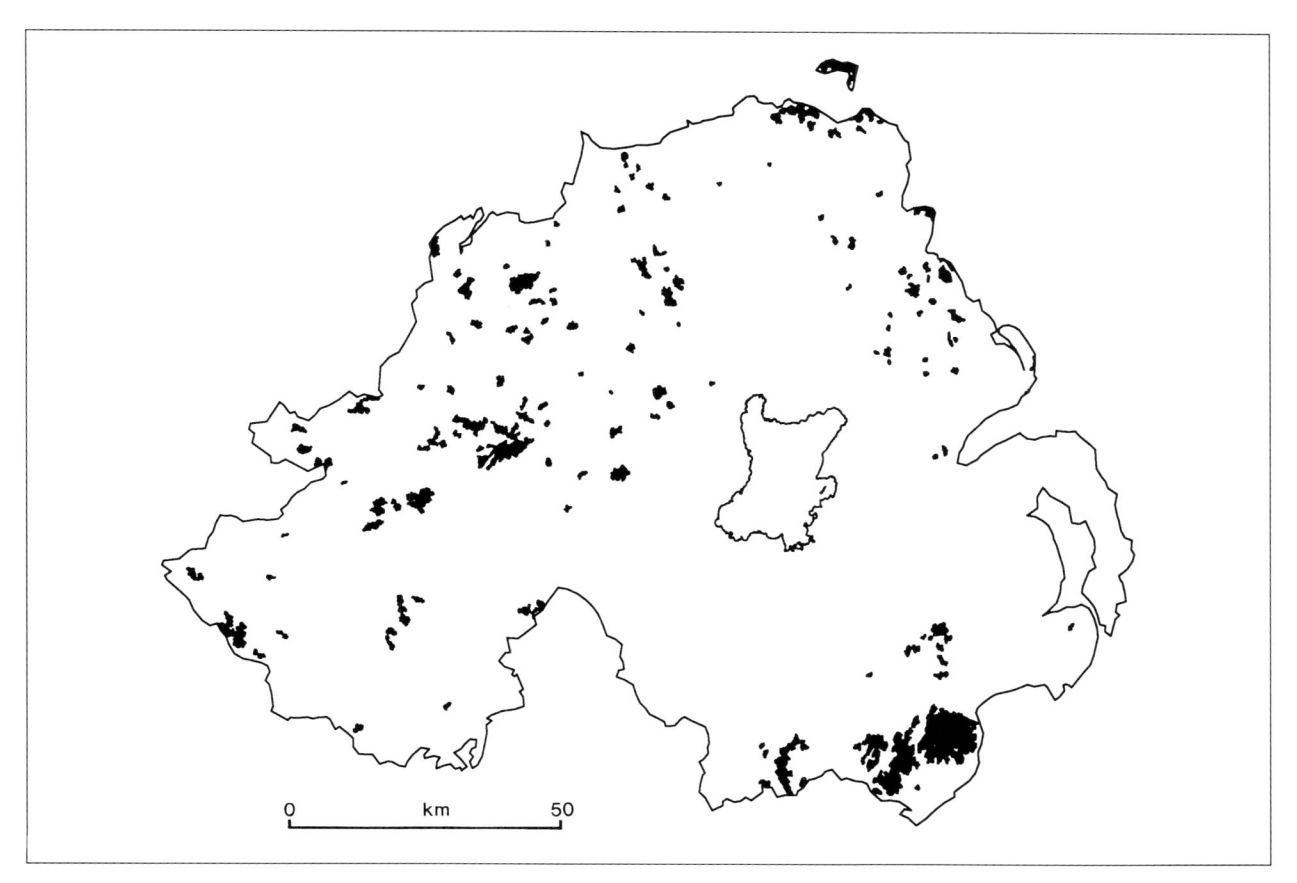

Figure 6.12b Moors and heathland, CORINE 3.2.2

Conclusion: the need to protect the diversity of land cover in Northern Ireland

Within a small land area Northern Ireland has many land cover types, each with complex spatial distributions. In part, this complexity has been caused by the range of soil types which in turn have developed from the diversity of parent materials, the topography and the prevailing moist, temperate climate. The particular combinations of parent materials, including many glacial tills, with the moist climate, has ensured that pastures dominate the agricultural land – in which the majority of soils are poorly drained. However, socio-economic factors are important in helping to determine land cover. Northern Ireland has a dominance of small family farm businesses with low incomes, so that pasture fields are often less improved than in GB, hedges between the small fields have not been removed and scrub has survived and even expanded. The physical and socio-economic factors have led almost two-thirds of Northern Ireland to be classed as Less Favoured Area from the viewpoint of agricultural production. In contrast, from the perspective of wildlife conservation, landscape or tourism, these factors are advantageous, especially when taken together with the open expanses of peatbog and other semi-natural vegetation on the uplands. The land cover which has evolved is now an asset or an environmental good, which must be protected and managed so that economic and other benefits may be gained by all and not least by the rural population.

The CORINE land cover programme has established a base-line against which change in land cover may be monitored. It presents also opportunities for management of the landscape, for example through designation of ESAs and AONBs, for drainage basin management and for calculation of carbon pools and fluxes as part of studies on the effects and amelioration of climatic change (Cruickshank and Tomlinson, 1994). It is an important first component of any attempt to manage the Northern Ireland countryside.

Acknowledgements: the author thanks Paula Devine and Stephen Trew for their assistance in extracting and mapping CORINE data and Maura Pringle for other cartography. Margaret Cruickshank was co-interpreter for the CORINE project in Northern Ireland. Digitisation of the CORINE maps was done by the Ordnance Survey of Northern Ireland (OSNI).

Notes: The maps of the distribution of CORINE classes in Chapter 6 are much reduced from the originals, and consequently some boundaries have coalesced. In some cases, distributions seem to overlap.

References

Brown, L.T.(1968) *A survey of turf working in Co. Down.* Unpubl. MSc Thesis, Queen's University of Belfast.

Cmd Paper 550 (1970) Forestry in Northern Ireland. HMSO, Belfast.

Cruickshank, J.G. and Cruickshank, M.M. (1981) The development of humus-iron podsol profiles, linked by radiocarbon dating and pollen analysis to vegetation history. *Oikos* 36, 238–53

Cruickshank, M.M. and Tomlinson, R. W. (1990a) An evaluation of LANDSAT TM and SPOT imagery for monitoring changes in peatland. *Proceedings of the Royal Irish Academy*, 90B, 109–25

Cruickshank, M.M. and Tomlinson, R.W. (1990b) Peatland in Northern Ireland: inventory and prospect. *Irish Geography* 23, 17–30

Cruickshank, M.M. and Tomlinson, R.W. (1994) First analysis of CORINE land cover as an indicative framework for environmental policy in the North of Ireland. *in* Bond, D. *et al*, (ed) *GIS and Public Policy '94*. Economic and Social Research Council/ University of Ulster.

Cruickshank, M.M., Tomlinson, R.W., Bond, D., Devine, P.M. and Edwards, C.J.W. (1995), Peat extraction, conservation and the rural economy in Northern Ireland. *Applied Geography*, 15, 365–83

Curtis, D.J., Bignall, E.M., Curtis, M.A. (1991) *Birds and pastoral agriculture in Europe. Proceedings 2nd European Forum on birds and pastoralism.* Scottish Chough Study Group/Joint Committee for Nature Conservation. Paisley.

DANI. (1993) *Afforestation – the DANI statement on environmental policy.* Forest Service, Department of Agriculture for Northern Ireland, Belfast.

Guyer, C.F. and Edwards, C.J.W. (1989) The role of farm woodland in Northern Ireland: an appraisal. *Irish Geography* 22, 78-85

Hammond, R.F. (1979) *The peatlands of Ireland.* Soil Survey Bulletin 35. An Foras Taluntais, Dublin.

Mather, A.S. (1986) *Land Use.* Longmans, 286pp

McCracken, D.M. (1993) *Extensive farming systems in Europe: an initial assessment of the situation in the UK and Ireland.* Joint Committee for Nature Conservation. Paisley.

Morrison, M.E.S. (1954) *The ecology and post-glacial history of a Co. Tyrone bog.* Unpubl. MSc thesis, Queen's University of Belfast.

O'Sullivan, G. *et al* (1994) *CORINE Land Cover Project (Ireland).* Project Report Ordnance Survey of Ireland and Ordnance Survey Northern Ireland for CEC, DG XI and DG XIV

CHAPTER 7

THE ECONOMIC AND POLICY ENVIRONMENT FOR FARMING

DR. JOAN E. MOSS, CENTRE FOR RURAL STUDIES, THE QUEEN'S UNIVERSITY OF BELFAST

'The main problems in farm management with which every farmer is confronted are those determining first of all the branches of production which offer the greatest opportunity for profit under prevailing economic conditions and having regard to the physical geographical and climatic conditions of the farm; and secondly, the level of production which gives the greatest margin of profit.' The Inquiry into the Economic Position of Small Farms in Northern Ireland, co-funded by the Empire Enterprise Board and the Ministry of Agriculture for Northern Ireland (Government of Northern Ireland, 1928).

Agricultural Land Use

Almost 80 per cent of the 1.35 million hectares of land in Northern Ireland is used for agriculture and a further 6 per cent for forestry (DANI, 1995). In total 75 per cent of the land is covered by grass or rough grazing and less than 5 per cent, approximately 60,000 ha is under arable crops. Hence the Northern Ireland countryside is dominated by grassland and a visitor in the 1990s might be tempted to conclude that the very low incidence of arable crops was a direct consequence of the interactions of climate, land quality and topography.

Had the visitor been travelling through Northern Ireland countryside in the 1850s, however, a very different scene would have been observed. Instead of 60,000 ha of agricultural and horticultural crops there was 441,000 ha, with over half the arable area devoted to oats (Government of Northern Ireland, 1926). In this chapter the far ranging changes which have occurred in the Northern Ireland farming industry in the past 150 years in terms of farm structures, labour force, crop and livestock enterprises will be highlighted and the policy environment in which farming has operated in recent years will be considered. The historical data presented in this chapter were obtained from a series of annual and periodic reports and statistical reviews of agricultural statistics published by the Ministry of Agriculture for the Government of Northern Ireland over the period 1925 to 1970 and subsequent statistical reviews and periodic reports published by the Department of Agriculture for Northern Ireland. Since much of the information was obtained from time series, the year(s) for which data are cited may not always correspond with the year of the reports or reviews. Particular attention will be paid to the past 25 years of membership of the European Community, and latterly the European Union, during which time the Common Agricultural Policy (CAP) has dictated the terms and conditions under which farm businesses operate.

In Figure 7.1 the areas of principal crops are presented for four periods[1]: the 1850s for which the earliest agricultural statistics exist; 1925 when the effect of the United Kingdom's Government Compulsory Tillage Order for the First World War ceased to impact on farmers' decisions; 1961 when the tractor had taken over from draft animals; and 1995. In the 1850s oats accounted for over half the total area cultivated. The oat crop was used for human consumption and to feed the 100,000+ horses then on the farms. The second most important crop was potatoes (97,000 ha) followed by flax, wheat and turnips. By 1925, the oats and potatoes had almost halved in area and all other crops were also in decline; wheat and barley cultivated was negligible and flax, the non-food crop had also reduced by a third. Only 160,000 ha of crops were cultivated in 1961. Oats and potatoes had undergone a further halving of area and there were negligible areas of flax and turnips. The only crop to increase in area planted was barley (45,000 ha). Thirty four years later in 1995, the long term trend of declining arable area was still evident. Potato and oat cultivation had declined to 9,000 and 3,000 ha respectively and the area of barley had reduced to 33,000 ha.

---

[1] Agricultural statistics have been collected regularly for Ireland since 1847, nineteen years before such information was first obtained in Great Britain, and provide a rich source of information on the trends in land use over the past 150 years.

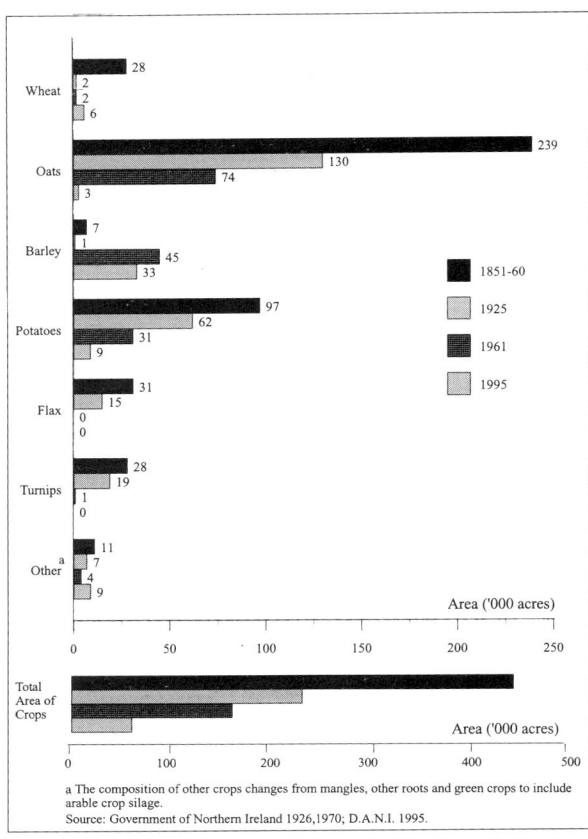

Figure 7.1. Areas of Principal Crops in Northern Ireland, 1851-60 to 1995

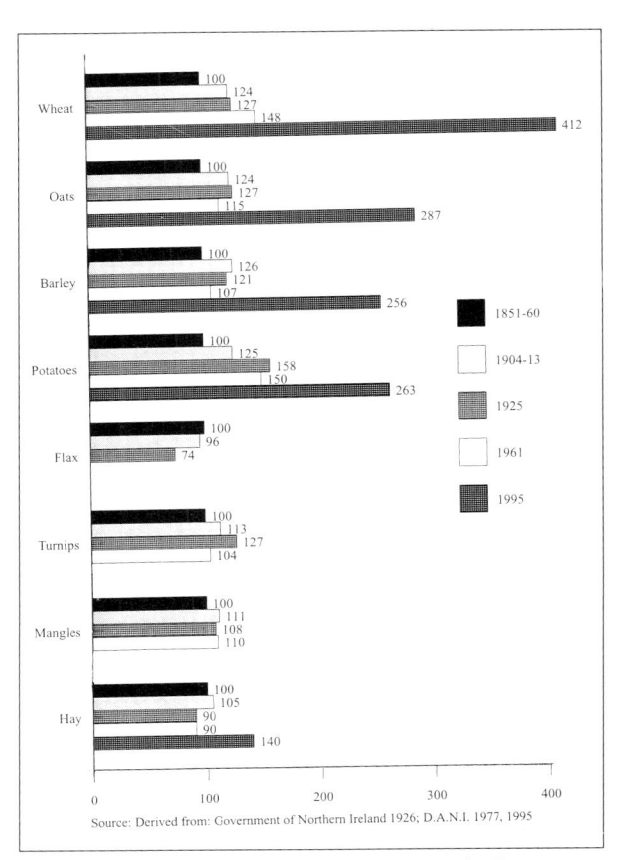

Figure 7.2. Indices of Crop Yields relative to the Base Period 1851-60

Agricultural Yields

Crop output is determined by area cultivated and yield. Over the 145 years for which there are detailed agricultural statistics for Northern Ireland, yield estimates have been made (Figure 7.2). Until 1961, there was only modest improvement in crop yields but in the last 34 years wheat, oats, barley and potatoes have all experienced substantial improvements in yield as a result of scientific plant breeding. Agri-chemicals have been applied for both plant nutrition and control of pests and plant diseases; precision cultivation was facilitated by mechanisation. Wheat has benefited most from crop husbandry research reflecting its dominant role in world agriculture and the international research effort it has attracted culminating in the Blueprint for cultivation. In line with improvements in crop yields, the milk yield of dairy cows has also increased. In 1925 the typical dairy cow produced, on average, 1,900 l of milk per year. By 1973, average milk yield had risen to 3,900 l and in 1995 it was 5,250 l (Figure 7.3).

Livestock Numbers

While crop production has been in long term decline, the area of land devoted to grass production for grazing livestock has increased, resulting in a gradual build up of the cattle and sheep populations from the mid 1920s through to the 1990s (Figure 7.4). The populations of non grazing livestock, pigs and poultry have exhibited different trends. Pig production peaked in 1971 when numerous relatively small pig units were to be found on farms. Thereafter the higher cereal prices under the Common Agricultural Policy eroded the profitability of pigmeat production and resulted in a loss of confidence among producers (DANI, 1974). In 1995, the pig population was approximately half the 1961 level. The total poultry flock of both broilers and layers has maintained a continuous upward expansion path to the present day and now stands at just under 15 million birds.

The other animal, once present in large numbers on Northern Ireland farms but now absent, was the draft horse (Figure 7.5). From the mid 1850s through to the mid 1920s the working horse population remained fairly constant at just under 100,000. From then onwards the number of draft horses declined in line with reductions in the area of arable cultivation until the early 1950s. During the 1950s mechanisation in the form of tractors gained widespread popularity on Northern Ireland farms and by 1961 only 11,000 were recorded. The reduced number of draft horses required less oats cultivation for feed. There are still approximately 11,000 horses and ponies on Northern Ireland farms but the animals are kept for recreational and amusement purposes.

Agricultural Holdings

The number of agricultural holdings on which the

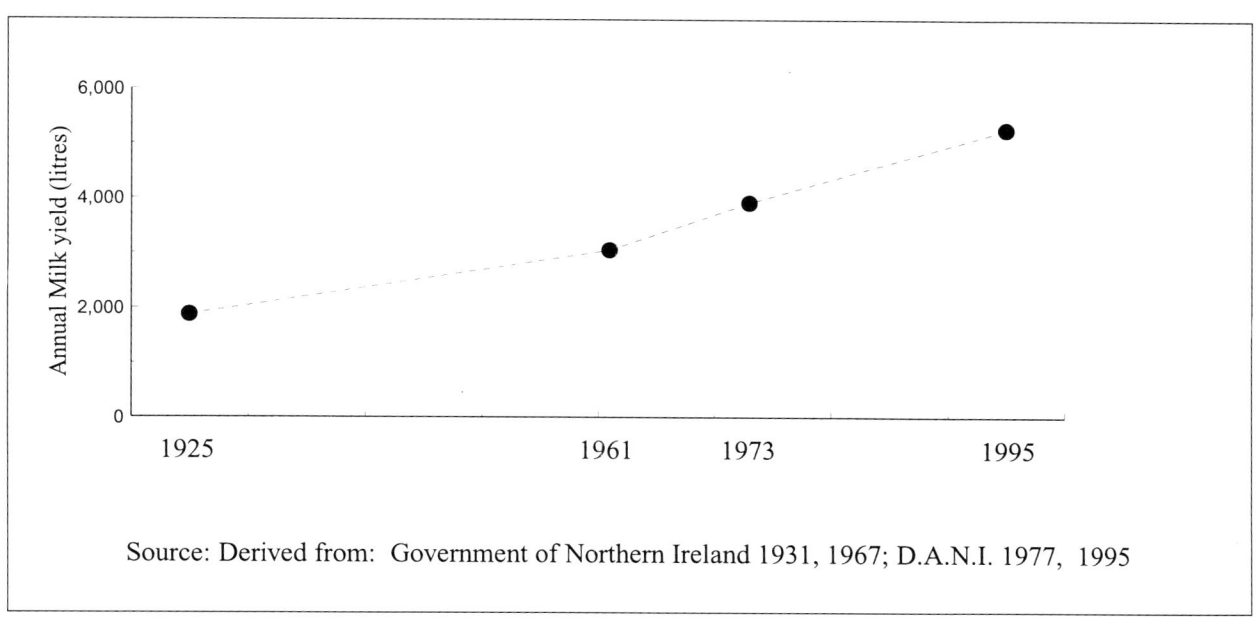

Figure 7.3. Average Annual Milk Yields (litres), 1925 to 1995

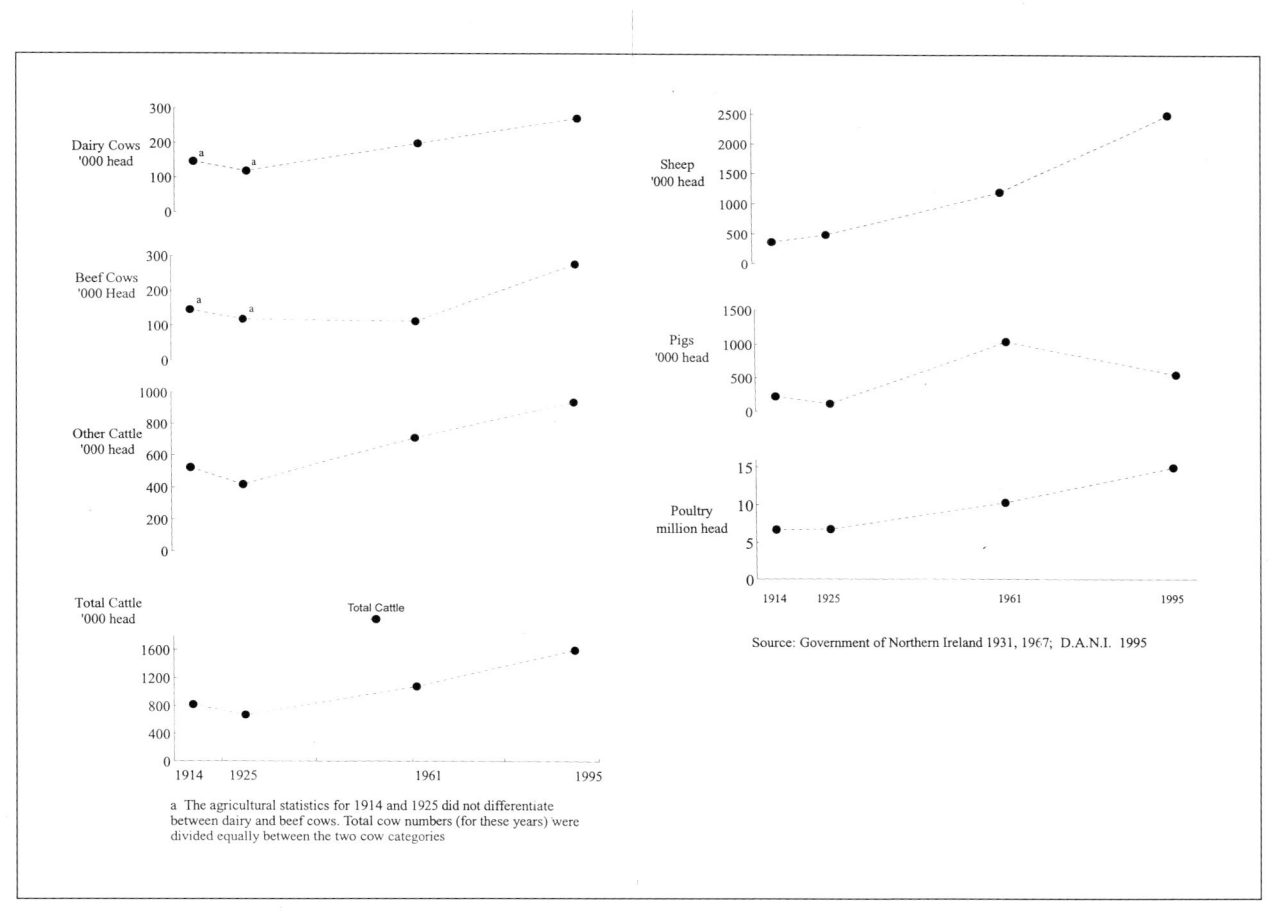

Figure 7.4. Livestock Numbers, 1914 to 1995

crops were grown and livestock reared has contracted steadily since the first records were kept (Table 7.1). Originally the number of agricultural holdings enumerated was approximately the same as the number of farms, however small, which supported a farming family. There were 138,000 such holdings in 1851. A succession of Irish Land Acts from 1870 to 1925 transferred to tenants the ownership of the land they had formerly rented from landlords. This process culminated in the Northern Ireland Act of 1925 which consolidated the system of owner-occupation over nearly all the farmland in the Province. In that year,

121

104,000 holdings over 1 acre of crops and grass and a further 75,500 holdings under 1 acre were recorded. It was noted that the holdings under 1 acre 'do not make any appreciable contribution to the agricultural output of the country' (Government of Northern Ireland, 1926). The average size of holdings increased as owners tended to purchase neighbouring holdings for amalgamation and thereafter to make a single return for the combined holding. By 1961, there were 71,000 holdings of at least one acre of crops and grass.

The number of farmers declined much more rapidly than the number of holdings. As elderly owners gave up farming or the heirs to farmland did not wish to continue farming on their own account, they were able to let most or all of their land under the conacre system of 11 month letting, while retaining their rights to their land as owner occupiers. Of the 71,000 holdings in 1961, it was estimated that there were only 50,000 full-time and part-time farm businesses.

By 1995, the number of active farm businesses with significant agricultural activity had declined to 28,000. It should be noted that the methodologies used to estimate the number of active farm business units recorded in the Northern Ireland Agricultural census, has undergone a number of changes since the early 1950s and the estimated numbers of farm businesses are not strictly comparable, but are presented to illustrate a general trend.

Agricultural Labour Force

The size of the agricultural labour force has declined as the number of farm businesses contracted. The earliest recorded full labour force, in 1912, enumerated a total of 214,000 persons (Figure 7.6). By the early 1960s this had declined to 115,000 and the following 30 years saw a further reduction to 57,000 by 1995, of which 33,000 were returned as either part-time or casual workers. The farm labour force in 1995 accounted for approximately 6 per cent of the total Northern Ireland workforce. Only 24,000 were recorded as working full time on Northern Ireland farms and of these 2,000 were hired labour. Given the number of active farm businesses, it can be seen that the vast majority of farm workers are farm owners.

Scale of Farm Enterprises

As the number of active farm businesses declined, the area of land farmed per farm has increased and the number of enterprises per farm has also fallen. The cumulative impact of the two long term trends of concentration and specialisation has been a significant increase in scale of enterprises. Since the accession of the United Kingdom to the European Community in 1973, there has been an acceleration in enterprise scale.

Dairying has undergone the greatest change in average size of enterprise. In 1973, almost a quarter of dairy cows were in herds of fewer than 15 cows and almost half were in herds of more than 30 cows. By 1995, only 3 per cent of dairy cows were in small herds and 85 percent were in herds of 30 cows plus (Figure 7.7). The trend towards larger herds of beef cows has been much less dramatic, with only half the beef cows in herds of 30 or more by 1995. Likewise, sheep flocks have risen in size with a negligible proportion now in very small flocks of below 50 ewes and over three quarters in flocks above 200 ewes, compared with less than half in larger flocks in 1973.

Despite the long term decline in the aggregate cereal acreage, the proportion of cereal production grown in areas of at least 20 ha (almost 50 acres) per farm has risen from one quarter in 1973 to almost one half in 1995 (Figure 7.8). Even greater changes have occurred in the average area of potatoes grown per farm, with over a third now grown in areas of at least 20 ha, compared with less than a tenth grown on such a scale in 1973.

Consideration has to be given to the factors which have led to these marked changes over time in the numbers of people gaining a livelihood from farming, the crop and livestock yields obtained and the scale of enterprises on Northern Ireland farms. At the beginning of this chapter a definition of the main problems in farm management, taken from an economic enquiry of Northern Ireland farming in 1928, was quoted. That definition of the economic objectives of farmers and the constraints under which they operate still holds true almost 70 years later.

Structural Change

The changes in employment in farming, the number of farm businesses, trends in specialisation and size of enterprises, reflect the resources deployed in agriculture adjusting to what is commonly called the 'Farm Income Problem'. A key characteristic of agricultural products is that they face relatively static demand. As society becomes more prosperous, as a consequence of economic growth, individuals have more money to spend but they do not increase their expenditure on basic food products in line with their increasing disposable incomes. More money may be spent on convenience foods such as ready prepared meals or restaurant meals but the amount of money spent on food 'raw materials' does not change significantly, unlike expenditure on transport, consumer goods, clothing and leisure services to name a few. Consequently, over time, the proportion of total consumer expenditure on household food declines.

At the same time as demand is fairly static, advances in modern production methods arising from technological advances in the fields of mechanisation, agro-chemicals and biological knowledge have greatly enhanced the supply of agricultural products onto the market. When market forces operate untrammelled, the combined effect of static demand and increasing supply causes a downward pressure on farm product prices. This in turn exerts a downward pressure on the sectral income of the agricultural industry.

120,000

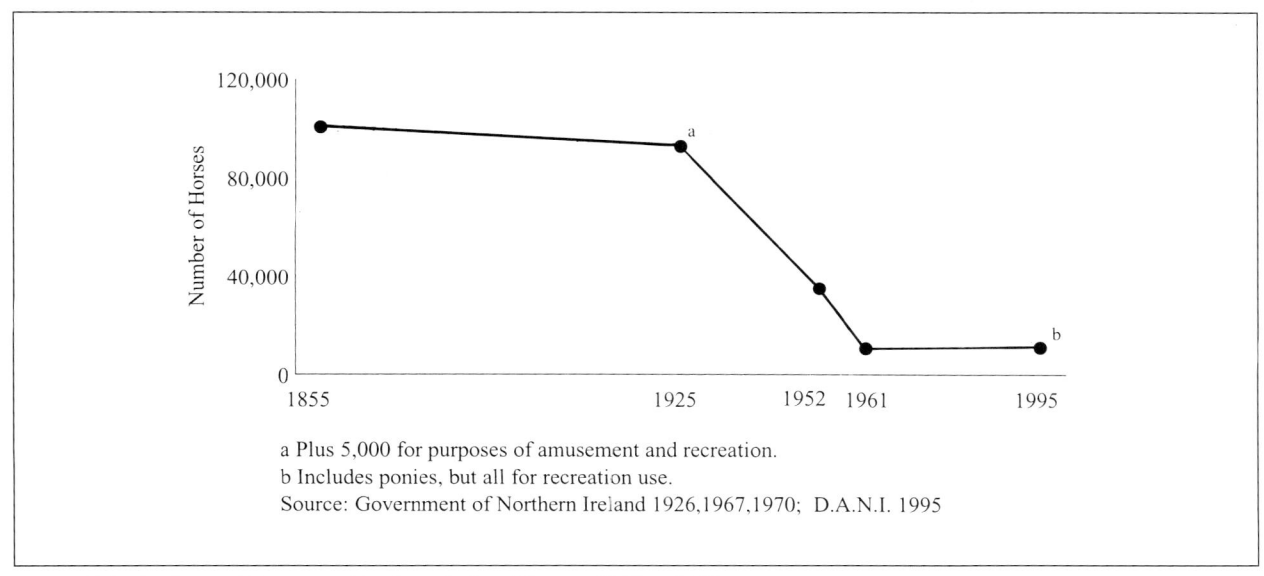

80,000

Number of Horses

40,000

0

1855　　　　　　　　　1925　　1952　1961　　　　　1995

a Plus 5,000 for purposes of amusement and recreation.
b Includes ponies, but all for recreation use.
Source: Government of Northern Ireland 1926,1967,1970; D.A.N.I. 1995

Figure 7.5. Numbers of Horses on Northern Ireland Farms, 1855 to 1995

The extent to which the pressure on sectoral income translates into pressure on individual farm incomes is determined by productivity, ie the relationship between the quantity of goods produced and the resources required for their production, in particular, the partial measure of labour productivity, i.e. output per worker. The productivity of labour is influenced by the input of all other resources used in the farming process.

Agriculture has a proud record of productivity growth mainly arising from technological developments and organisational changes in farming. These organisational changes include specialisation. Many technical developments require a certain scale of production if the investment is to be justified e.g. automatic cluster removers in milking parlours, combine harvesters, potato planters. When, as a result of technological advances, supply starts to 'run ahead' of demand, there are two possible responses. The first is that price drops, acting as a brake on supply and a stimulant to demand (whether in the domestic or export markets); and farm incomes then come under pressure, because the increase in volume of sales does not compensate for the reduction in price. A second possibility is that the government intervenes to put a brake on price reductions and surpluses accumulate.

In the face of downward pressure on product prices, farmers have the incentive to introduce even more advanced technology to increase their agricultural output. This tendency, when applied across the whole farm sector, further increases supply

Table 7.1
Number of Agricultural Holdings and Estimated Number of Farm Businesses, 1851 to 1995

Year	Holdings	Farm Businesses
	('000)	
1851	138	138[1]
1901	116	116[1]
1925	104[2]	60[3]
1961	71[2]	50[4]
1995	40[5]	28[6]

1. All holdings considered active farm businesses, irrespective of size.
2. Area 1 acre or greater.
3. Estimate derived from 1926 Northern Ireland Census of Population which enumerated 59,513 farmers.
4. Estimate derived from special sample survey June 1961 conducted with Agricultural Census.
5. Holdings defined as ownership units.
6. Estimate based on number of active main holdings.

Source: Derived from Government of Northern Ireland 1926, 1967; D.A.N.I. 1977, 1995.

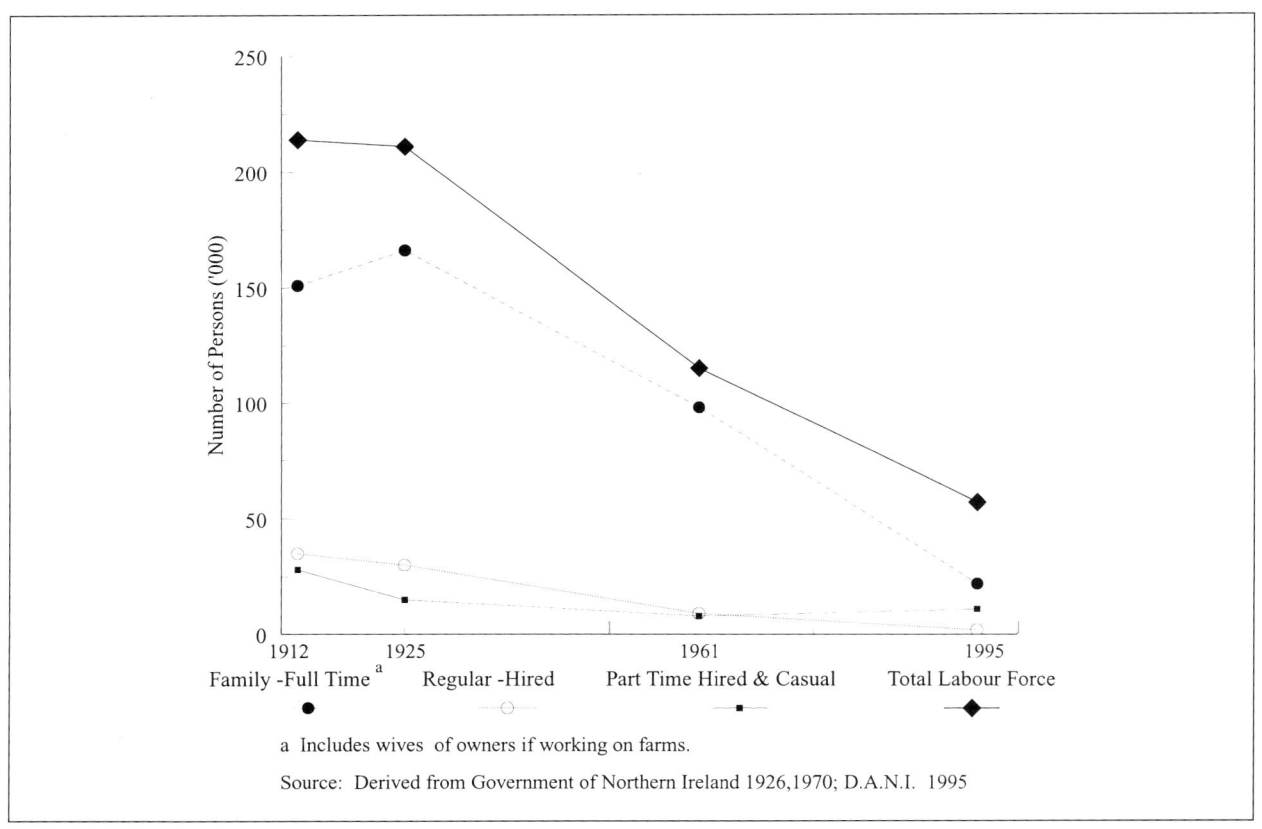

Figure 7.6. Agricultural Labour Force, 1912 to 1925

which in turn places even greater downward pressure on prices. This phenomenon has been described by Willard Cochrane as the Technology Treadmill.

'The average farmer is on a treadmill with respect to technical advance. In the quest for increased returns ... which he hopes to achieve through the adoption of some new technology he runs faster and faster on the treadmill. But by running faster he does not reach the goal of increased returns, the treadmill turns over faster ... and ... grinds out more and more farm products for consumers'.

(William Cochrane, 1952, *Farm Prices: Myth and Reality*)

If there is an outflow of labour from agriculture and this outflow matches or exceeds the increase in labour productivity, then individual farm incomes can be maintained, even though sectoral income may be in decline (relative to the rest of the economy). An individual farmer by reducing the labour force, investing in plant and machinery, specialising in one or two key enterprises and farming a larger area of land can earn a living considered acceptable, in terms of incomes in society at large.

It has long been recognised that the outflow of labour from the agricultural sector has not matched the increases in labour productivity. Governments intervene in the operation of agricultural markets, in response to pressure on farming incomes. Other reasons for intervention include:–

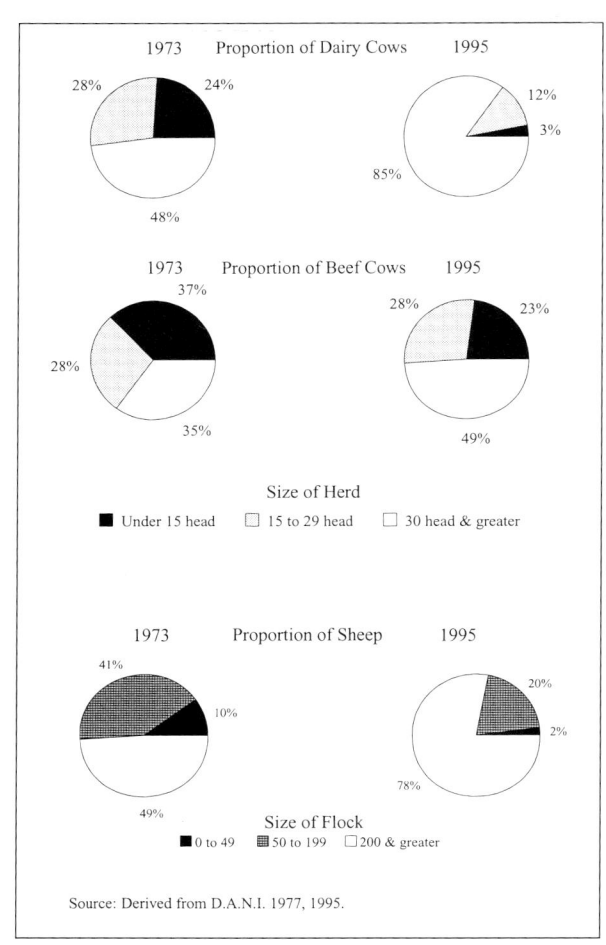

Figure 7.7. Changing Size of Livestock Enterprises, 1973 to 1995

1. Market price instability – a high degree of variability in the volume of agricultural output from year to year causes price instability. The main reasons for output variability are weather, incidence of animal and plant pests and diseases and the tendency for concerted production decisions by a large number of farmers to lead to over and under supply. (The potato sector in Northern Ireland exhibited such price instability over many years.)

2. Security and stability of supply – governments wish to ensure that in times of international emergency there are adequate and secure stocks of food. (The experiences of food shortages in Western Europe during the Second World War led to great importance being attached to a high degree of self-sufficiency in food production. Until recently this was a dominant consideration in the formulation of the Common Agricultural Policy).

3. Price of food – a relatively high proportion of consumer expenditure (typically between 10-25 per cent) goes on food and beverages, depending on the income of consumers. At very low levels of income, an even higher proportion of disposable income is spent on food. There are, consequently, strong welfare arguments for keeping down the cost of food. (These arguments had a strong influence on the design of UK agricultural measures prior to access in to the European Community).

4. Consumer protection – governments acknowledge the importance of ensuring that consumers are protected by specifying standards for food quality.

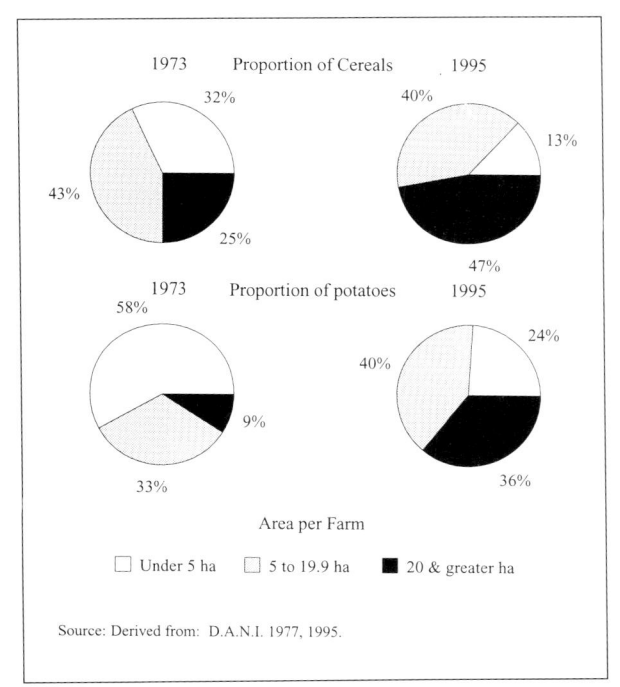

Figure 7.8. Changing Size of Crop Enterprises, 1973 to 1995

Expansion of Production under the Common Agricultural Policy[2]

The major expansion of output in Northern Ireland's dairying, beef and sheep sectors has occurred since accession to the European Community in 1973. From that date the Northern Ireland farming sector has operated subject to the CAP which was established by The Treaty of Rome in 1957. Article 39.1 of the Treaty identified the objectives for 'a common policy in the sphere of agriculture':

(a) *to increase agricultural productivity by promoting technical progress and by ensuring the rational development of agricultural production and the optimum utilisation of the factors of production, in particular labour;*

(b) *thus, to ensure a fair standard of living for the agricultural community, in particular by increasing the individual earnings of persons enjoyed in agriculture;*

(c) *to stabilise markets;*

(d) *to assure the availability of supplies; and*

(e) *to ensure that supplies reach consumers at reasonable prices. (Treaty of Rome 1957 Article 39.1).*

There are two main elements in the CAP, structural measures and price and market support. Structural measures are designed to tackle the fundamental structural problems of the agricultural sectors of the Community. The price and market policies are concerned with the market alleviation of the adverse consequences of the structural problems, mainly low prices and increases. Structural measures have attracted a relatively small proportion of CAP funding, 5 per cent in 1973 rising to 8.5 per cent in 1993. The recent increase in structures funding has arisen because of the 1992 reform of the CAP, which is discussed later (Commission of the European Communities, 1994).

Prior to reform, the price and market policies of the CAP operated whereby a producer's income had to be determined principally by the price produce fetched in the market, rather than by direct income supplements or allowances determined by the area of crops grown or livestock numbers. Prices were established by the interaction of supply and demand in the EC market within the framework of the CAP commodity regimes, at a level which was designed to ensure that even relatively high cost producers could earn a living.

This was achieved by deriving annually a desired market price for Community produce termed a Target, Base or Guide price depending on the Commodity. This was a reference point and from it two additional prices were determined for each commodity:

---

[2] For a more detailed presentation of the Common Agricultural Policy see Moss, 1994.

(a) Threshold price – the minimum entry price chargeable by importers from third countries (this sets a price culling);

(b) Intervention (or buying-in) price – the floor price at which national intervention agencies were obliged to purchase Community produce offered to them. The intervention price was set significantly higher than the world price. Consequently market prices were kept above world prices.

This high price was maintained by controlling the flow of imports into and subsidising the flow of exports out of the Community. When the Community was a net importer of a commodity, the EC price was maintained above the level of world prices by imposing taxes (levies) on imports, whereby the difference between the prices had to be forfeited by the importers. When the Community was a net exporter of a commodity (ie the commodity was in surplus in the EC market), intervention buying was the primary internal market price support mechanism. The intervention stocks could not be held indefinitely and much had to be exported on to the world market, where they fetched the (lower) world price. Export refunds (restitutions) financed from Community funds financed the subsidised exports.

Because of the high cost of maintaining intervention stocks, particularly of perishable commodities such as meat and dairy produce, it was cheaper to subsidise traders to export directly onto the world market, than to go through intervention. This price support mechanism has been described in some detail, although avoiding the full complexity of the various relevant EC Commodity regimes, because of its importance to Northern Ireland agriculture. Since accession to the EC Northern Ireland has been a net exporter of dairy produce, beef and sheepmeat. In 1994, for example, 80 per cent of Northern Ireland's agricultural produce measured in value terms, after adjusting for the value of subsidies, was exported, whether to Great Britain or further afield (DANI, 1995). Throughout the 1980s there was a shift in support away from intervention buying to the payment of various premia to livestock farmers. In addition to price support measures and payment of premia, farmers in the disadvantaged Less Favoured Areas (which account for approximately 75 per cent of farm land) received headage payments (Figure 7.9).

Northern Ireland farmers also, indirectly benefited financially from the weakness of the UK economy relative to the rest of the European Community. EC prices and premia are set in European Currency Units (ECU). Each time Sterling was devalued against the other major European Currencies, the target prices and premia rose in value when expressed in pounds Sterling.

With price support mechanisms in place, the onward march of technology and corresponding increases in agricultural productivity, steadily increased supply and the cost of support. The build up of dairy produce surpluses, which were very expensive to store, led to the introduction of dairy producer quotas in 1984 and subsequent quota reductions. By the late 1980s the cost of supporting the other major commodities had reached such a level that the Commission of the European Communities was forced to reassess price support as part of a wide scale reform of the CAP. The negative impact on world trade of disposing subsidised exports on the world market also had to be considered in the protracted Uruguay round of GATT negotiations.

Reform of CAP

The reform of the cereals regime in 1992 involved a reduction in support price levels, supply control via set-aside and arable compensation payments to farmers. The initial obligation to set-aside 15 per cent of the eligible area, for which compensation could be claimed, was restricted to those farmers cultivating, on average, at least 19.5 ha of cereals. When the measure was introduced only 127 of Northern Ireland farmers were subject to this restriction (DANI, 1993).

The milk regime was little affected by reform and production quotas remained the cornerstone of the milk regime. The other two major livestock enterprises, beef and sheep meat production, however, were subjected to far-reaching reforms involving a significant move from price support to direct support payments. The direct (compensatory) payments for beef production were designed to compensate farmers for progressive reductions in intervention price and the trigger price for safety-net intervention. They were also designed to even out the flow of cattle slaughtering throughout the year and reduce stocking rates. Entitlement to compensatory premia was restricted to the number of cattle existing on the farms in the base year with a maximum of 90 eligible cattle in each age group per producer per year for the Beef Special Premium. It has been estimated (Moss and Phelan, 1994) that for the small scale beef producers in the Less Favoured Areas, the direct payments may account for more than 100 per cent of the income derived from their beef enterprises.

The sheep meat regime was also subjected to individual producer quotas on the number of annual live premia which could be paid (with 1990 the base year) and ceilings for producers fixed at 1,000 head in the LFA and 500 head elsewhere. Only 50 per cent of the premium is payable for ewes held in excess of these ceilings. There were no stocking rate restrictions introduced on the ewe premium payments to sheep only farmers, but where cattle and sheep are mixed, ewes are taken into account when computing the stocking rate for the beef entesificatia premia.

A set of accompanying measures were also introduced to promote the use of agricultural land for

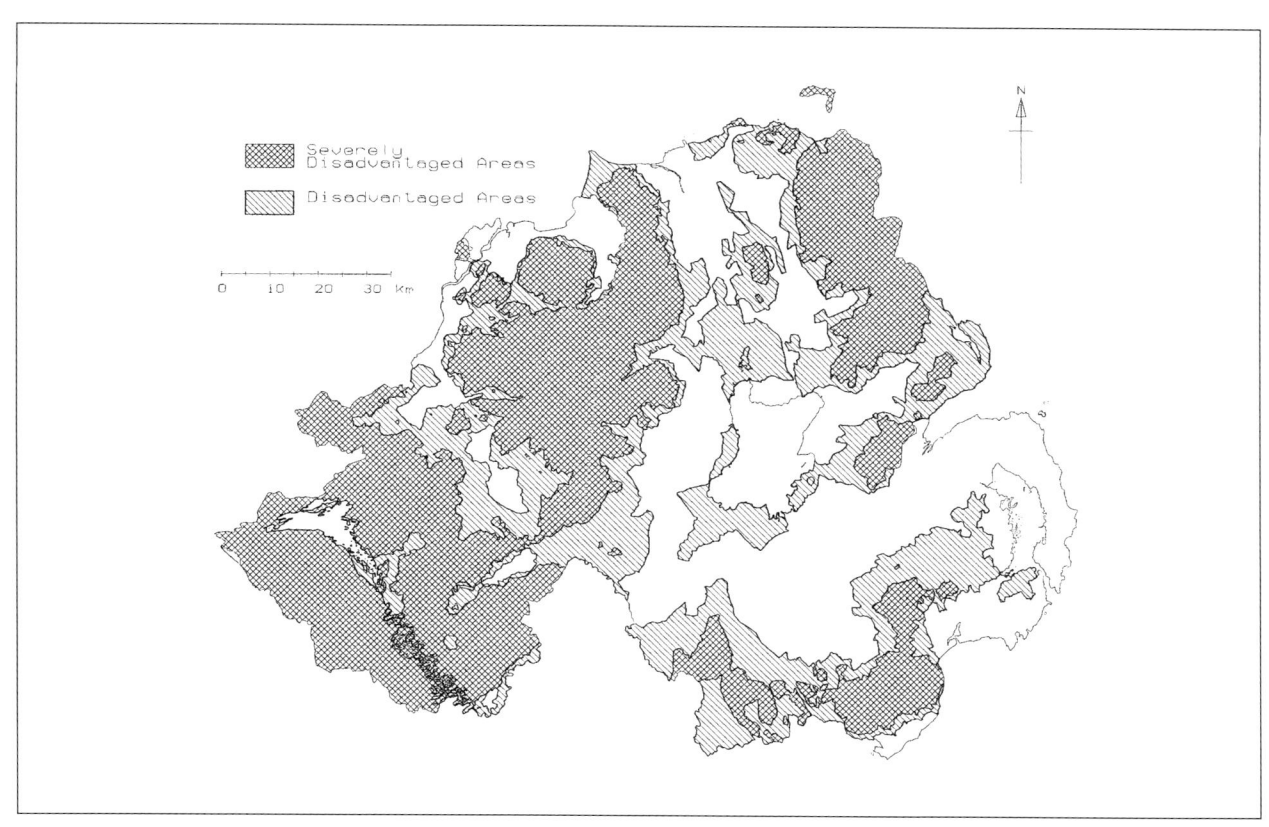

Figure 7.9. Northern Ireland Less Favoured Areas

forestry and to grant aid the introduction or maintenance of production techniques which encourage the protection of the environment, the landscape and natural resources (Commission of the European Communities, 1992).

The combined effect of the CAP reforms on the beef and sheep enterprises has been to halt the expansion of suckler cow and ewe numbers. The compensatory payments were designed, as their name suggests, to compensate for reductions in guarantee prices. On 17 September 1992, however, as a result of a change in monetary policy, the United Kingdom Government decided to leave the Exchange Rate Mechanism, within which Sterling had been shadowing the Deutschmark, at an exchange rate which had become increasingly difficult to sustain. The consequence was an immediate devaluation of Sterling vis a vis the main EC currencies. This, in turn, led to a devaluation of the Green Rate – the rate at which support measures expressed in European Currency Units are converted into Sterling and an immediate rise in support prices and compensatory premia when expressed in Sterling. By the end of 1992, the Green Rate for Sterling had been devalued by 11.4 per cent, with corresponding rises in value of support measures.

Subsequently, Northern Ireland farmers have had the double benefit of higher prices and direct payments to compensate for price reductions which have not materialised. The earlier outline of government intervention in agriculture emphasised the importance of agricultural policy when farmers are planning their farm businesses, in terms of enterprise choice and scale. The financial remuneration associated with the major land-using enterprises is highly dependent on the portfolio of agricultural policies in place at any one time.

There is research evidence (Lund and Hill, 1979) that the land using enterprise with the greatest economies of scale (i.e. where the larger the enterprise, the greater the financial return, per unit of enterprise) is cereal production. This arises mainly from the high capital outlay associated with mechanisation, which can only be profitably utilised when a sufficiently large area is cultivated. The relatively small scale of holdings in Northern Ireland, reflecting the settlement patterns and land tenure practices of earlier centuries, associated with a reluctance for farmland owning families to sell land, even when they do not wish to farm it, has inhibited the establishment of significant numbers of holdings large enough to specialise in cereal production. The cereals enterprises in England and Wales averaged 41.2 ha in 1994, compared with 9.5 ha in Northern Ireland (DANI, 1995). Apart from any climatic influences which may exist, a further factor which may also have inhibited large scale cereal production has been the Northern Ireland topography typified by small rolling hills, called drumlins.

The predominance of grazing livestock enterprises, with dairying being the most profitable, is consistent

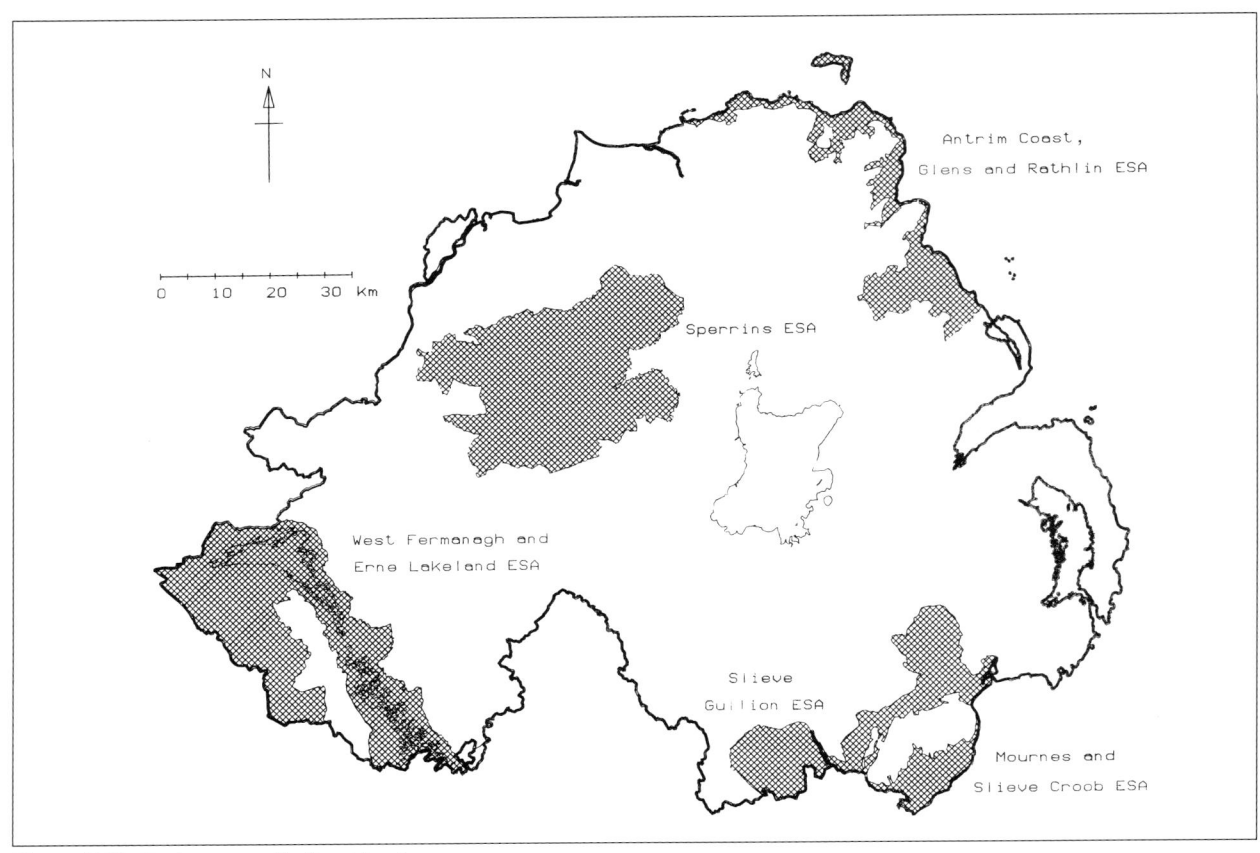

Figure 7.10. Environmentally Sensitive Areas

with the size distribution of the farm holdings in Northern Ireland and the high incidence of part-time farming, where farming income is augmented by income from the employment off-farm of either the farmer and/or the farmer's spouse.

Agricultural Structural Policy

In addition to discussion of the reform of the price and market support measures in CAP and the subsequent impact on farmers decision making, consideration must be given to the other arm of CAP, namely, structural policy. The two main elements of structural policy are:

1. measures directed at decreasing the numbers of people employed in agriculture;
2. measures directed at increasing farm size.

The early CAP structural measures were designed with the objective of improving the structure of the agricultural sector by the modernising and enlarging of holdings; training labour and encouraging the exit of excess, particularly elderly, labour from agriculture; the strengthening of processing and marketing structures; the reduction of structural handicaps and the improvement of infrastructure in specific regions.

A structural measure which has had a significant impact on farming incomes was the differentially larger capital grants and headage payments for breeding cattle and sheep in the designated Less Favoured Areas (Figure 7.9). Most farmers consider the headage payments, Hill Livestock Compensatory Allowances, to be price support policy but they are an important regional element of structural policy. They were designed as direct income payments to help farmers to remain in the Less Favoured Areas, despite their production and marketing disadvantages associated with their location.

The reasons for reforming the price support measures have already been discussed but by the mid 1980s it was also recognised that there still existed a serious low income problem for smaller farmers, despite the expensive support measures and burgeoning surpluses. As support was directly linked to production the larger farmers, while only accounting for 20 per cent of the holdings in the European Community, gained 80 per cent of support (Commission of the European Communities, 1990).

It could no longer be argued that supporting agricultural prices would guarantee all farmers an acceptable income. It was envisaged that, in future, market forces would have to be more important in determining levels of farm output and price restraint was introduced, as already discussed. In the face of surpluses, it was no longer considered feasible to provide assistance to farmers to improve their farming incomes by modernising their farm business, if this entailed expanding output of those products already in surplus. From 1985 onwards, a number of structural measures were introduced which sought to: improve working conditions on farms; reduce costs; improve the quality of farm produce; and protect, and where possible enhance, the rural environment without increasing farm output.

It was recognised that the problems faced by the farming community could not be resolved solely within the agricultural sector. New economic activities were required, on and off the farms to stimulate the economies of rural areas. There was also the prospect of the Single European Act coming into force at the end of 1992 and the recognition that disadvantaged peripheral regions might require additional assistance.

The resultant reform of the EC Structural Policy which led to Northern Ireland being designated an Objective One region, the highest priority category of disadvantaged regions, resulted in a Rural Development initiative which highlighted the improvement of the structure of the agricultural sector as essential to the development of the rural areas.

There have subsequently been two Operational Programmes relating to the agricultural sector.

1. The Agricultural Development Operational Programme (ADOP) for the period 1991 to 1994;
2. The Sub Programme for Agricultural and Rural Development (SPARD), 1995-1999.

Both sub programmes have provided capital grants to 'assist the agricultural sector and related industries to met the challenges and take full advantage of market opportunities and protect the natural environment.' With regard to agricultural production, the emphasis has been on improving quality, as opposed to increasing agricultural outputs. Measures have also been funded to enhance the rural environment, reduce pollution incidents and conserve the ecological and amenity value of the rural landscape.

In recognition of the failure of agriculture to generate a standard of living for farmers commensurate with that enjoyed by those employed in the non-farm sectors, funds have been made available to encourage farmers in Northern Ireland to diversify and develop alternative enterprises, both agricultural and non-agricultural in nature.

Part-time Farming

The farmers in Northern Ireland can be divided into three almost equal groups: those above the state retirement age of 65 years; full-time farmers below retirement age who obtain all their income from their farming activity; and farmers below retirement age whose household income includes income from the non-farming employment of the farmer and/or the farmer's spouse (Moss, 1992).

Where the farmer has off-farm employment the major farm enterprises are usually beef cattle and/or sheep. These enterprises, with their low daily labour requirements, can be undertaken while leaving time for the farmer to engage in work off the farm. The growth in part-time farming, otherwise known as pluriactivity, reflects the inability of relatively small land holdings to generate an income which will sustain a standard of living acceptable for the farming family. By combining farming with off-farm work, a higher income is generated and the farmer can retain involvement with agriculture, an activity which carries a high status in rural communities.

Environmental Objectives in Agricultural Policy

The incorporation of environmental objectives within agricultural policy further influences how farmers husband their land e.g. one element of CAP reform was the introduction of annual extensification premia, paid if the stocking rate is held below a specified level, which over time is set to decline. This achieves the two fold objective of restricting output and reducing the level of intensity of farming, with the subsequent impact on utilisation of grazing land.

A major agri-environmental initiative has been the introduction of Environmentally Sensitive Area Schemes, see Figure 7.10.

The first area thus designated was the Mourne Mountains and Slieve Croob in 1988, followed by the Glens of Antrim (now extended to include the Antrim Coast and Rathlin Island), West Fermanagh and Erne Lakeland, Sperrins and Slieve Gullion. In all, approximately 20 per cent of the agricultural land area of Northern Ireland was designated as environmentally sensitive by 1995. The main objective of the ESA Schemes is to 'help safeguard areas of the countryside where the landscape, wildlife or historic interest is of particular importance and where that interest would benefit through farmers continuing with, or engaging in, environmentally sensitive farming practices' (DANI, undated).

Farmers participating in the schemes volunteer to enter into 10 year management agreements in return for which they receive an annual payment for each hectare of land covered by the scheme. The ESA Schemes have 3 tiers, the higher the tier, the more restrictive the agreement, the greater the environmental benefits and in return, the higher the per hectare payment. The farmers, in addition, are subject to a range of environmental enhancing prescriptions which cover stocking rates, rates of application of organic and inorganic fertilisers, land reclamation or improvement, extension of drainage systems and application of agrichemicals.

Summary

While there has been a move towards market determination of agricultural product prices in the European Union and the opening up of such markets to international competition, the existence of direct payments in the form of various premia and compensatory amounts, leaves the Northern Ireland farmers still heavily dependent on government support for their livelihoods. Consequently, when farmers are determining *'the branches of production which offer the greatest opportunity for profit and the level of production which gives the greatest margin of profit'* their range of options is heavily curtailed by policies which restrict expansion of production, unless additional quota or compensatory premia entitlements are purchased. While individual farmers can expand their scale of enterprises, production of the main grazing livestock enterprises of dairying, beef cattle and sheep is capped at the regional level.

The limited opportunity to purchase additional land and restrictions on the intensity of farming further inhibit the farmer from taking full advantage of whatever economies of scale modern agricultural technology permits. The outflow of labour from the land usually starts with hired labour who cease to secure full-time paid employment, followed by farm family members who are encouraged in their youth to look off-farm for employment. The last to leave the land, usually at retirement or death, are farmers who usually have little experience of off-farm employment. Consequently, structural change in farming, now that the vast majority of farm labour is owners, is much slower than in other primary sector industries facing long term decline and subsequent restructuring e.g. fishing and mining. Many of the smaller full-time farmers are hoping to see out their days on their land, augmenting their meagre low farm income with welfare supplements and anticipating the day when they qualify for a state Retirement Pension (Moss and Phelan, 1994).

Such changes as are likely to occur to the mix of land-using enterprises and their scale in the foreseeable time will be dictated by prevailing agricultural policies. Unless the regional limitations placed on production of cereals, dairying produce, beef and sheepmeat are lifted, the pace of change will be slower than in the past decade, although the underlying forces for structural change in agriculture will persist.

References

Cochrane, W.W., 1952. *Farm Prices: Myth and Reality*, University of Minnesota Press, Minneapolis, USA.

Commission of the European Communities, 1990. *The Agricultural Situation in the Community*, CM58/90, Brussels.

Commission of the European Communities, 1992. Directorate General for Agriculture. *The Reform of the Common Agricultural Policy*. CAP Working Notes, Brussels.

Commission of the European Communities, 1994. *The Agricultural Situation in the European Union*, Brussels.

Department of Agriculture for Northern Ireland, undated. *Environmentally Sensitive Areas*, Explanatory Booklet (ESA/1). HMSO Belfast.

Department of Agriculture for Northern Ireland, 1977. *Ninth Report on the Agricultural Statistics of Northern Ireland 1966–67 to 1973–74*, H.M.S.O., Belfast.

Department of Agriculture for Northern Ireland, 1993. Memorandum, October 19, 1993.

Department of Agriculture for Northern Ireland, 1995. *Statistical Review of Northern Ireland Agriculture*, Economics and Statistics Division, A Government Statistical Publication, Belfast.

Government of Northern Ireland, 1926. Ministry of Agriculture. *First Annual Report upon the Agricultural Statistics of Northern Ireland 1925*, Cmd. 64, H.M.S.O., Belfast.

Government of Northern Ireland, 1928. Ministry of Agriculture. *The Inquiry into the Economic Position of Small Farms in Northern Ireland*, Belfast.

Government of Northern Ireland, 1931. Ministry of Agriculture. *Fifth Annual Report upon the Agricultural Statistics of Northern Ireland 1929–30*, Cmd. 133, H.M.S.O., Belfast.

Government of Northern Ireland, 1967. Ministry of Agriculture. *Seventh Report on the Agricultural Statistics of Northern Ireland 1952–1961*, Cmd.508, H.M.S.O., Belfast.

Government of Northern Ireland, 1970. Ministry of Agriculture. *Eighth Report on the Agricultural Statistics of Northern Ireland 1961/62 – 1966/67*, H.M.S.O., Belfast.

Lund, P.J. and Hill, P.G., 1979. 'Farm Size, Efficiency and Economics of Size', *Journal of Agricultural Economics*, Vol 30.

Moss, J.E., 1992. The Role of Pluri-Active Family Farms in the Development of the Rural Economy of a Marginal Area of the European Community in *Spatial Dynamics of Highland and High Latitude Environments*. Occasional Papers in Geography and Planning, Volume 4, Appalachian State University, Boone, North Carolina.

Moss, J.E., 1994. Agriculture in *Political Issues in Ireland Today* edited by N. Collins, Manchester University Press, Manchester.

Moss, J.E. and Phelan, J.F., 1994. *Farm Household Incomes: Prospects in a Changing Policy Environment,* Co-Operation North, Third Study Series, Report No. 4. Dublin.

CHAPTER 8

THE SOIL MAPS

This book, 'Soil and Environment: Northern Ireland', serves as a report on the work of the DANI Soil Survey 1987–1995, and is also the only written memoir to accompany the publication of 17 soil maps at 1:50,000. It is therefore both desirable and appropriate that one chapter of the book is devoted to descriptions of the distributions of the main soil series on each of the 17 soil maps. The other chapters in the book are either background material for these distributions on the soil maps, or describe the nature and properties of the main soil series, or show how new maps can be derived from soil maps. A knowledge of the 17 soil maps is basic to each and every other chapter in the book. Accordingly, an extract from the published colour soil maps is reproduced in Colour Plate No. 9.

All, but one, of the 17 soil maps is illustrated with a simplified sketch map to show the distribution of the main soil series. The exception is Soil Map 4, which overlaps on the soil maps around (5, 7 and 8) and has only a small area of about 160 km^2, between Coleraine, Binevenagh and Magilligan Strand, that is exclusive from the surrounding soil maps. This part of Soil Map 4 is described in the description for Soil Map 7. Each soil map described is provided with its own key, as the approach adopted in the map description is slightly different in each case. For example, a regional approach (dividing the area of the map into different soil regions) is adopted for the north coast and Fermanagh maps, as soil landscapes of these maps can be divided naturally into regions.

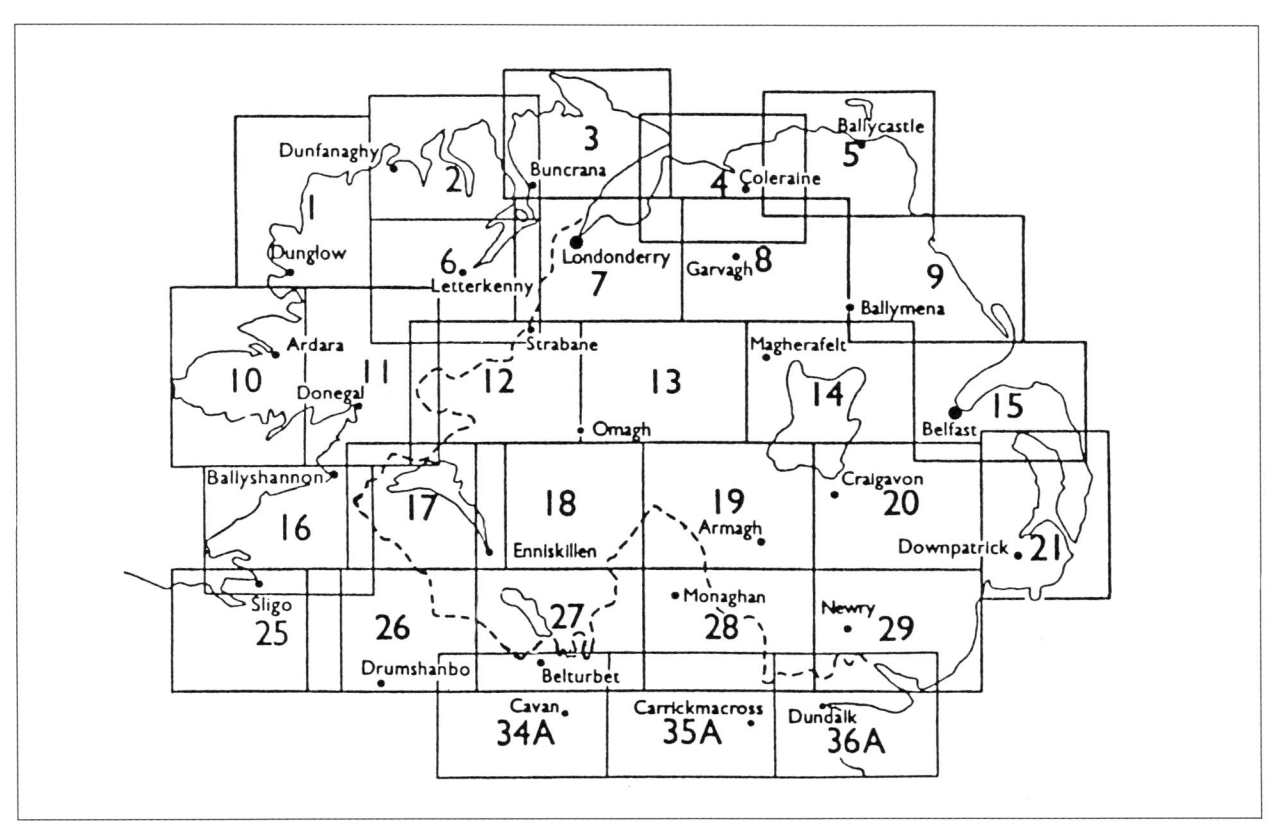

Figure 8.1 Framework of 1:50,000 Soil Maps

How to use a Soil Map

The lines on a soil map are the boundaries of soil map units, or soil polygons where each is uniform in soil profile (or vertical section) and uniform in soil parent material. The polygons are parts of a distinctive **Soil Series**, and in Northern Ireland, there are about 300 Soil Series. They are described in Chapter 2.

Each soil map unit, or polygon, bears the colour chosen to show its soil series, as well as a code of letters indicating the soil profile. The user should start using the soil map by reading this soil profile code (eg BR, BE, SWGl etc), and then refer to the table giving the full description of soil profile types. The second step is to move to the main key or legend colours, where each colour represents a soil series. The user moves to the headings of this key, and for a particular soil profile of interest, eg ranker, brown earth etc, the user works vertically down that column under ranker, brown earth etc, until the matching colour is found. At that point, the second part of the soil series character, the soil parent material, can be read from the column on left. Colour comparisons are relevant only within one vertical soil profile column, belonging to the soil profile of interest, and should not be read horizontally across the key. See Colour Plate No. 9.

The 1:50,000 soil maps for Northern Ireland are overlays on OSNI topographic maps, so valuable altitude and topographic information is available for each soil map unit from the topographic maps. If this is combined with observation of the location or site within Northern Ireland, climatic conditions may be estimated approximately, and combined with soil to provide an agricultural land quality classification (ALC) of each soil polygon.

Agricultural land classification

The agricultural land classification (ALC) of England and Wales is adopted here, and will be the scheme on which future land classifications are developed by DANI in land research work. The full details and definitions of the scheme, and how it is being developed, are reported in Chapter 9. It is a scheme of only 5 main classes or grades, but does depend on climatic and topographic, as well as soil, criteria. Class 1 is the best quality agricultural land, of which there is no extensive area in Northern Ireland. It has no physical limitations of ground surface, climate or soil, that would limit agricultural use or productivity. There are some small area examples of Class 1 land in the Province, but they have been subsumed in Class 2 land. The latter is still regarded as very good quality agricultural land, with only slight physical limitations on use or productivity. Class 3 is divided into 3A and 3B, so creating an extra class, and most agricultural land in the Province is in Class 3A or 3B, divided often between sandy textures and better drainage of 3A, and clayey textures and worse drainage of 3B. Class 4 is very limited in use, either by very poor drainage, extremes of climate, or by both in situations of droughtiness in shallow soil. Class 5 has almost no agricultural potential in lowland swamps, coastal marshes, or humic rankers of mountain tops. These classes of ALC are used frequently in the soil map descriptions which follow.

Soil map databases

It should be noted also that all the soil map information is available in digital form in two databases held by The Ordnance Survey of Northern Ireland (OSNI). These are the graphical database holding digital information on the boundary lines of the soil polygons, and the textual database holding information about the soil profile and soil parent material character of each soil polygon.

Soil polygon boundaries were captured digitally at OSNI by scanning outline soil map overlays. The data were then manipulated using graphical editing techniques to transform the data on to the Irish Grid. 12 figure georeferences are automatically generated for every soil polygon, and these georeferences provide the main link between the graphical and textual databases. The textual database contains separate records for each soil polygon (the soil profile type, soil parent material, georeference, Irish Grid sheet number and the area of the polygon in square metres being the main information stored).

The soil maps described here will be taken in order from the north coast, Soil Maps 4 and 5, progressively to the south, to the Mournes, in Soil Map 29.

1A
Soil map of the Braid Valley, east of Ballymena in Co. Antrim, showing brown earths (BE) in orange.

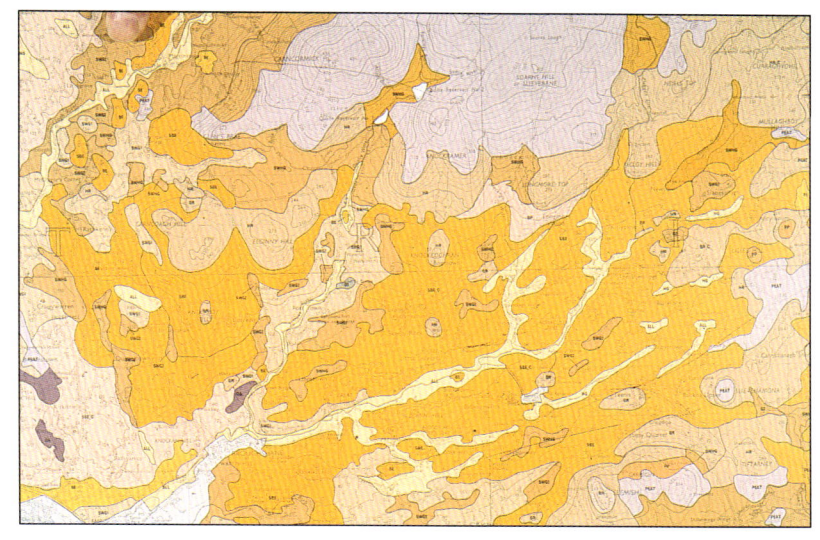

1B
Soil map of the north coast of Co. Antrim showing peat in purple and brown earths (BE) in orange.

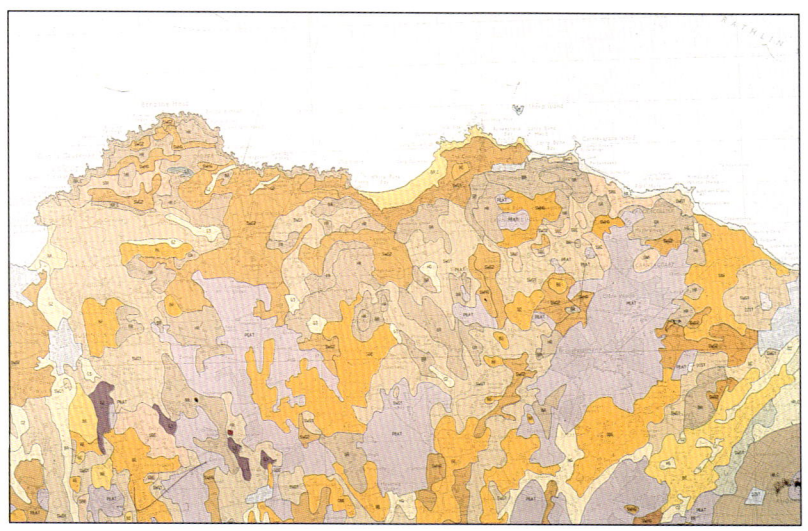

1C
A key for a Northern Ireland 1:50,000 Soil Map

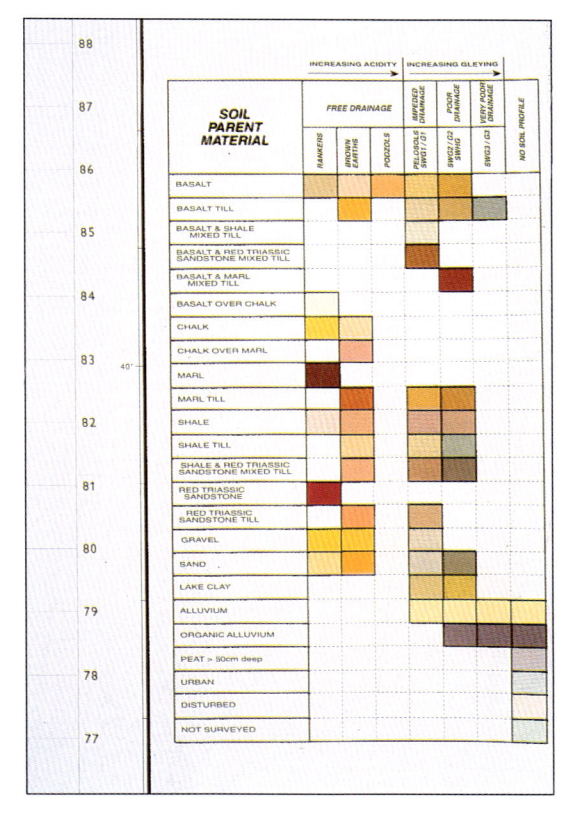

2A *The estuary of the River Roe, showing the Myroe Level, or polder, behind the sea wall. The ploughed surface reveals white, shelly sand*

2B *The west end of the polders, Donnybrewer and Longfield, where the soil has a peaty surface. Photographs by Esler Crawford*

2C *Soil map of the Myroe polder, Roe estuary, Magilligan sand spit and Binevenagh Mountain, all features in the photographs.*

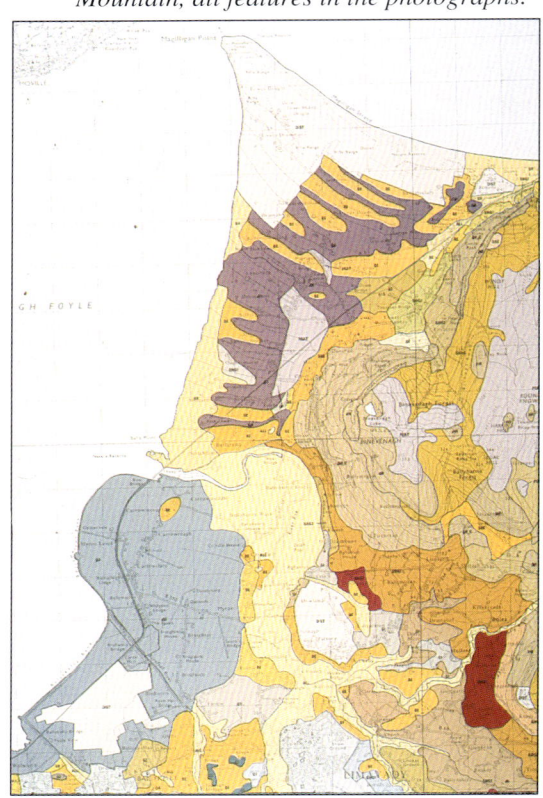

3A & B *Mixed farming in the Glenelly Valley of the Sperrins, Co. Tyrone, where the soils are SWG2 gleys on mica-schist till. Valley farms are centres for extensive hill sheep production*

3C & D *Diatomite soil, close to the Lower Bann River and Lough Beg near Toome in Co. Antrim, behaves like gleyed fine sand with a wide range of crop use, and which has been commercially exploited until recently.*

4A
Excavating a brown earth (BE) soil profile on basalt till for sampling at Bellair, 180m above Glenarm in east Co. Antrim. A chalk outcrop can be seen in the distance at Glenarm harbour.

4B
Preparing to sample a SWG2 on Calp till, near Donagh and Kilmacbrack in Co. Fermanagh.

4C
Sampling an undrained SWG2 on Calp till, near Derrylin in Co. Fermanagh, west of upper Lough Erne.

5A *Fields on the south-west slopes below Slemish Mountain in east Co. Antrim, showing peat, humic rankers (HR) and SWG2 on basalt till. Photograph by Esler Crawford.*

5B & C
Fields and stone walls of the Braid Valley, east of Ballymena, where soils are either BE or SWG1 on basalt till.

6A *Scrabo Hill, which is a dolerite plug with shallow rankers, and south-facing slopes of arable land on brown earths (BE) or gleys (SWG1) on Red Triassic Sandstone till.*

6B *Drumena Cashel is a soil landscape of rankers and shallow brown earths (SBE) on shale and granite rocks, near Castlewellan. Both photographs by Roy Anderson.*

7A *Slieve Commedagh in the Mournes, where the soils are rankers and podzols on granite, viewed from near Kilcoo, where soils on lower ground are brown podzolics (BP) on granite till.*

7B *Knockmore Cliff, where the soils are rankers on limestone and strongly gleyed SWG2 and SWG3 on Calp till in the foreground. Both photographs by Roy Anderson.*

8A
Traditional hand-cut peat on a cutting through blanket peat showing variable degree of decomposition.

8B
Lough Fea, in Co. Tyrone, showing the sausage-shape extraction of blanket peat. This machine exploitation of peat creates an undesirable internal drainage system.

8C
Derryleakagh, in south-west Co. Down, near Rathfriland, showing basin peat (mapped as organic alluvium) between ridges and drumlins of SWG1 on granite till. Photographs by Roy Anderson.

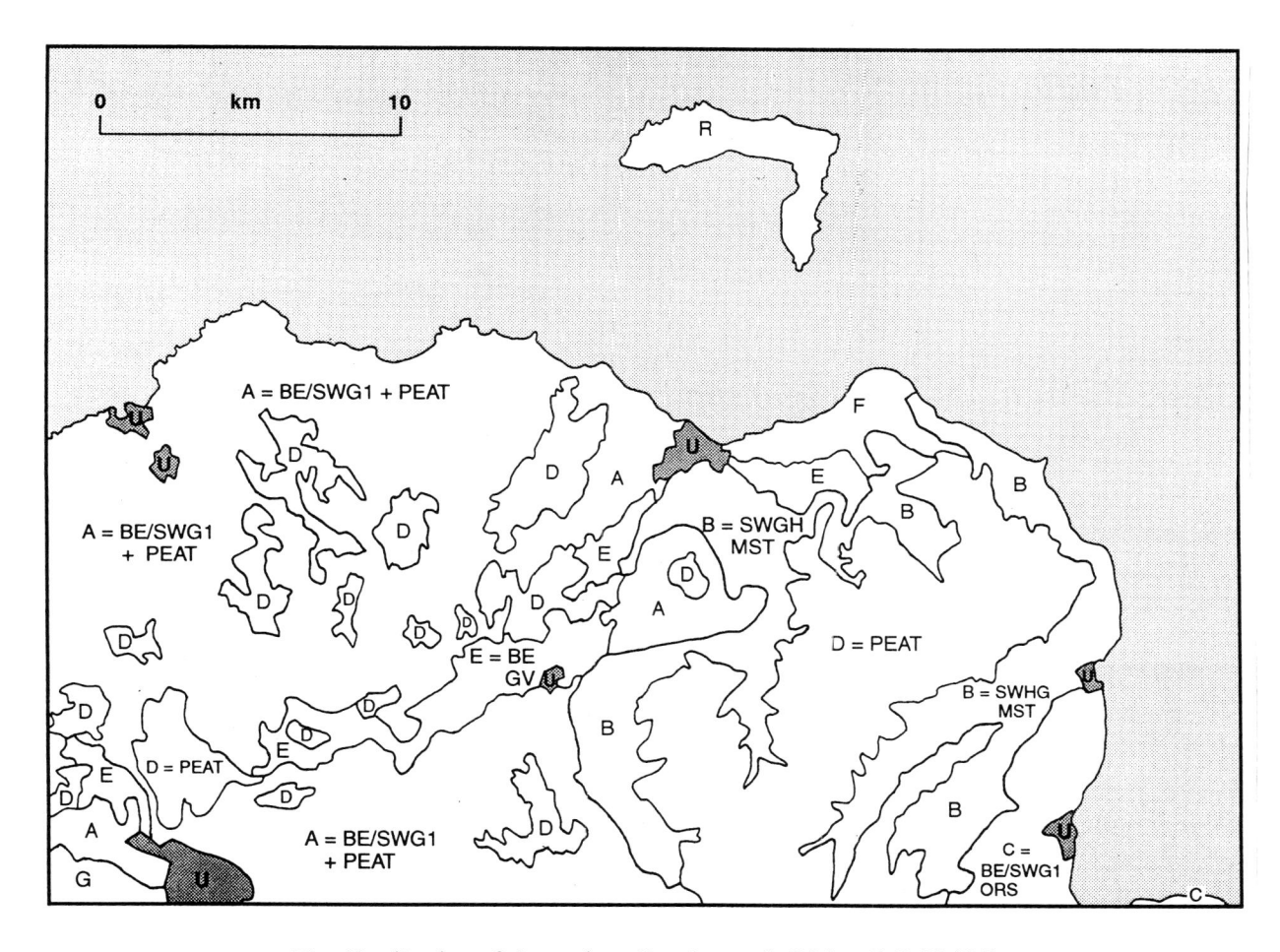

The distribution of the main soil series on Soil Map 5 (1:50,000)

Key for soil map sketch 5

Apart from Rathlin Island, where the soils are mainly shallow humic and brown rankers on basalt and brown rankers on chalk rock, the soils of north-east Antrim divide into seven main regional groupings:

A   the first is the Ballycastle-Ballymoney-Bushmills triangle of well-drained soils on thin drift over basalt, which is mixed with basin peat in the hollows and brown earths on the freely draining rises. There are SWG1 and SWG2 basalt till soils present, but in a minority proportion.

B   the second is the province of Upper Dalradian mica-schists in the north-east corner, where blanket peat covers the level plateau surface tops and mineral and mixed organo-mineral gleys are on the valley slopes.

C   a small upland landscape rising over 200 m above Cushendall of Old Red Sandstone and Triassic Red Sandstone soils, where slope contributes to good drainage, and soils are brown earths or SWG1.

D   extensive areas of blanket and lowland raised peatbog.

E   brown earths on glacial sand and gravel deposits.

F   mixed soils on chalk, dolerite and sandstone near Fair Head.

G   a small area of SWG1 on stone-free basalt till.

It must be mentioned that the alluvium on **Soil Map 5** has been classified by profile, partly because so much is humic gley (not organic enough to be organic alluvium). Alluvium comprises 4 per cent of the area of the map.

A: The area of Ballycastle-Ballymoney-Bushmills triangle is a very distinctive soil landscape developed on basalt till. Soils are well-drained – rankers, shallow brown earths and SWG1 – on basalt till. There are no drumlins or drumlinoid features, and hollows are in-filled with peat or humic alluvium. It has the appearance of a rock-scoured surface, and is bounded on its south-east margin by the 'Armoy Moraine'. This important landscape feature of the last glaciation stretches from Ballycastle to Armoy and on farther west to the Garry Bog. Soil mapping revealed that the Armoy Moraine was not composed entirely of gravel, but included areas of brown earth and SWG1 on basalt till – particularly east of Ballymoney, to Kilraghts and over to Stranocum. The whole area of this triangle is good quality agricultural land. Most of the brown earths, of Class 2 quality, which are 16 per cent of the area of the map, are found here.

B: Mica-schist valleys cut down through the upland block in the north-east that supports the blanket peat, and in these valleys, upper slopes around the peat are SWHG on mica-schist till, while SWG1 or SWG2 are found on lower till slopes. From Glenaan around to Glendun, and on to Tor Head and Fair Head, there are shallow humic rankers on the steep, rocky slopes of mica-schist.

C: No further information.

D: These are the organic soils, covering 174 km^2 or just over 26 per cent of the land area on Soil Map 5. In the east, the large unit of upland blanket peat from Ballypatrick to Slieveanorra is 107 km^2, and joins-up with a similar size of blanket peat bog on Soil Map 9 to the south. It is an extremely wet upland with over 1600 mm rain annually. Of the lowland basin peat in the western half of Soil Map 5, the Garry Bog is the largest at 10 km^2, and when combined with other similar smaller bogs, the total lowland peat area is 67 km^2.

Of interest is the considerable height variation of the edge of the upland blanket peat. On Soil Map 9 to the south, the peat edge around Douglas Top (10 miles east of Ballymena) can be as high as 350 mm above sea level, and co-incides with 1300 mm annual rainfall. On Soil Map 5, the relationship with 1300 mm of rainfall remains, but the height of the peat edge is about 300 m in Glendun, falling to 250 m on the west and north-facing slopes, and to even 200 m and lower in Glenshesk. The climate of north-east Antrim is very wet, (with annual rainfall up to 1800 mm), which explains the extent of the upland blanket peat. The peat edge can have annual rainfall from 1250 mm upwards.

E: These are various sand and gravel deposits, including the gravel of the Armoy Moraine, where the soil profile is brown earth. The formation starts at Ballycastle, and extends to Armoy and Dervock.

F: Mixed soils close to Fair Head are dominated by rankers on dolerite, but also include rankers, and shallow podzols, apparently on chalk. The latter is covered by a skin of mica-schist drift. The important case study of podzol development is at Goodland (Proudfoot, 1958).

Not falling into any one of the six major landscape divisions of soils, is the prominent Knocklayd Mountain – a dome of basalt rising to over 500 m between the valleys of the Tow and Glenshesk. The soils on its steep slopes are humic rankers on basalt, and on its south side, are some SBE on chalk, and mixed basalt and chalk SWG2 soils below. These alkaline soils lie immediately adjacent to Breen Wood (D125335), an area of acid and podzolised soils from which a pedological history was constructed (Cruickshank and Cruickshank, 1981). Radio-carbon methods were used to date assemblages of pollen in both surface and humic B horizons.

It was mentioned earlier that peat comprises 26 per cent of the area of the map, and to this should be added 6 per cent of humic rankers (mainly on mica-schist rock) and 9 per cent of peaty or humic gleys. All the organic soils collectively make-up 41 per cent of the map. Other types of ranker comprise 6 per cent, and brown earths 16 per cent. Both are found mainly on basalt and gravel parent materials on the western and lowland part of Soil Map 5. These are good soils for agriculture and would be classed as Class 2 or Class 3A. The gleyed mineral soils are widely distributed on basalt till, with SWG1 more common around Dervock and Ballymoney, comprising 13 per cent land area. These gleyed soils are drained artificially and are regarded as good quality soils for pasture farming. They would be in Class 3A or Class 3B, with the majority going into 3A.

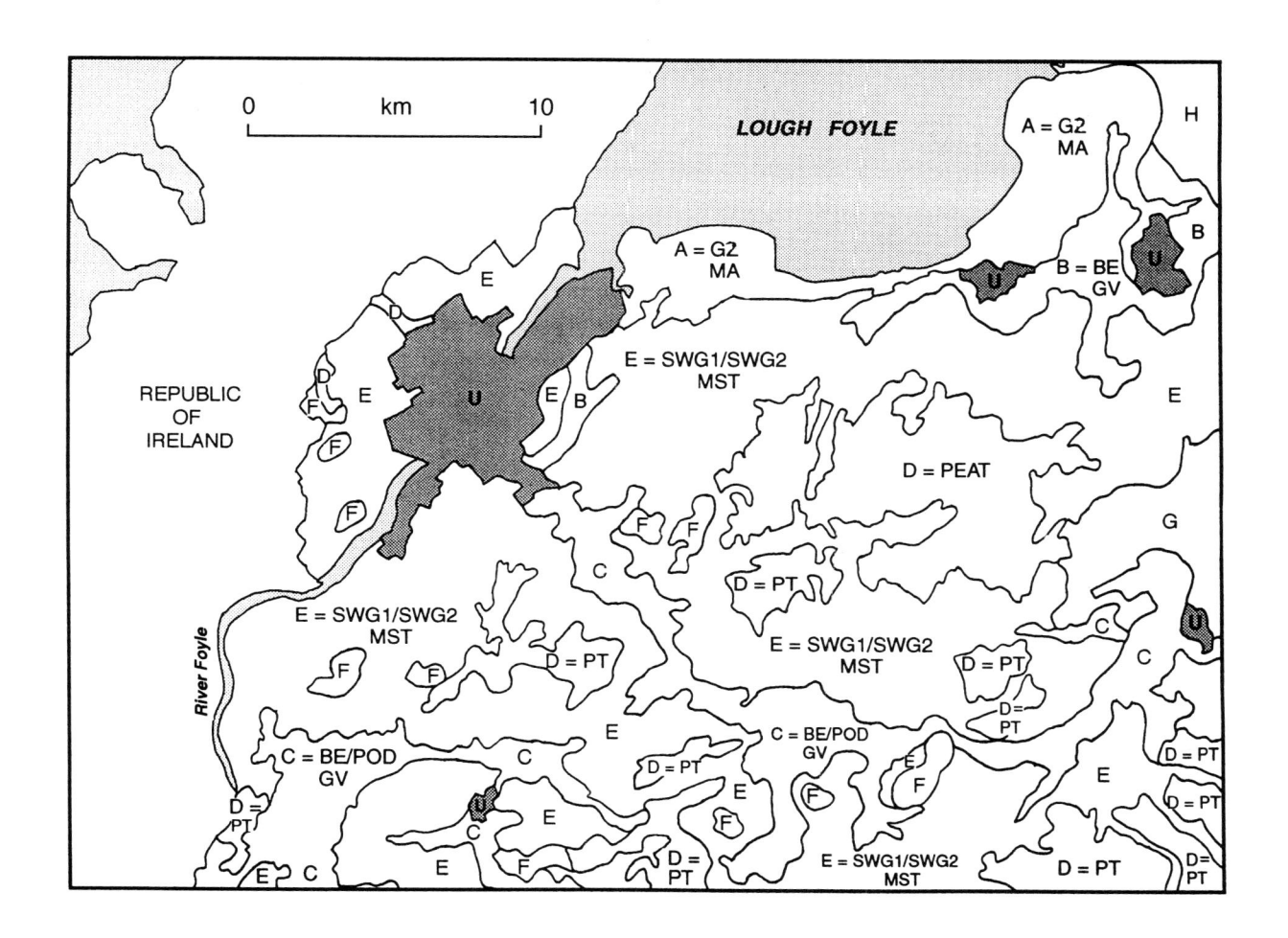

The distribution of the main soil series on Soil Map 7 (1:50,000), plus the north coast extension to Coleraine on Soil Map 4

Key for soil map sketch 7

The most evident features of the soil series distribution on **Soil Map 7** are:

A   reclaimed marine alluvium (shown in blue-grey on the soil map) with its artificial embankment enclosing it from the sea, and where pumping maintains control of the soil drainage. Soils still appear to be gleyed G2 and G3.

B   level outwash gravels around Limavady and in the Foyle Valley.

C   freely draining and steep-sided gravel formations filling the Sperrin Valleys, ranging from brown earths to peat podzols.

D   extensive blanket peat on the gently rounded tops of the North Sperrin summits.

E   gleyed soils developed on mica-schist till where most of the SWG1 soils (shown in grey-green) are on the outermost slopes of the uplands, while the SWG2 and SWHG soils (shown as pale yellow-green) lie above the gravels in the internal valleys.

F   mica-schist rankers identify the smaller, rocky summits in the Sperrins.

These, and some other soils on **Soil Map 7**, will be discussed in more detail under the above headings.

A: The coastal flatland, reclaimed from the sea in the mid-19th century and enclosed by a sea wall with pumping stations, is a form of 'polder' (see Chapter 5). The area is 31 km^2 in size, and about 80 per cent of that area is divided among 20 farms. The landward edge of the polder, locally called the Levels or the Carse, is at 10 metres above sea level, but the seaward edge is at or below sea level. The Levels merge into, and hence may include a further 10 km^2 of the wide estuary of the Rive Roe, north of Limavady. All the soils are of sandy texture, often coarse sand, because the landward uplands draining down into the Levels are of quartz-rich, sandy composition, being mainly mica-schists, quartzite and sandstones. The Levels, and the adjacent estuary of the River Roe, are underlain by Carboniferous Sandstone rock (basal clastics). Before the 19th Century drainage and the construction of a sea wall, there was no continuity of farming in this area. The Levels were frequently flooded by the sea, and farm habitation was limited to a few elevated gravel mounds and ridges. The area may have been grazed by flocks of sheep, but arable farming was not possible. Today, soil moisture is controlled by pumping drainage water through the sea wall, so that now the soils are still grey in colour, gleyed G2, with coarse shelly sand below 45-50 cm. At the western end near Eglinton, the soils of the Black Brae (C 500230), are peaty on the surface, and at Longfield (C 550220), silty clay overlies the shelly sand. The stratification of marine deposits is highly variable in the Levels.

B: and C: Brown earths and podzols on sands and gravels deposited by meltwater at stages of deglaciation in the Sperrins. The deposits range in altitude from 10 m at the edge of the marine alluvium to over 250 m above sea level, adjacent to the peat edge. The areal units divide between restricted and steep-sided ridges on valley sides, and level outwash formation, with deep kettle holes, at low altitudes. In soil profile type, they are almost always freely draining, brown earths being replaced by podzols at about 80–100 m, and in turn replaced by peat podzols at about 200 m above sea level, beside areas of blanket peat. These soils are a dominant element on Soil Map 7, comprising about 60-62 km^2, or 8 per cent of land area.

D: The blanket peat of the North Sperrins covers the gently-sloping, domed summits and plateau surfaces, but the outer edges of the peat vary in altitude, due partly to 19th century peat cutting around the edges. 250 m above sea level might be considered a working figure for the height of the edge, and although it is rarely higher, the edge comes down as low as 150 m in places. The Loughermore blanket peat unit is 58 km^2, and within it, a 500 hectare bog at Altahullion (C 625115) has been cut commercially for over a century. There was no activity there in 1995–96, but past exploitation gives an indication of great depth of peat on Loughermore. 9 km^2 of blanket peat is tree-planted. There are nine other small summit areas of blanket peat, plus a small amount of lowland bog within the area of Soil Map 7.

There are several visible signs that the blanket peat has been cut away at its edges over the past two centuries. The actual edge of the peat (over 50 cm deep) is straight in many parts due to cutting and so does not have a close relationship with altitude or contours. The former natural extent of blanket peat was probably over what is shown as surface-water humic gley soils (SWHG), which surround most of the blanket peat on Loughermore and where the Ap ploughed horizon is very peaty, but less than 50 cm thick. In such areas, and the area around Glack (C 615175) on the north side is typical, field size is quite large and regular in shape. This is a further indication from the man-made landscape of late-19th Century reclamation of the agricultural area from blanket peat. These reclaimed areas are above 150 m, and in an acid, wet and cold environment, with over 1200 mm annual rainfall. Such SWHG or SWG2 soils on mica-schist till would probably be regarded as ALC Class 4.

E: The most extensive and significant soils on Soil Map 7 are developed on mica-schist and mica-schist till. Brown earths are to be found on moderately steep slopes on the north-western edge of the uplands, above Artigarvan, Dunnamanagh, New Buildings, Londonderry, Eglinton and Greysteel. Some of these areas are at low altitude (below 100 m), but steepness of slope and higher altitude, in examples which are found as high as 170 m above sea level, detract from the agricultural potential of these soils, usually depressing their classification from Class 2 to Class 3A. More extensive than the various types of brown earths, are the SWG1 and SWG2, the gleyed soils of mica-schist till. They can be seen on Soil Map 7 in two shades of green, and their colour dominance on the map reflects the regional character of the mica-schist till gleyed soils. Perhaps surprisingly for acid and sandy material, there are few map units of podzols on this parent material. And although most of these few units of podzol are found at about 150 m above sea level, there is no direct relationship with altitude in their distribution on Soil Map 7.

The main soil profile types (brown earths, SWG1 and SWG2) on mica-schist till show very slight differences in soil analytical data (as seen in the tables in Chapter 2). The brown earths are slightly more acid, and do reveal indications of leaching. All three types usually have a sandy loam texture, and a profile range of pH between 5.0 and 6.5 (Natural subsoil being pH 5.0-5.6). Total iron and oxidised iron values are low, and this probably explains a lack of much visible sign of podzolisation. All the exchangeable cations are low in value, and available K is usually low as well. The tendency of mica-schist soils to become compact, and even indurated in B and C horizons, possibly reduces the effects of leaching in these soils. There is a striking contrast between the much greater degree of podzolisation in sands and gravels in the Sperrin valleys (where they are usually podzols or peat podzols), than on adjacent mica-schist till soils at the same altitude. However, it should be noted that some polygons of podzols on mica-schist till do exist (11 or 12 of them, shown in dark purple on the soil map), but collectively do not exceed 5 km^2. They are on the Sperrin valley slopes, not at the highest altitude.

F: The smaller rocky summits, on mica-schist rock, in the Sperrins, have no agricultural value and are limited in total area. There are 24 summits, usually between 250 m and 300 m, where the mapped soil is humic ranker, which means that peat cover is less than 40 cm deep. They are individually always less than 1 km^2 in area, and the total area involved on Soil Map 7 is about 10†km^2. In addition, there are 12 soil polygons of brown ranker on mica-schist, on rocky slopes at lower elevation and which may have slightly more agricultural value for grazing. In some cases, the humic rankers rise above areas of blanket peat, but more commonly, the summit humic rankers are surrounded by SWG1 or SWG2 on mica-schist till.

Outside these 6 soil groups, there remains one soil parent material of significance and reasonable areal extent, namely the Carboniferous Sandstone (otherwise called basal clastics). The grey-coloured sandstone underlies the areas around Dungiven in the Roe valley, Limavady, the Myroe Levels and along the shore at Culmore on the west side of the Foyle estuary. At the last site, the grey and some red colour from the sandstone imparts that colour to the local glacial till around Culmore (C 470240), where there is a poorly drained SWG2 on Carboniferous Sandstone till - with sandy clay loam texture. A more extensive area of both SWG1 and SWG2 on the same till is found in the upper part of the Roe Valley between Dungiven and Bovevagh (C 663143). The interesting point is that, although the Carboniferous Sandstone till does continue east a short distance on to Soil Map 8, it extends beyond the limits of its underlying derivative rock. This is despite the direction of the last glaciation northward down the Roe Valley (Colhoun 1970, 1971 and 1972). This till has its northern boundary defined by the moraine across the valley at Bovevagh, which is coincident with the solid geological boundary. There are no drumlins in the Roe 'Valley', which has the appearance of a glacial till plain. Modal SWG1 and SWG2 profiles are provided in Chapter 2.

Reference to the Geological Map of Northern Ireland (1:250,000) will show two further rock types in the area of Soil Map 7. These are also metamorphic rocks of Dalradian age, as is mica-schist. One is the very quartz-rich Quartzite, extending from west of Dungiven to the River Foyle at Strabane, and the other is Dungiven Limestone in a band on either side of the Quartzite. It has proved very difficult to identify soils of these two rocks, that are distinctively different from mica-schist soils. The northern band of Dungiven Limestone does not show its presence at all, but the other area, running south-west from Feeny (C 630053) has SWG2 soils that may be different from mica-schist soils by the till parent material being dark grey in colour. There is no chemical or fertility or land use difference from the surrounding mica-schist soils, and even the 'limestone' area around Lough Ash (C 482002) has not raised the pH of the Lough water. The soils developed from Quartzite parent material sometimes appear as almost white, ash-like sandy loam, but generally it is not possible to separate them from mica-schist soils.

The distribution of the main soil series on part of Soil Map 4 (1:50,000)

Only a small part of this Soil Map (north of Northings 300 through Binevenagh, and west of Eastings 900 through Portrush) is exclusive from the adjacent Soil Maps 5, 7 and 8. It is a soil landscape with several distinctive features of its own, but it continues eastward, the general soil distribution described for Soil Map 5. The glacial till cover is thin, and there are no drumlins. On the upland of Binevenagh and the Round Knowe, the drift cover is very thin, and shallow humic and brown rankers are developing from basalt rock. On lower ground, there is a general alignment of ranker and shallow brown earth ridges on basalt till from NNW to SSE. In the hollows, there are parallel strips of peat and alluvium, and notably, there are two encircled former lakes, the first at the polygon of organic alluvium (C 825320), just west of Coleraine, and the second on flat land of clay deposits east of Articlave (C 795345).

The organic soils on this part of Soil Map 4 include several basin sites of organic alluvium, which may be what survives from cut-over peat. On Magilligan Point, the parallel beach bars are separated by organic alluvium. The mineral soils of the ridges are either brown earth if the site is stable, or brown ranker if not. The blanket peat on the high ground of Binevenagh Mountain and Round Knowe comprises about 9 km², but with a particularly high elevation of 250 m – 300 m for the peat edge, due to lower rainfall and higher base status of the underlying basalt soil.

Finally on Soil Map 4, the unstable soils of the north and north-west facing slopes of Binevenagh Mountain are mapped as a complex of brown rankers (BRc). The area has become uneven and unstable because of underlying Triassic clays, and has no agricultural value. It is mostly tree-planted.

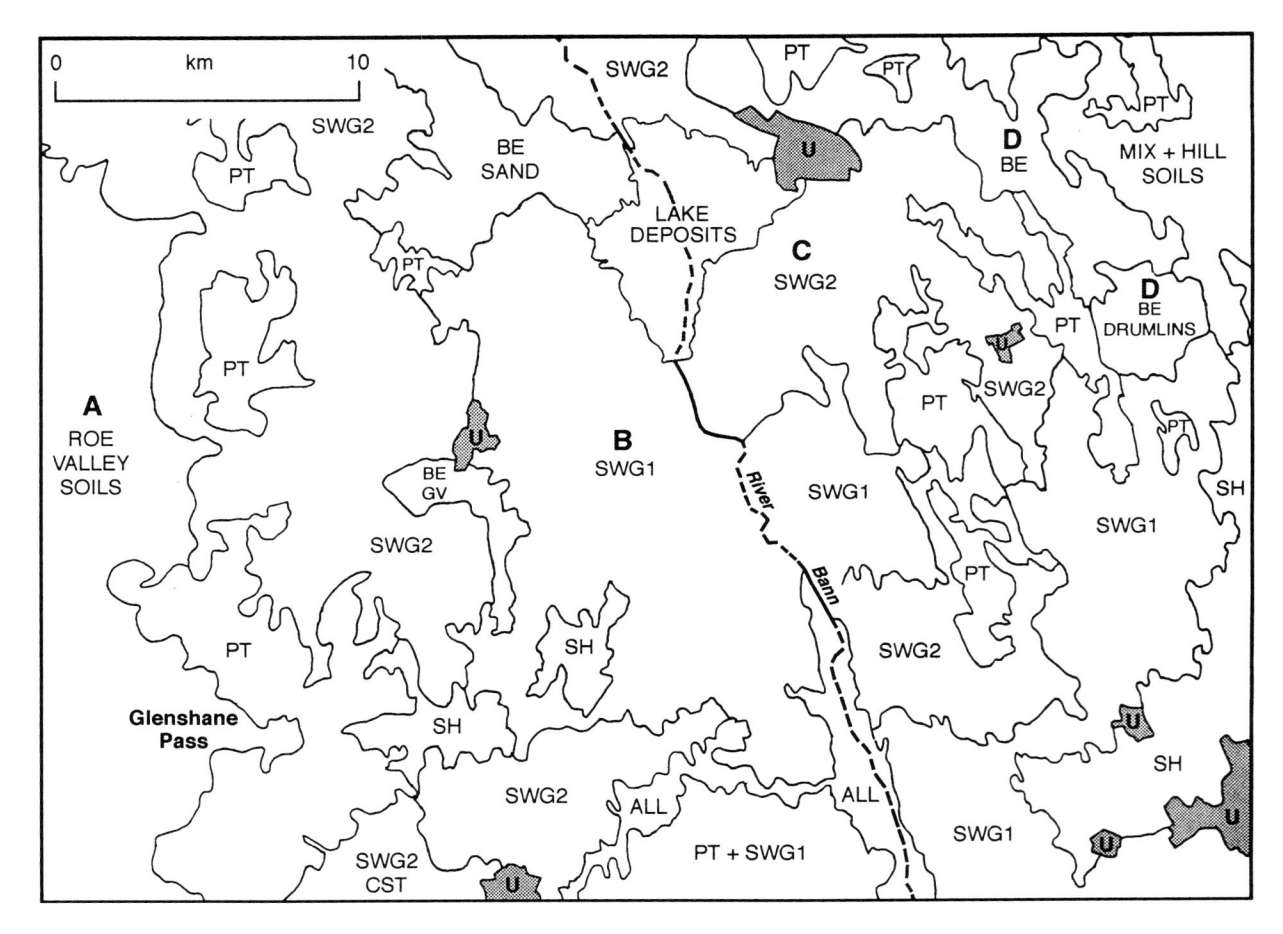

The distribution of the main soil series on Soil Map 8 (1:50,000)

Key for soil map sketch 8

With the exception of Roe valley soils (which include Carboniferous Sandstone (CST) and mica-schist (MST) soils), on **Soil Map 8** all mineral soils are derived from basalt or basalt till, gravel (GV) and lake deposits (also derived from basalt). Only profile codes are used in Soil Map 8, and can be found in Chapter 2. SH is here used to designate shallow soils, including shallow brown earths and rankers.

PT means PEAT.

ALL means ALLUVIUM (without profile subdivision).

Soil Map 8

Basalt rock underlies the soil on most of **Soil Map 8**. The only exceptions are on the western edge, north and south of Drumsurn in the Roe Valley, and the south west corner around the Glenshane Pass, west of Maghera. Here the soils are developed on Dalradian mica-schist rocks, Carboniferous Sandstone and Triassic Sandstone and Mudstones, and are described for Soil Map 7.

Basalt soils, developed on both weathered rock and till, dominate the rest of the map. The pattern of soils does not relate to the occurrence of Upper Basalt flows lying to the north of the Garvagh-Ballymoney line, or with Lower Basalts elsewhere. Rather there is a 'grain' to the soil landscape, aligned with the direction line of the River Lower Bann, north-north-west to south-south-east. This alignment has its origin in a similar pattern of minor faults and intrusions, and major downfaulting to south west, in the underlying basalt rocks, and this has been exploited further by glaciation. There are only a few, about twenty, drumlins on Soil Map 8, and these (north of Clough Mills at D070210) are also aligned north-north-west to south-south-east. The main soil divisions are aligned in the same way, and four of these can be identified, from west to east across the map.

A: The first region, with the Roe valley and Glenshane Pass on its western edge, is high ground formed on mica-schist in the south and upper basalts in the north. Summits exceed 300 m, and a large proportion is covered with blanket peat. The height of the western edge of the blanket peat is controlled topographically, but on the gently-sloping east, the peat edge comes down to 250 m in many places. Peaty gleys (SWHG) are extensive on the lower slopes on the east side of this first region, and SWG2 is the main mineral soil profile – statements applying to both the mica-schist and basalt parent materials.

B: Between the high ground of region A and the River Lower Bann, the majority of soils are in the well-drained end of the drainage spectrum – brown earths, but also some rankers and SWG1 on basalt and basalt till, and extensive areas of brown earth on gravels along the River Bann. Many units of lowland basin peat and organic alluvium infill the hollows among the ridges. These till and gravel ridges are again orientated NNW-SSE, reflecting the 'grain' of the country, but there are no true drumlins.

C: This region is parallel to the others and between the Rivers Lower Bann and Main, almost as one ridge, called Long Mountain (C995160), rising 200 m above sea level and covered with blanket peat. There are about 21 km^2 of peat on Long Mountain, which is more like a plateau with rock (basalt) cliff edges (rankers) in places. The mineral soils are mainly SWG2 on basalt till, but there are some areas of better drainage around Rasharkin, and in the surrounds of Ahoghill, Cullybackey and Ballymena. On both sides of the River Lower Bann, between Ballymoney and Agivey, there are three unusual stone-free deposits, each at its own level. The lowest level (below 10 m) of the Agivey plain is clearly a G2 lake clay deposit, as there is evidence of stratification in the material. The two higher levels lie between 15 and 30 m, and seem to have unstratified stone-free deposits (possibly a variant of basalt till) divided between G1 and G2.

D: In the extreme east is the lowland before the land rises eastward to the Antrim plateau and the Upper Basalts. It is effectively the 'valley' of the southward-flowing Main and Clough Rivers. There is a mixture of brown earths on gravel (shown in yellow), brown earths and shallow brown earths on the thin till cover on the ridges, and brown earths of the basalt till drumlins north of Clough Mills. There is very extensive basin peat around these drumlins and along the channel of the River Main.

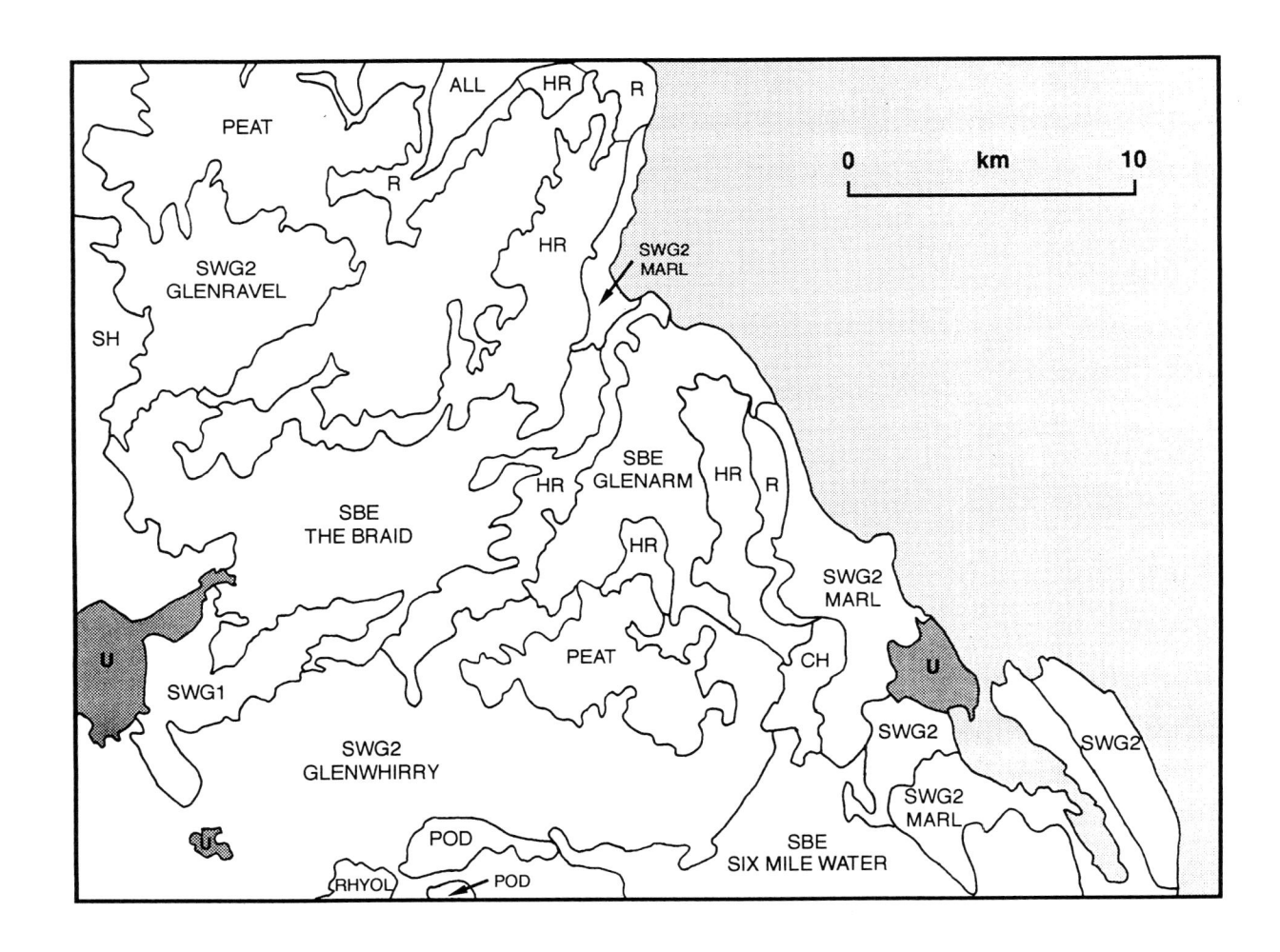

The distribution of the main soil series on Soil Map 9 (1:50,000)

Key for soil map sketch 9

Basalt is the underlying rock for most of **Soil Map 9**, and where no lower or parent material codes are used, basalt can be assumed. There are also areas of CH = Chalk, RHYOL = Rhyolite, and MARL = Triassic Mercia mudstones. All soil profile codes can be found in Chapter 2. SH is here used for shallow soils and R for all rankers except HR. ALL = Alluvium. Geographical place-names for the Antrim valleys are given.

The first impression of the soil distribution on **Soil Map 9** is of two very extensive areas of upland blanket peat, one in the north of over 100 km^2 and the other in the centre-south measuring 29 km^2, which are surrounded by over twenty small peat units on lower platforms, making an overall total of 151 km^2 peat. The lower edge of peat in the large units is just above 300 m in height, which coincides with an annual rainfall of about 1300 mm. The peat edge is higher in the southern unit, at about 350 m, on south and south west facing slopes, such as around Douglas Top (D245030) and Glenhead (D262024). The boundaries of the units of blanket peat are controlled by topography such as the steep slopes marking the edges of the basalt lava flows (for example the edge of the upper basalt at Agnews Hill – D330015), as well as by climatic wetness. Blanket peat is surrounded mainly by peaty or humic gleys, as well as some humic rankers. In the north, the pattern is reversed with humic rankers dominating on steeper, rocky slopes on the north, east and south of the blanket peat. Here again, the blanket peat overlies lava flows of the Upper Basalt, but the significance is simply because of their greater height rather than chemistry or mineralogy of the rock.

Below the blanket peat and generally below 300 m, the mainly mineral soils are associated with four south-west flowing river systems and the coastal lowlands of Trias Marl-derived till, around Carnlough and more extensively around Larne. All four river systems have formed valleys into the Lower Basalts, and each has a distinctive pattern of soils, controlled by the thickness of clay-loam basalt-derived till. From north to south, first the Glenravel Water cuts through deep deposits of clay loam till with strongly gleyed SWG2 and SWHG profiles, second, the Braid River (and including connecting Glenarm to the east) is in a valley of thin drift cover and well-drained shallow soils, mainly shallow brown earths. The 'Braid' is renowned for the high quality of its agricultural land, because the shallow soils are developed on a thin cover of clay-loam, they do not dry-out quickly in summer. An extension of slightly gleyed SWG1 runs from Racavan to Ballymena. The appearance of the landscape is marked by the stone dykes which enclose almost every field. The sequence is continued to the south with the Glenwhirry Valley being very strongly gleyed clay loam till, while the fourth and last, the valley of the Six Mile Water returns to high quality land (Class 2) of brown earths and shallow brown earths on a relatively thin cover of basalt till. The brown earths of the Six Mile Water extend south-west on to Soil Map 15, in the direction of Ballymena and Ballyclare.

The most common and widespread of all basalt till derived soils is the clay-loam SWG2, which comprises 25 per cent of Soil Map 9 and 35 per cent if the closely related humic gley (SWHG) is added, a good example here is Glenwhirry. Only 6.4 per cent of Soil Map 9 is SWG1 on basalt till – the Racavan to Ballymena area and 5 other polygons. More common, on Soil Map 9, are brown earth and shallow brown earth on basalt till which make-up 20 per cent of the soil area. The special example given in Chapter 2 is a high-level site at Bellair (D298151), 105 m above Glenarm.

South Antrim is predominantly a grassland district, but with some cropping of potatoes. Maize at one farm near Crumlin is a new development of the 1990s. The land of south Antrim is difficult to classify for agricultural land quality, but is mostly in Class 3B. Soils are mostly clay loam (or even sometimes clay) in texture, and well-gleyed SWG2, except for the areas of shallow brown earths. Annual rainfall varies from 900 mm at the east coast to over 1300 mm on the plateau surface. Growing season ranges from 280 days at the coast to 220 days on the plateau, so land classification is mainly controlled by climate, lowering to Class 4 on the plateau.

There are several soil groupings outside basalt and basalt till. The first is the acid igneous intrusion of rhyolite at Tardree Hill (J200955) and associated with place-names like Sandy Braes. This rock is volcanic, but there ends the similarity with basalt. The soils are sandy in texture, acid in reaction, with podzols and peat podzols as the main profile types. It is shown on Soil Map 9 in a dark purple colour, and is more fully discussed for the overlapping Soil Map 14. Other exceptional soils are found along the east coast of County Antrim, from Glenariff, to Carnlough, Glenarm, Larne and the Larne Lough area, where rocks of Cretaceous, Jurassic and Triassic age, normally capped by basalt, emerge along the coast. Soils are usually calcareous, base-rich, have a high clay content, and are often found in an unstable state due to the nature of clay minerals present, wetness and slope.

The highest in sequence are the chalk soils. There are a few places in coastal sections, such as just south of Glenarm where the ancient, red, early Tertiary, tropical soil can be found sealed below the first basalt flows. This is only of academic interest, and elsewhere (inland from Larne, between Kilwaughter and Blairmount – D365035) there is a chalk 'platform', between 100 m and 150 m above sea level. The soils of this chalky till are calcareous brown earths (CBE) and shallow brown earths (SBE), mixed with pelosols (PEL). Pelosols are calcareous, reddish in colour, and clayey in texture, mapped in 3 polygons, covering 302 hectares. They are gleyed, but gley features are not easily seen against the red colour of the soil. They are considered to be SWG1 soils, but could be SWG2. Calcareous pelosols have pH values above 6.0.

Some of the adjacent Triassic Marl or Mercia Mudstone till soils may also be pelosols by their pH and red colour, as evidence of gleying is difficult to detect. However, these soils are not quite so calcareous, though mixed with chalky material, and pH is not quite as high. They are strongly red in colour, and red colour masks the features of gleying. The clay content of between 30 and 40 per cent, gives the soils a clay texture. No analytical data are supplied for pelosols, but the Red Marl soils are so similar, that for all practical purposes of soil management, they may be taken as representing both soil groupings. The Triassic Marl SWG2 soils cover 4690 hectares, or 5.8 per cent of the land area on Soil Map 9.

This soil map also has a larger than usual percentage of rankers (15.6 per cent), and most rankers are on high-level, mountain summits or cliffs. There are some areas of unstable soils, developed on swelling clays (near Glenarm). All this group would be Class 4 or 5 for agricultural quality.

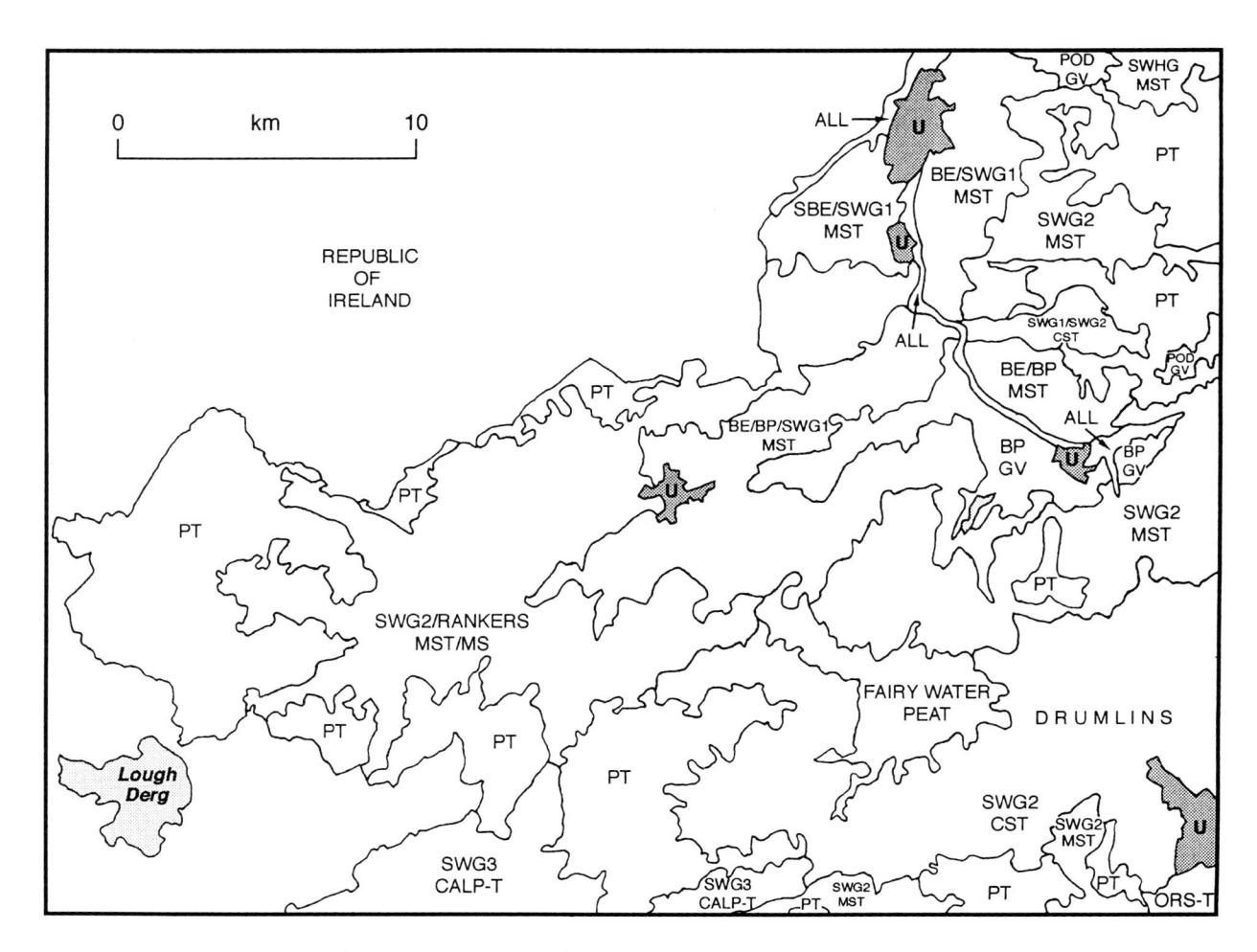

The distribution of the main soil series on Soil Map 12 (1:50,000)

Key for soil map sketch 12

The underlying rock in **Soil Map 12** is mostly mica-schist, MS, with exceptions for Carboniferous Sandstone = CS and Carboniferous Calp = Calp, and Old Red Sandstone = ORS. T is added for till, ie ORST and CST. PT = PEAT and ALL = Alluvium. The upper or soil profile codes can be found in Chapter 2.

The land area of **Soil Map 12** is dominated by the well-developed river systems of the Mourne-Strule, and the tributaries of the River Derg, Owenkillew and Drumquin – Fairy Water. The landscape is one of mainly smooth and gentle slopes, such as that of shallow brown earths and shallow brown podzolics, north and south of Castlederg (H265850). There are large areas of fluvio-glacial deposits in valley floors, such as the spread of brown podzolics on gravel running from Castlederg to Newtownstewart, but there is only one drumlin field in the lowland between Drumquin and Omagh. The orientation of these drumlins is WNW to ESE, reflecting the movement of the last ice outward from Donegal, and they are partly submerged by peat of the lowland bog systems of the Fairy Water River.

The mineral soils belong to either the Dalradian mica-schist and quartzite formation, or Carboniferous Sandstone, with some developed from the Calp shales, sandstones and limestones of the Middle Carboniferous Series (in the south-west corner of the map, running south into the Kesh-Boa Island area of Lower Lough Erne). Peat is the single dominant soil on Soil Map 12, being found in six main units plus countless more smaller blocks. The most western block, west of Castlederg and running over the political Border beyond Mullyfa and Cross Hill, has an area of 72 km^2 and a very low lower edge between 100 and 200 m above sea level. The south-central block around Lough Bradan (H260715) covers 27 km^2, and has a lower edge between 200 and 250 m, largely because of topographical breaks in the

144

landscape, and 3 km² at Bessy Bell (H390820), and 7 km² at Pollnalaght (H370710) – both with very variable elevation for the peat margins. There is at least 17 km² of lowland bog in the Fairy Water bogs. In the Sperrins, in the north-east corner of the soil map, there is about 25 km² of blanket peat, divided between Owenreagh Hill with a lower edge around 250 m, and a southern part where the lower edge is commonly around 100 m. North of Castlederg on the political Border, Moneygal Bog (H240880) extends down to 140 m above sea level.

Mineral soils on Soil Map 12 cover a wide range. The freely draining soils include 22 km² of brown podzolics on gravel in the valley systems centred on Newtownstewart (H400860). Particular areas, close to Newtownstewart, are being worked for gravel extraction, and the whole gravel formation is steep-sided and hummocky throughout. Wet inclusions of organic alluvium are commonly found, all of which reduces the current agricultural value of the gravel. In the past, the dry sites of the freely draining gravel must have been more highly valued as many archaeological sites occur within it, including Harry Avery's Castle (H392852) and parts of the demesne land of Barons Court (H363830) estate, around Lough Catherine.

The best quality land for agriculture on Soil Map 12, is found back from valley floors on higher ground, from Strabane through Sion Mills, south to Clady Hill (H340890) and across west to Castlederg. Indeed, the land on both sides of the Castlederg valley is moderate to good quality on shallow brown earths, shallow brown podzolics (shown in dark maroon colour), and SWG1 soils, all on mica-schist till. Drainage of some areas of these soils has been artificially improved, and in other parts by a relatively shallow depth of less than 1 m on even, gentle slopes. They would all be classed as Class 3A (and even as Class 2, especially if they had been in a drier or more easterly location). Surprisingly for an area of annual rainfall 1200–1500 mm, this is an important area for winter barley.

The mica-schist and quartzite till soils are intermixed in this area, and difficult to differentiate morphologically. No soil analysis data are available for these soils from the Castlederg area. Soil/land quality deteriorates rapidly to the west in the upper part of the Castlederg valley, west of Killen (H240800), where a mixture of SWG2 and shallow rocky rankers is found. Most of the land west of Killen, and all land west of Killeter (H205800) would be in Class 4 because rankers are humic and SWG2 soils are almost SWG3. Annual rainfall exceeds 1500 mm.

The last remaining major grouping of soils is developed on Carboniferous Sandstone till, and till mixes, east to west through Victoria Bridge (H350903) and Douglas Bridge (H372900) and again in the Drumquin to Omagh lowlands. SWG2 dominates in the drumlins of the Drumquin basin, which is about 120 km² in size, including the inter-drumlin and river valley peat. However, the upper parts of individual drumlins are the better SWG1 on Carboniferous Sandstone till. There is a morphological difference between the two but little difference in the soil chemical analysis data. A profile table for SWG2 is provided in Chapter 2.

To the west of Drumquin, there is small area of limestone till SWG2 below 200 m elevation (shown in a grey colour), and even more gleyed and wetter soils farther west are developed on 'Calp' rocks and 'Calp' till of mixed shales, and sandy limestone. They are very poor quality, Class 4 soils, SWHG or humic gleys and SWG3. These poor soils are high in elevation in an area of high annual rainfall (1600–1700 mm), which is adjacent to the political Border, and runs south-west onto Soil Map 17 in Co Fermanagh. The assemblage of soil profiles on Old Red Sandstone till, found south of Omagh in the south east corner of the map, is described fully for Soil Map 18.

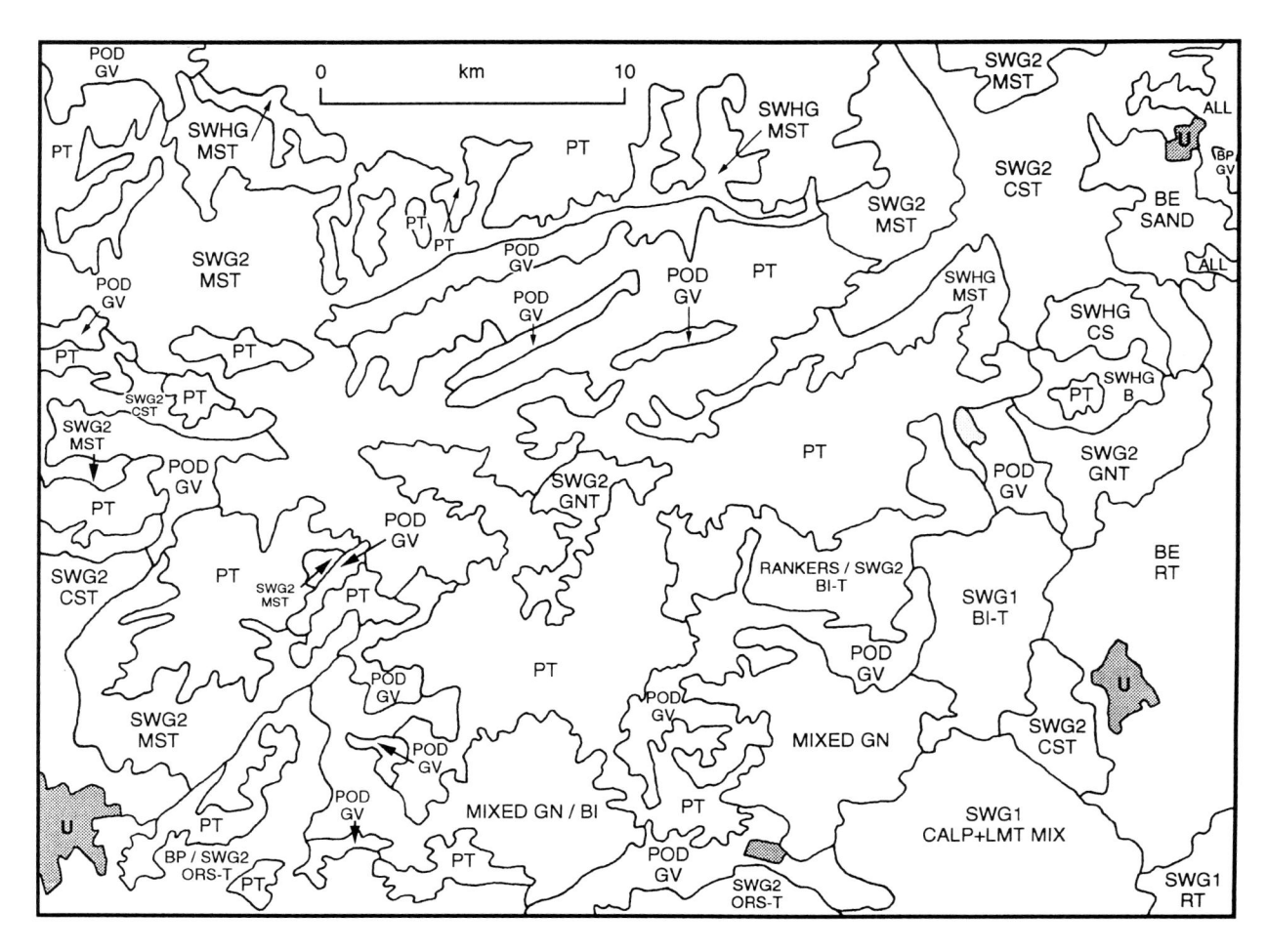

The distribution of the main soil series on Soil Map 13 (1:50.000)

Key for soil map sketch 13

The area of **Soil Map 13** overlies highly complex geology. The main rock types and their codes are:

B	=	Basalt	R = Red Triassic Sandstone
GN	=	Granite	ORS = Old Red Sandstone
BI	=	Basic Igneous rocks	CALP = Carboniferous CALP rocks
LM	=	Carboniferous limestone	MS = Mica-schist
GV	=	Gravel	PT = Peat

T is added for till

The upper or soil profile codes are provided in Chapter 2.

The striking feature about the distribution of soil series on **Soil Map 13** is the division, controlled by a geological fault line, the Omagh Fault, between ancient Dalradian mica-schist rocks of the Sperrin Mountains in the north-west, and to the south-east of the Fault, a mixture of volcanic rocks, merging into soft sediments of limestone and sandstone in the eastern lowlands. The latter are around Cookstown (H810780), lying between 50 and 100 m above sea level, and take the form of a drumlinised lowland of good quality Class 2 farmland. This is the only such area on Soil Map 13.

The Omagh Fault runs from Omagh town to the River Moyola, in an area of Carboniferous sandstone, near Draperstown (H785945). On the north-west, the rounded uplands of the Sperrins rise to over 500 m on many ridges, and to over 600 m elevation on the E-W ridge of Dart-Sawel and Meenard Mountains, north of Glenelly. From above 250 m on valley slopes, the Sperrins are covered with blanket peat, but in some

146

cases, high summits like Sawel, rise above the blanket peat. About 35 per cent of the land area of Soil Map 13 is in blanket peat, plus a further 3 per cent in basin peat. Against that background, it is surprising that this Sperrins district has a very small area in planted forest (less than 40 km^2 or about 3 per cent), and only three large forests on blanket peat, at Davagh (H710860), Gortin Glen (H490810) and Goles (H680930). The agricultural value of the hills is low, in Class 4 or 5, and they are used mainly for low density hill sheep production.

Sheep farming is based on home farms and infield land in the valleys. The uplands have been dissected by ice-melt and river erosion into steeply sloping valleys in which water-laid gravels have developed podzolised soils. These are found in the Owenreagh River, especially around Gortin (H495860), and in the valleys of the Glenelly and Glenlark Rivers, and Coneyglen Burn. On the valley slopes between the gravels and the blanket peat edge, the soils are mostly a mix of humic gleys and SWG2 on mica-schist till. A modal profile of a SWG2 from the Sperrins is provided as Profile 9 in Chapter 2. Occasional polygons of better drained SWG1 do occur, but are confined to steeper slopes. Humic rankers (HR) on mica-schist, and podzol profiles on mica-schist till (the latter shown in purple) are also found in the Sperrins.

Two units of Carboniferous Sandstone till project into the map area from the west, from Soil Map 12, between Gortin and Omagh. Surface water gleys SWG2 are the commonly found soils, as they are also, on parent material till mixes with mica-schist on the north side of Omagh, and with Old Red Sandstone till on the east side. More information can be found in the description of Soil Map 12, and likewise for the soils of Old Red Sandstone till (on the southern edge of Soil Map 13) in the description of Soil Map 18. Again, SWG2 is the common soil of Old Red Sandstone till, and the main features are sandy clay loam texture, moderate to poor drainage, slightly acid in reaction. In quality, these soils are in Class 3B.

On the south-east side of the Omagh Fault, some upland remains, and there is a marked annual rainfall gradient from over 1100 mm in the hills to 750 mm at Lough Neagh. Climate, therefore, contributes in large measure to land quality classification, but the soil variety and complexity also plays an important part. Almost all of the 108 soil series shown on the key (the largest for any soil map in the Province) are represented in this small area of 500 km^2. Most of the 'western' soils of mica-schist, Carboniferous Sandstone, and Old Red Sandstone are repeated here, and so it becomes difficult to point to any dominant soil grouping. The range of character and of agricultural quality of the soils is very wide, ranging from acid soils of granite and mica-schist, to intermediate reactions of soils on shales and sandstones, to basic or alkaline soils on limestones and basic igneous rocks. The last mentioned rocks are similar to basalt, but in this area they are found at high elevation, with high rainfall in the east Sperrins (shown in bright green colours on the map), and are often mixed with acid, sandy, granite-derived materials (shown in strong blues). Thus, the base-rich chemical character of the parent material is modified considerably. These soils, and the mixed variations in till, are almost equally divided between SWG1 and SWG2, and neither is extensive in area. They are usually in small polygons surrounded by blanket peat in the central area of the map, at elevations over 150 m. Here, their agricultural value would be Class 4, but in the N-S zone closer to Cookstown in the east, the elevation is around 100 m to 140 m, with the possibility of Class 3B.

There is a small grouping of soils of shale till, plus a mix with granite till, immediately east of Pomeroy (H690723), with both SWG1 and SWG2 soil profiles. There are also extensive areas of podzolised sands and gravels around Pomeroy, and in the valley of the Ballinderry River (H675790). Within the area of Soil Map 13, 10 per cent of the soils are freely draining, podzolised to some degree, and 6 per cent either podzols or peat podzols, which is the highest proportion of any soil map. Most of these podzols are found on the sand and gravel formations of the eastern Sperrins.

The upland of Slieve Gallion (H815896) has two summits at 496 m and 528 m, and is shown in very similar colours, to be divided between SWHG on Carboniferous Sandstone in the north, and SWHG on basalt (or basic igneous rock) in the south. Despite the height, there is very little blanket peat, probably because of steep slopes. There is a large spread of SWG2 on Carboniferous Sandstone till around Draperstown, as well as the contrasting good drainage of brown earths on sand, extending east to Tobermore (H830965). Extensive areas of sandy alluvium also exist in the north-east.

The area around Cookstown in the south-east, is notable for very good quality agricultural land (Class 2) of brown earths on Red Trias Sandstone till, which is further favoured in areas where it is mixed with limestone till. The landscape is slightly drumlinoid in landform, but steepness of slope is not limiting, and climate also is favourable to farming. Poorer quality (Class 3B) soils are found in the SWG2 on Carboniferous Sandstone till, and also mixes with Old Red Sandstone till west of Cookstown. Around Sandholes (H790725), there are sand deposits in various states of drainage, and extensive SWG1 soil on till mixes with Old Red Sandstone till. This lowland area is climatically favoured as one of the lowest rainfall areas of the Province (around 750 mm annually).

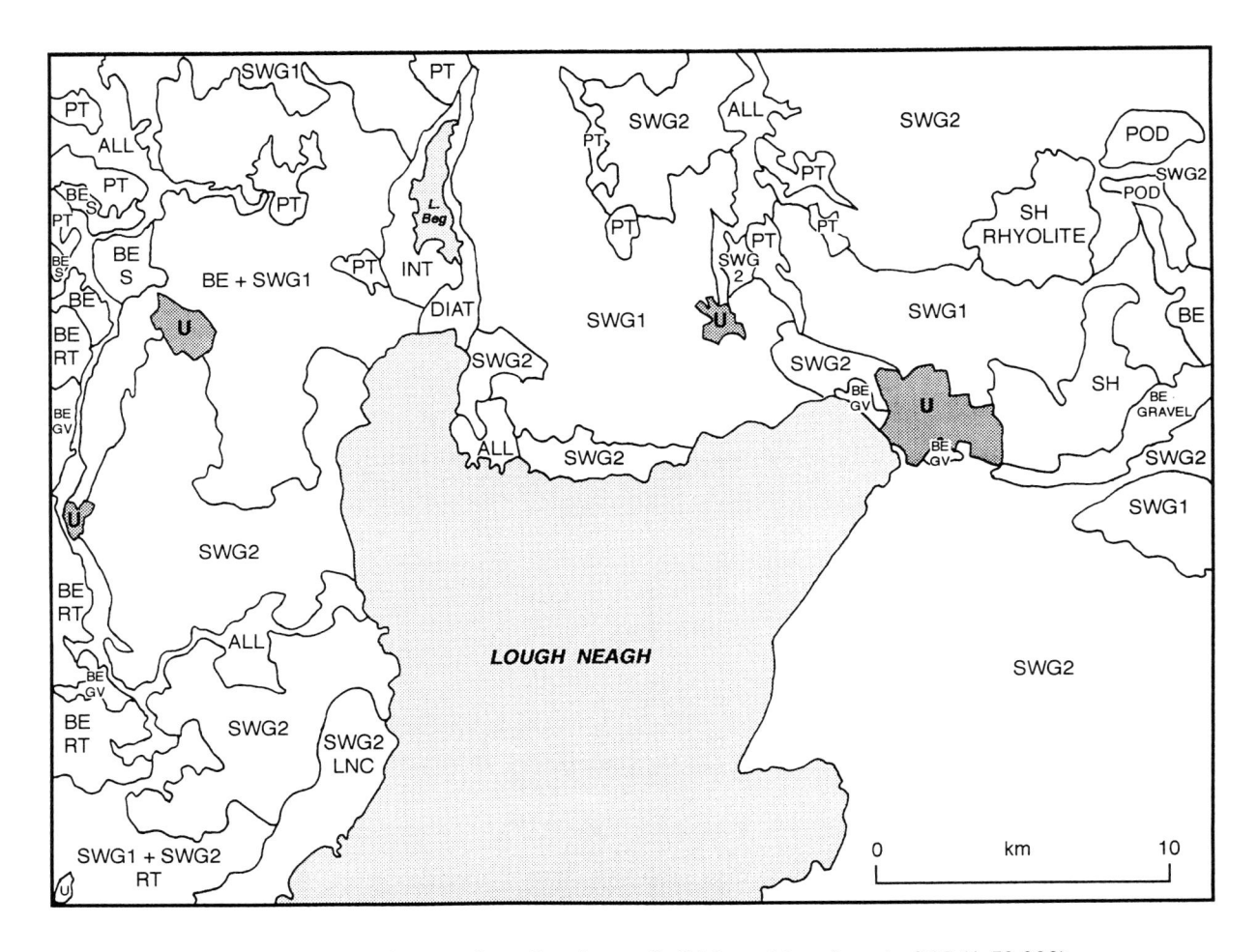

The distribution of the main soil series on Soil Maps 14 and part of 15 (1:50,000)

Key for soil map sketch 14

In the area of **Soil Map 14**, the main rock is basalt and the main soil parent material is basalt till. In such cases, no lower or parent material code is given. Other parent materials present are:

RT	=	Red Trias Sandstone till	LNC	=	Lough Neagh Clay
PT	=	Peat	GV	=	Gravel
S	=	Sand	INT	=	Intake lake alluvium
DIAT	=	Diatomite alluvium	ALL	=	Alluvium

Soil profile codes are provided in Chapter 2, except SH = Shallow soil profiles.

There is a continuity and cohesion in the land area and in the main soil series from **Soil Map 14 into 15**, as far as the edge of the basalt plateau and the area around the city of Belfast, are concerned. These are soil series developed on basalt, basalt till or mixes with them. The remainder of Soil Map 15 comprises North Down and the northern part of The Ards, and naturally belongs to the soil series groupings of Soil Map 21. The description may be found there.

West of Lough Neagh, the basalt till soil series lie between the lough shore and a very sharp physical boundary of the river valley through Moneymore. This channel carries the Ballmully River south to join the Ballinderry River. The river system is the boundary between the Triassic and Carboniferous rocks on the west and basalt on the east, and is a channel with an underlying geological control but over-deepened by glacial meltwater. There seems to have been little carry-over of glacial till from their derivative rocks in this area, and there is only one small example of basalt till (H855825) found on the west side of the channel, opposite Moneymore, and just 2 km west of the basalt rock boundary. There is some mixing of basalt and Red Triassic Sandstone till, east of Stewartstown, which indicates a short southward carriage of Red Sandstone till. However, immediately west of Stewartstown, the summit of Crewe Hill (H853715), showing shallow brown earths on weathered basalt rock, appears not to have been glaciated. It is thought that this area around Stewartstown was a divide between ice moving north and ice moving south (McCabe, 1992). To the immediate east and south-east of Magherafelt is a drumlin field where the alignment of drumlins is NNW-SSE, as elsewhere in lowland, central County Antrim (see Soil Map 8). This natural alignment or 'grain' of the landscape, is also reinforced by the road system, following the drier rises or ridges.

SWG1 and SWG2 soils on basalt till are clearly separated in discrete blocks in the area of Soil Map 14. SWG1 soils are in the block from Moneymore to Ballyronan on the loughshore, and over to Magherafelt. In this unit, better drainage of SWG1 soils is extended to areas of brown earth on basalt till in drumlins, and also by areas of red sand north-west of Magherafelt. The other unit extends from around Randalstown over to Antrim, where there is further good quality land, again of brown earths on basalt till and brown earths on gravel. In both areas, there is considerable river alluvium deposition in the Moyola and Main valleys, as well as infills of lowland, basin peat. Around Lough Beg, evidence of a larger lough at the end of the glacial period is evident in the lake deposits of "Intake" and "Diatomite". The latter is a deposition in fresh water of silica-encased diatoms, and the deposit (diatomite) behaves like a sandy soil, with a wide range of possible land uses.

The more strongly gleyed SWG2 on basalt till is found in the south-west corner of Soil Map 14, east of Coagh, and there joins with the wetter strongly gleyed SWG2 on Lough Neagh Clay till on level land along the lough shore. A finger-like extension of SWG2 basalt till reaches Magherafelt and includes many drumlins aligned NNW-SSE. So, it is the case that drumlins in the area around Magherafelt can have the full range of brown earth, SWG1 and SWG2 soils on basalt till.

On the east side of Soil Map 14, lie the two other blocks of SWG2 soils on clay-loam basalt till, that continue off this map northwards in the direction of Ballymena, and eastwards to the Glengormley side of Belfast, on Soil Map 15. There is a drumlin field between Randalstown and Kells, on either side of the River Main alluvium, where the drumlins are aligned NNW-SSE and also NW-SE, in the direction of the movement of the last glaciation. In marked contrast, the adjacent area to the south, around Crumlin, Glenavy and the two airfields, is a large, uniform, almost level, till plain. The overall dimensions are 20 km by 20 km, or an area of 400 km^2, which, with four exceptions of SWG3, makes it a very large and uniform unit of SWG2 on clay-loam basalt till. It is the most extensive example of one soil series anywhere in the Province, i.e. this south Antrim till plain.

Tardree Mountain (J192943) at over 240 m above sea level, on rhyolite rock, deserves special mention. Although only about 13 km^2, soils such as acid brown earths and peat podzols are found on the acid, and cinder-like, weathered rhyolite rock, and SWG2/SWHG soils are found on the sandy rhyolite till, on surrounding slopes. A place-name like Sandy Braes (J203958) reveals the texture of this soil, which in all physical and chemical properties, is a marked contrast from the basalt soils. The hill, a prominent landmark, is used mainly for forestry, and would be Class 4 in agricultural value.

In contrast, the area around Parkgate (J230880) is regarded as high quality land for a wide range of agricultural uses, and is in Class 2. The brown earths are on both basalt till and gravel terraces on the north side of the Six Mile Water. There is an arc of this high quality land running from north of Antrim town through Parkgate, to Doagh and Ballyclare on Soil Map 15. It does include some areas of SWG1 on basalt till, and the best land is below 120 m in altitude. Most is freely draining, and most is on level or almost level land. The disadvantage is that it is not very large in size, and is less than 30 km^2.

The soils on Soil Map 15 continue this general pattern for the basalt parent materials on the north side of Belfast Lough. SWG2 is still the dominant soil, to the extent of 121 km^2 or 81 per cent, on basalt till on the north side. Shallow brown earths and rankers on basalt rock are found on the hills, which are of little agriculture value. However, the area connecting the two soil maps (14 and 15), around and south of Templepatrick (J230855) is good quality land of SWG1 and brown earths on basalt till. With increasing altitude and closeness to the 'interbasaltic' iron-rich layer, the brown earths become ferric brown earths FBE and ferric rankers FR, at Lyles Hill (J246826) and Boghill (J270812). These hills have summits at 227 m and 283 m respectively, so altitude, exposure, and very shallow and very iron rich soils are of low agricultural value class, of Class 4. There is a significant area of SWG2 on Triassic Marl or Mercia Mudstone till along the north shore of Belfast Lough.

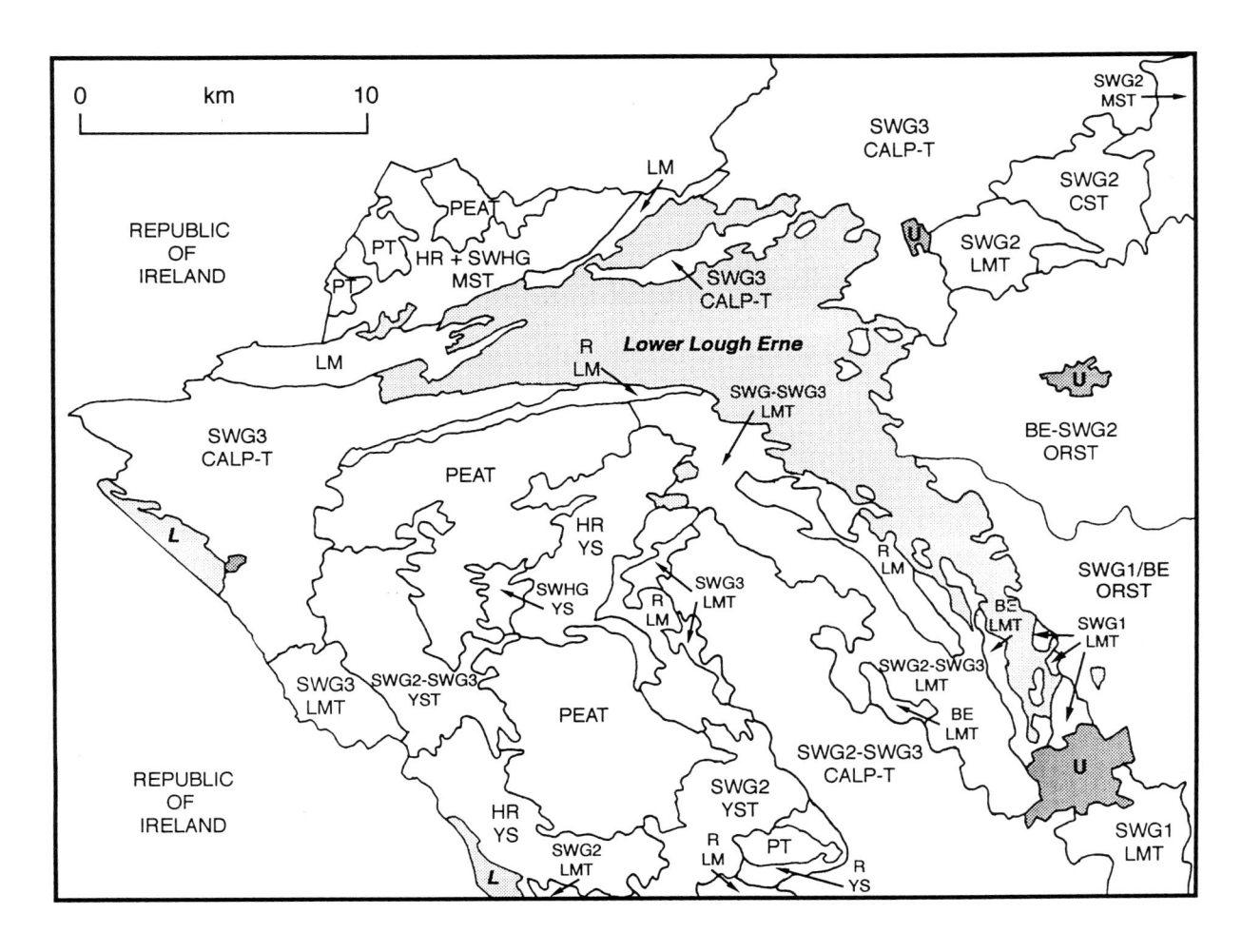

The distribution of the main soil series on Soil Map 17 + 26 (1:50,000)

Key for soil map sketch 17

In the area of **Soil Map 17**, the following rock types are present:

YS	=	Yoredale Sandstone of the Carboniferous	CALP	=	Calp rocks of the Carboniferous
LM	=	Limestone of the Carboniferous	CS	=	Carboniferous Sandstone
ORS	=	Old Red Sandstone	MS	=	Mica-schist

T is added for till

The codes for uppers or soil profile types are given in Chapter 2.

R   =   All rankers, except HR Humic Rankers.

PT =   Peat

Soil Map 17

Soil Map 17 covers the area around Lower Lough Erne, climatically one of the wettest areas in the Province, and where blanket peat can be found almost at sea level in the farthest west. This is particularly the case in the Garrison lowlands, a block of land of clay-rich Calp till SWG3, which is extremely difficult to drain or improve, and which includes very low level blanket peat in its 56 km^2 area. Similarly, the area of Lower Dalradian Moinian mica-schist along the north-west Border (30 km^2 in size) is divided between low level blanket peat, with an edge below 100 m above sea level, on an ice-scoured mica-schist rock surface. These are two areas of the very lowest potential for agriculture, and would be at the lowest level of Class 4 (or even Class 5 for rock outcrop) in land classification. Annual precipitation on the lowland is more than 1200 mm, combined with a high frequency of rainfall.

Elsewhere on higher ground, blanket peat is extensive on the upland from Lough Navar to Ballintempo, covering about 66 km^2. In many parts, the peat edge is defined by limestone cliffs, but in the west, the natural, climatic edge is between 120 m and 180 m above sea level, on gently-sloping Yoredale Sandstone till. Summits and ridges of coarse grained, yellow brown Yoredale Sandstone (at the top of the Carboniferous sequence) rise above the blanket peat at heights of around 300 m, with their raw rankers and humic rankers. The eastern slopes of the west Fermanagh upland from Church Hill to Knockmore and Boho are brown ranker complex, with much rock outcrop and solution landscape features on limestone. A very high proportion of the upland blanket peat, considerably more than half and almost all the deep peat of the western and northern parts, has been tree planted.

The soils of the Border area, adjacent to County Leitrim, are extremely poorly draining SWG3 soils on Carboniferous Upper Limestone and Yoredale Sandstone. These are very low value soils of Class 4 (ALC) which merge in a complex of humic rankers on Yoredale Sandstones, beside Upper Lough Macnean. These soils have their counterparts across the Border in County Leitrim, notably in the Kiltyclogher series, which is described as strongly gleyed, with a shallow peaty surface, on sandy clay overlying massive, sticky clay (c.38-40 per cent clay). Land Class is E, or very poor, in the classification used in the Republic of Ireland.

Moving from the Belmore to Lough Navar upland, towards the shore of Lower Lough Erne, there are three parallel zones of distinctive soils: a) Upper Limestone of Knockmore limestone cliffs and rankers, and very high clay content SWG3 till (shown in blue colour) and merging into, b) SWG3 soils on Calp till in the 'valley' of the Sillees river, mixed with extensive G3 alluvium, and c) mixed soils from rankers to SWG3 on the Lower Limestones on the Derrygonnelly to Enniskillen ridge. The mix of soils in the two limestone zones is similar, and both have great variety over short distances. A point worth making is that the polygons shown in yellow are brown earths on limestone till. There is no gravel in this area, although the topography of limestone till often resembles the form of gravel (the same feature is found on the Colebrooke estate, near Brookeborough).

The soil map units of single dominant soil series that lie to the east of Lower Lough Erne, such as the SWG3 on Calp till on the north-eastern edge, that runs on to Boa Island and south to Kesh, are soil units where a glacial movement from east to west is obvious. The best example is the Carboniferous Sandstone till, south of Ederney, which is moved about 2 km west from the boundary of the derivative rock. These soils, east of the Lower Lough Erne, are described in more detail for the over-lapping Soil Map 18.

Soil Map 17 naturally merges into **Soil Map 26** along it southern edge. The SWG2 on Calp till (shown in dark brownish grey), the dominant soil west of Enniskillen, continues into Soil Map 26 around Lower Lough Macnean. It is crossed by extensive lowland peat and alluvium of the Arney River lowlands. Facing each other across this lowland, are the ranker limestone slopes of Belmore to the north, and Cuilcagh in the south (in the area of Marble Arch). The lowland SWG3 on Calp till, around Florencecourt extends into Soil Map 27, and the soil description is continued there.

Sketch of Soil Map 26 appears on page 162

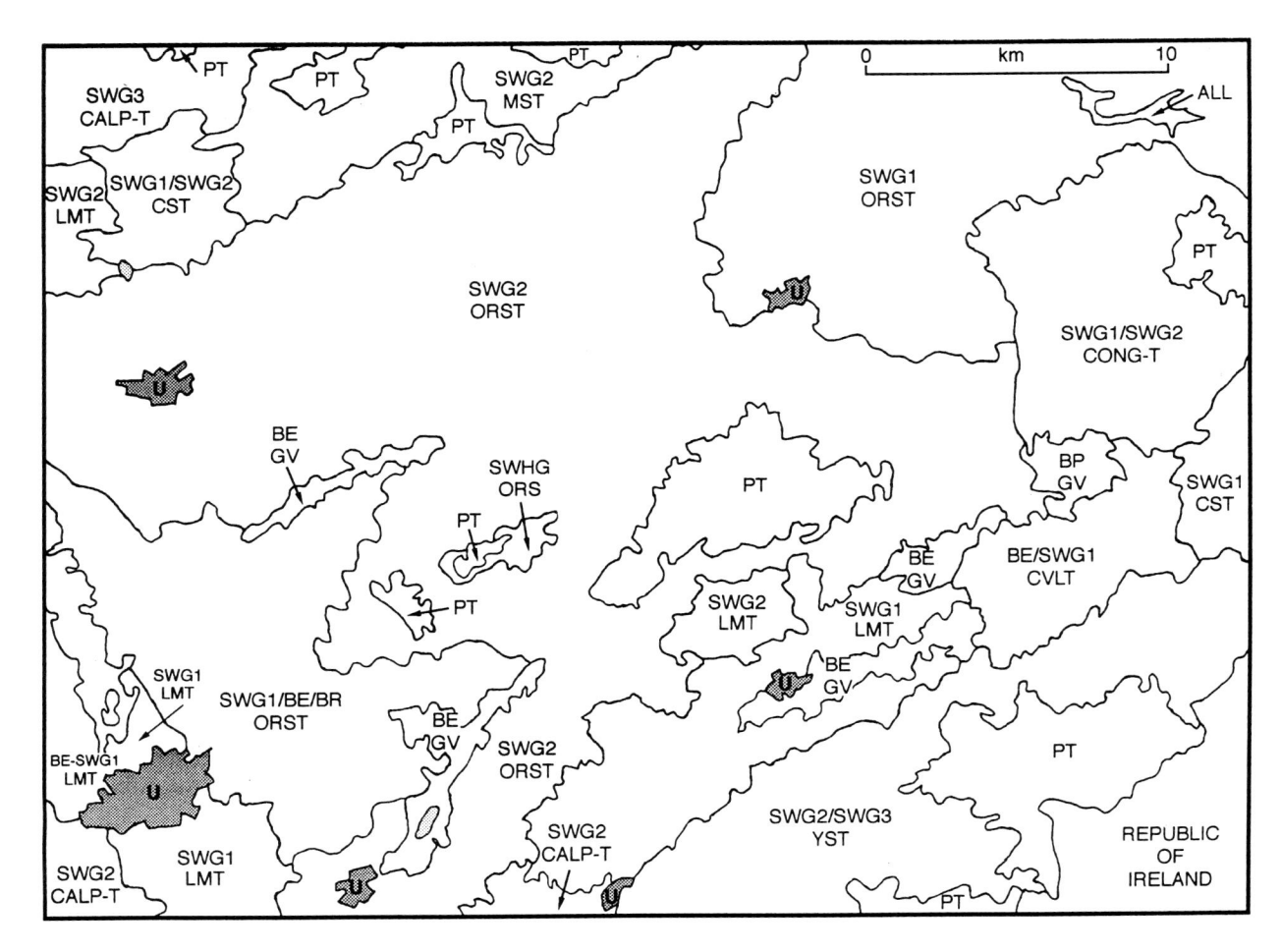

The distribution of the main soil series on Soil Map 18 (1:50,000)

Key for soil map sketch 18

In the area of **Soil Map 18**, the main rock type is Devonian Old Red Sandstone (ORS).
Other rock types present are:

CONG	=	ORS Conglomerate
YS	=	Yoredale Sandstone of the Carboniferous
CALP	=	CALP rocks of the Carboniferous
LM	=	Limestone rocks of the Carboniferous
CVL	=	Clogher Valley limestone of the Carboniferous
CS	=	Carboniferous Sandstone
MS	=	Mica-schist
GV	=	Gravel

T is added for till. Soil profile codes are provided in Chapter 2.

Soil Map 18 is primarily the province of Old Red Sandstone (Devonian) and ORS Conglomerate soils. It is also an area of four major concentrations of gravel, and three upland units of blanket peat. The red colours chosen for the ORS soils contrast clearly with the blue and yellow chosen for the limestone till soils of the Clogher Valley. These are mostly brown earths around Clogher, and SWG1 soil around Fivemiletown, both are good quality agricultural soils in Class 2 or 3A. They abut on to the poor SWHG and SWG3 of Yoredale Sandstone till on the south-east, with marked contrast of agricultural quality. The upland of Slieve Beagh runs east, over the political Border, mainly with peat and peaty gleys, and is partly tree-planted.

There are three large units of blanket peat on Soil Map 18. The peat on Slieve Beagh is about 32 km² in area, with a lower edge at about 200 m elevation and rising to over 380 m on the summit. The other two units of upland peat are on Brougher Mountain (H350528) extending to Ballyness Mountain (H456535) with 21 km^2, and at Slieve More (H595615) with 6 km^2. The total area of all blanket peat on Soil Map 18 is about 90 km^2, as there are many isolated small polygons. When combined with 40 km^2 of raised or basin peat, this soil map has an 11 per cent cover of peat. This is to be expected in the move west, into wetter climate and annual precipitation of just over 1100 mm on the uplands. The topographic edge of blanket peat usually coincides with this rainfall value, at about 200 m above sea level.

In the north-west corner of the soil map, around the villages of Ederny (H220650) and Lack (H275665), there are four distinctive soil units. In sequence from the north-west corner, first is the SWG3 on Calp till (shown in dark grey-brown colour) which continues south-west on to Soil Map 17. The very clayey, or sandy clay, soil textures retain moisture, and make this SWG3 Calp till very difficult to drain. The district also contains a large proportion of basin peat and peaty gleys (SWHG and HG), so that its agricultural value is poor, possibly Class 4. The second unit is the area to the south, on either side of the Glendurragh River, where slightly better SWG2 soils are found, on Carboniferous Limestone and Sandstone till. Slope helps to improve drainage locally, and south of Ederny, there are four polygons of limestone brown rankers, which were regarded as valuable soils in the past, in pre-drainage time, as is evident from the number of archaeological sites. There is evidence of an east to west movement by glaciation of Carboniferous Sandstone till, of about 2-3 km as far as Parkhill Lough (H227625). The third soil unit is SWG2, well-gleyed soils on mica-schist till, on rising ground in the centre-north of the map. This unit includes a large proportion of SWHG (peaty or humic gleys) and blanket peat above 240 m. These soils extend northward into the Sperrins on Soil Map 12, and are fully described for that map.

The Old Red Sandstone till is remarkable for the degree and extent of gleying. The dominant soil on the Old Red Sandstone till is SWG2, and comprises about 40 per cent of this map. The ORS SWG2 extends in an arc from around Irvinestown to Dromore, on to Fintona and into the Clogher Valley. The topography is undulating, drumlinoid and between 130 and 150 m above sea level. There are many small units of peat, both basin and blanket peat within the area of these soils, where mean annual precipitation is between 1100 and 1300mm. Soil textures are usually clay loam, with clay content in the 33 to 38 per cent range, despite the source of the parent material being Old Red 'Sandstone'. There is very little difference in appearance and properties between the SWG1 and SWG2, but the former occur in two separate unified blocks, one lying north of Fintona (H440610) and the other between Ballinamallard and Enniskillen. The latter is particularly mixed and includes brown earths (the only examples on the Old Red Sandstone till), SWG2, brown earths on gravel, and basin peat. The glacial carriage of the ORS till passes by Tamlaght and Lisbellaw (H300410) on to Soil Map 27, beyond the southern edge of this map, and is probably the greatest distance of till carriage in the Province (McCabe, 1969).

The four concentrations of gravel, mentioned in the introduction, are in the Ballinamallard, and Tempo Valleys, the valleys south of Fintona, and the Clogher Valley. The gravel formations have freely draining brown earth soils, but also include basin peat in 'kettle hole' depressions. The topography is hummocky, and although the soils are freely draining, variability and angles of slope would lower the land quality to Class 3A or Class 3B.

In the Clogher Valley, the gravels are mixed with limestone till developing SWG1 and SWG2 soils, from Brookeborough to Fivemiletown. Even some of the SWG1 limestone till on the Colebrooke estate (H410445) is 'ridge-like' in form and may be mistaken for gravel. The SWG1 limestone till is classed as 3A, and possibly some of the more level areas of gravel east of Fivemiletown are in Class 2. The very good quality Class 2 land, west of Clogher, is fairly level, brown earth on more sandy Clogher Valley Limestone till, a sharp contrast with the very poor land on SWHG and SWG3 of Yoredale Sandstone till, on the adjacent Slieve Beagh upland.

On the north-eastern quarter of Soil Map 18, between Beragh (H542671) and Augher (H564537) lies the last extensive group of related soils, that of the rock and till of ORS Conglomerate sediments. The soils are similar in many property values to Old Red Sandstone soils, except that soil textures are more sandy (usually sandy loam) and soils are more stoney. As a result, these ORS Conglomerate soils are inclined to be acid and leach easily, on elevated sites up to 200 m with quite high annual rainfall. Drainage is improved by soil texture and slope, so that in this group of soils, SWG1 (shown in dark red) is the dominant soil profile type. It is a feature of these soils that good quality and well-managed farmland is found up to heights of over 200 m. Annual precipitation is around 1100 mm.

The remaining small soil units along the edges of Soil Map 18, are described for adjacent maps.

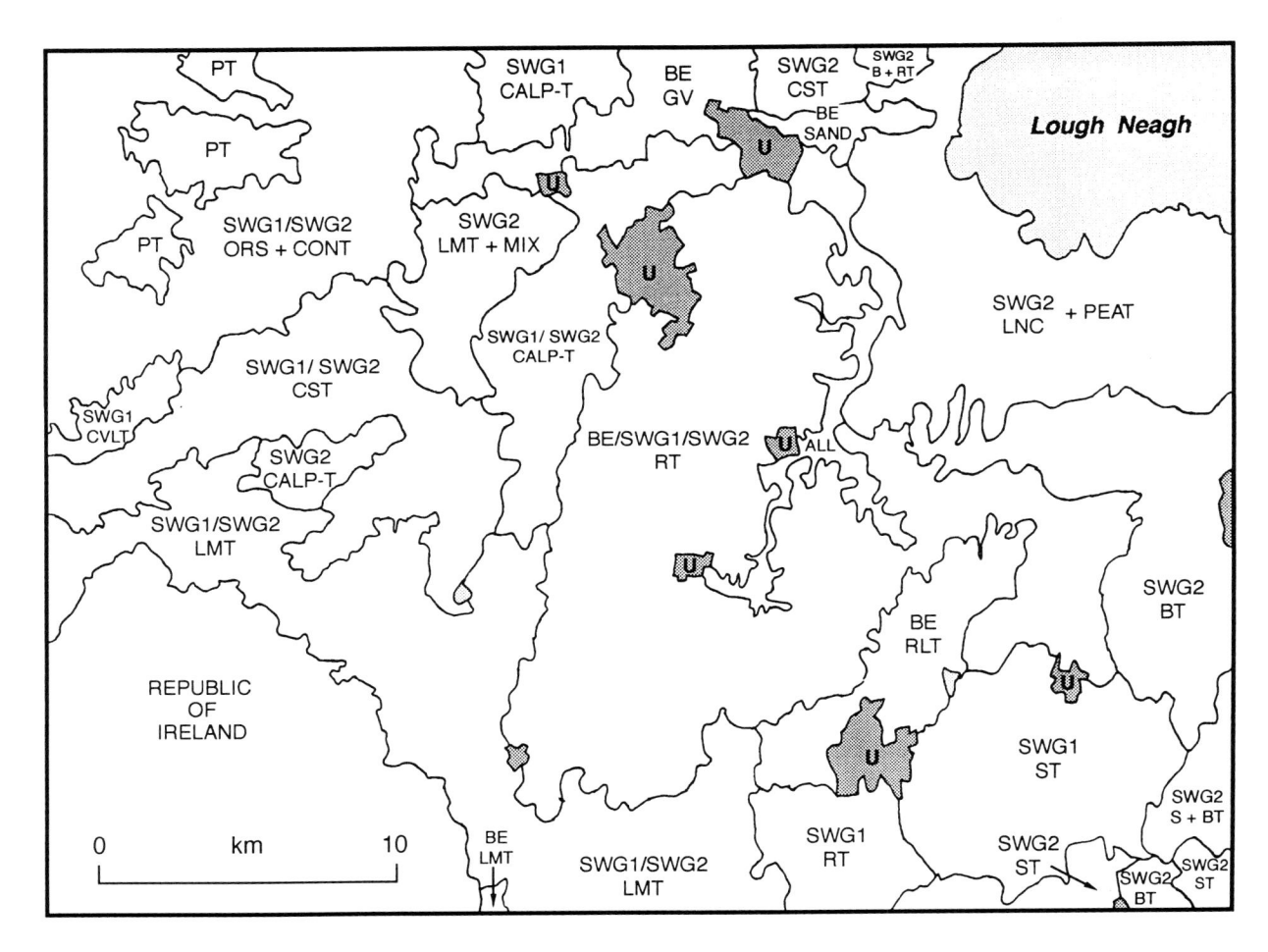

The distribution of the main soil series on Soil Map 19 (1:50,000)

Key for soil map sketch 19

In the area of **Soil Map 19**, a variety of rocks are found:

CVL	=	Clogher Valley Limestone	LM	= Limestone
RL	=	Red Limestone	CALP	= Calp rocks of Middle Carboniferous
CS	=	Carboniferous sandstone	S	= Shale
B	=	Basalt	LNC	= Lough Neagh Clay
GV	=	Gravel	ORS	= Old Red Sandstone
CONG	=	ORS Conglomerate		

T is added for till

The area of Counties Armagh and Tyrone, falling within the rectangle of **Soil Map 19**, includes a great variety of different soil series. 85 colour blocks are found in the legend of the map, and when adjustment is made to include subdivisions of rankers, brown earths and podzols, there are 102 soil series on Soil Map 19. The reason for this is the geological complexity, especially of Carboniferous series, in East Tyrone. Only the eight or nine main groupings of the soil series will be described here.

The first clearly defined soil landscape (of 79–80 km²) co-incides with the low-lying land around the southern shores of Lough Neagh. In this area, weakly formed drumlins composed of Lough Neagh Clay till, are aligned north to south. The table of soil analytical data for Lough Neagh Clay SWG2 (in Chapter 2) shows that the Ap surface horizon is usually deep (at 40–50 cm), due to a long period of farming improvement, and has a sandy clay loam texture. This sandy texture may be related to widespread sand deposits, often found at the south end of the individual drumlins, and released by meltwater at deglaciation. The lower B horizons of Lough Neagh Clay become clay loam or even clay. On the level landscape close to

Lough Neagh, drainage is poor, and so most of the surface of this landscape is peat-infilled. Peat here is nutrient receiving from surrounding base-rich tills, so is quite alkaline and is widely used for horticulture. It is probably the only peatland in the Province that deserves a higher land class than the usual 4, and should be considered Class 3B.

This first soil landscape is surrounded by SWG2 on clay-loam basalt till on the south side (an extension south-west from south Antrim and described for Soil Map 20), and by the very extensive alluvium of the Blackwater River to the west. The latter also bounds the region of the Red Triassic Sandstone till soils. extending from Dungannon southward to Armagh City, and to Killylea (H798450) and Caledon (H756455). This is the largest area of four in the Province, where soils are developed from Red Triassic Sandstone till. A wide range of soil profile types, shown in various shades of red (brown earths, SWG1 and SWG2) are found, and also mixed till with Carboniferous limestone and 'Calp' rocks. The landscape is weakly drumlinised, with stronger form appearing on drumlins in the western fringe.

The soil map unit shown in bright red, connecting Loughall to Armagh City and Navan Fort is apparently a brown earth, with no visible mottling and also bright red in natural colour, developed on till of Carboniferous limestone. It is here called Red Limestone till because of its strong red colour (Munsell colour 2.5 YR 4/4 reddish brown) in B and C horizons, but this colour may mask the features of gleying. The red colour may have come from the Red Triassic soils to the north, but the soil texture is not the same. The Red Limestone 'brown earths' have 30–45 per cent clay, high pH 7.0-8.3, high Ex. Ca and free carbonates at 80 cm. These soils may be slightly gleyed, and may be a type of pelosol.

There are some large spreads of sand and gravel around Dungannon. Brown earths on gravels spread-out from north of Donaghmore across eastwards to Coalisland (shown in yellow on the soil map), and broken-up with peat patches and organic alluvium in the hollows among the gravel ridges. To the immediate east and south of Dungannon, lie formations of red sand (shown in orange colour) which have been washed out from Red Triassic rock and till by meltwater at the stage of deglaciation. These sand and gravel deposits, along with a scatter of isolated gravel units elsewhere on the map, measure approximately 30 km^2 or 3 per cent of the area on Soil Map 19. Brown earths are otherwise found on the Red Triassic till and the Red Limestone till, around Loughall. Collectively they comprise 111 km^2 or 11.5 per cent of this soil map.

Moving west of the Blackwater River, the Carboniferous limestone, shales and sandstones extend continuously further westward through County Fermanagh to the Atlantic coast. The individual, named rock series may be a mix of these sedimentary rocks, and the named series may often be mixed through each other. Their influence on soils, and the identification of soil parent materials, can be difficult to establish in the field. Soils developed on limestone, normally grey or yellowish-grey in natural colour and with a clayey texture, are found south of Killylea (SWG1 strong blue in colour on the map) where the SWG1 soils are on well-developed drumlins with mineral alluvium in the hollows. Moving along the political Border, the light grey colour on the map represents SWG2 on limestone till on less well-developed drumlins, and now with organic alluvium in the hollows. Other units of SWG1 on limestone till, north of Aughnacloy, are mixed spatially with those of SWG2 (dark grey or blue grey), SWG1 (very light blue grey) and SWG2 (pale blue) on Carboniferous Sandstone till. The Clogher Valley Limestone SWG1 (yellow-green) emerges as a distinctive unit around Ballygawley, and then a sequence of ORS Conglomerate till soils, Old Red Sandstone till soils and blanket peat cover Cappagh Mountain (H680660) and the other hills, rising to over 250 m in the far north west of Soil Map 19.

Moving back to the south-eastern part of Soil Map 19, south of Richhill (H950480), shale till appears in a large unit of about 50 km^2 of SWG1. This is quite well drained land on strongly formed drumlins, from Richhill to Armagh City and almost to Markethill (shown in pale brown). At the south edge of the map and at the edge of rising ground, the dark green map-colour of SWG2 on shale till appears, and is far more extensive on Soil Map 28 to the south. To the west, the map unit of SWG1 on limestone till (shown in blue), is surrounded with units (in a pink colour) of SWG1 on mixed limestone and Red Trias Sandstone till, to the south of Killylea (H798450) and extending west to Tynan (H767430). The landscape is less strongly drumlinised, more gently-rolling in slopes, and with SWG1 soils on limestone till, is regarded as good quality Class 3A. It is better than the extensive areas of SWG2 on various tills to the west, as far as Aughnacloy and Ballygawley (H630577). The good quality is exceeded only by 1.5 km^2 of calcareous brown earth on limestone till, at Boneville House (H750410), which is Class 2. This soil might be regarded as grey-brown podzolic, but here called a calcareous brown earth CBE, with a clay-enriched B horizon.

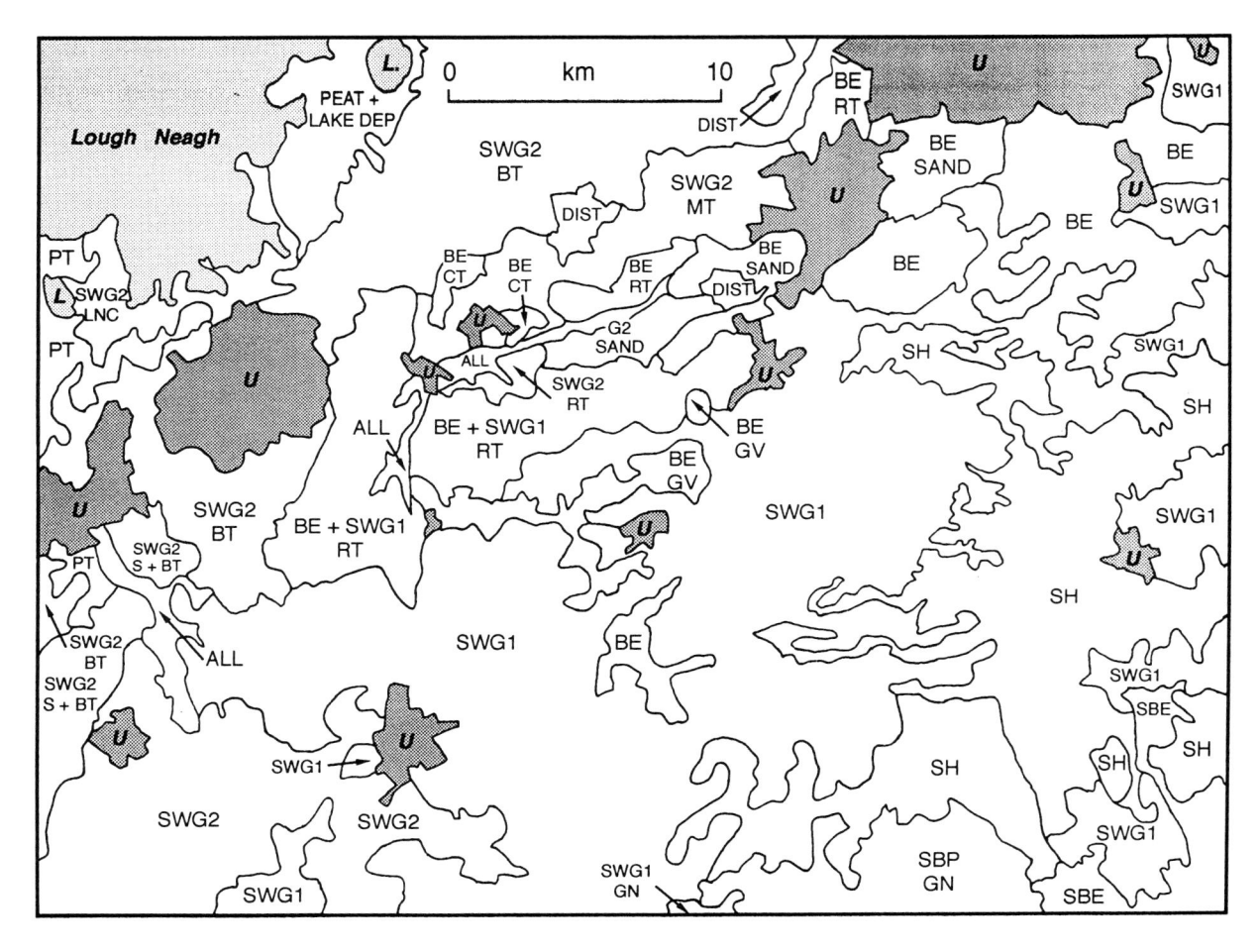

The distribution of the main soil series on Soil Map 20 (1:50.000)

Key for soil map sketch 20

In the area of **Soil Map 20**, the following rocks are found with provided soil profile codes:

LNC	=	Lough Neagh Clay	B =	Basalt
R	=	Red Trias Sandstone	C =	Chalk
M	=	Marl	GN =	Granite
GV	=	Gravel		

T is added for till

Where parent materials are absent (in the east), the underlying rock is shale. SH is used for shallow soils, and other soil profile codes are provided in Chapter 2.

Soil Map 20

Although the distribution of the main soil series units appears to be quite complex, particularly in the Lagan valley between Lurgan and Lisburn, there are only a few main subdivisions of soil groupings on **Soil Map 20**. Around the shores of Lough Neagh, there is a grouping of soils on Lough Neagh Clay till and basin peat (which is described for Sheet 19), and adjacent to the east and south, is the SWG2 on clay loam basalt till, which is an extension southward from Sheet 14, and is fully described there.

The mosaic of small soil polygons in the Lagan Valley can be grouped in the following way. First, there are the fluvial alluvial and fluvio-glacial deposits, laid-down on the floor of the valley at a late stage of deglaciation, when melting ice filled and blocked the drainage exit into Belfast Lough. These are the parent materials of the River Lagan G2 alluvium, the G2 Lake Sand and Lake Clay, and red sand and gravel – most of which lie on the south side of the River Lagan. Also mainly on the south side, are the brown earths, SWG1 and G1 of Red Trias Sandstone till, as well as mixes of that till with basalt, shale and chalky till. The soils are almost all SWG1, except for a polygon of SWG2 on Red Trias Sandstone plus basalt mixed till, east of Magheralin (J135585). In the floor of the Lagan valley, the topography is almost level, so drainage is largely dependent on soil texture. Sandy tills and sand deposits are favoured. Moving south, the land rises and general steepening of slope helps soil drainage. There are significant areas of brown earths on chalky till around Moira, Dollingstown and Waringstown (which probably explains the siting of these Plantation towns), and is the basis for a Class 2 land quality classification. Similarly, brown earths on the rising ground south of the Lagan, on Red Trias Sandstone till and on red sand and gravel are regarded as ALC Class 2 land. In the area around Blaris (J245630), and continuing into Belfast beside the M1 (J290655), brown earths on red sand are good Class 2 land. Unfortunately, there has been, and will be, considerable pressure on land in this area from urban and industrial development around the town of Lisburn.

The Lagan valley is also notable for the soil unit (shown in deep red on the soil map) that lies between the south edge of the basalt and the River Lagan – on its north side. This is a fairly level landscape of clay loam and clay textured soil, developing from a mixture of basalt and Red Trias Marl (or Mudstone) till. The soils are strongly gleyed, and soil texture and topography combine to make drainage improvement difficult. Small polygons of brown earth on gravel do exist, but make no difference to a land classification of Class 3B for this basalt-marl SWG2. There is a similar unit of SWG2 on a mixed till of basalt and Red Trias Sandstone on the south side of the valley at Magheralin.

In areal terms, the most extensive soils on Soil Map 20 are those developed on shale rock or shale till. For a distance of about 9–10 km beyond the built-up area of Belfast, the soils in County Down are either brown earths or SWG1 on shale till, and are of good quality for agriculture. Indeed the SWG1 on shale till in a drumlin landscape is the representative soil for most of County Down, and on Soil Map 20, extends south-west to Banbridge and Gilford. A modal soil profile of SWG1 on shale till is provided in Chapter 2. The more poorly drained SWG2 on shale till extends further south-westward from Tandragee and Banbridge towards Newry and across central and south County Armagh. It is one (more-or-less) cohesive unit, of drumlin country and silty-clay loam till. Because of relatively steep slopes on the drumlins, this SWG2 shale till is not as difficult to manage as the soil classification would suggest. Against the light grey colour of the shale till, the large rust mottles of SWG2 profiles are very clearly seen.

The remaining major soil units of this soil map are the shallow brown earth (complex) and brown rankers of shale rock, where till is either very shallow or absent. These soils are freely draining, but because slope is very irregular and the soils are extremely shallow in places, their agricultural value is lower, probably Class 4. Rock outcrop is quite common in these areas, and there is trouble with drying-out of these soils in dry summers. They are found in central County Down, in an arc around the town of Ballynahinch, where the landscape of these soils is more appropriate for gardens and demesne land of estates, rather than for agriculture. On Soil Map 20, they appear in a pale orange-brown colour, and would be in ALC Class 4.

The last group of soils is confined to the upland of Slieve Croob, where several summits are found around 500 m above sea level, and the rocks divide between shale and slate on the north side, and granite on the south side continuing south on to Soil Map 29. The group comprises brown podzolic and shallow brown podzolic soils on both rock types, and these soils fall in and between ALC Class 3 and 4. Around Lowtown (J258410), the shallow brown podzolics on weathered granite rock offer some quite good land in Class 3B, but climate at elevations around 150 m to 200 m will affect the classification. Above these soils, and also above some SWG1 soils on the highest drumlins at about 150 m, there are humic rankers on both rock types (granite and shale), where the Class is most likely to be Class 5, because of rock outcrop and heights above 300 m. Such high and rocky land has almost no agricultural value, except for very low density sheep grazing.

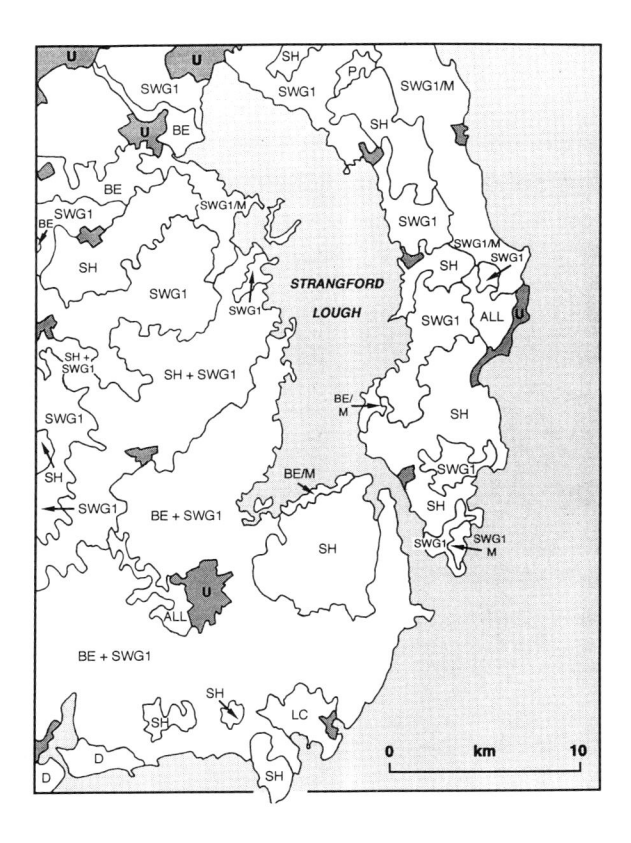

The distribution of main soil series on Soil Map 21 and Co Down part of Soil Map 15 (1:50,000)

Key for soil map sketch (Soil Map 21 - left)

The soil parent on Soil Map 21 is mostly shale or shale till, and only soil profile codes are used. The exceptions are the M = Marl, or marine-derived Marl pm, LC = Lake clay and D = Dunes. SH = Shallow soils.

Key for soil map sketch (Soil Map 15 - below)

On the north-west side of Belfast Lough (in Antrim), the soil parent material is assumed to be derived from basalt and only soil profile codes are used. On the south-east, the soil parent material is assumed to be shale, with exception for the marine-derived Marl = M on the east side of the Ards. SH = Shallow soils.

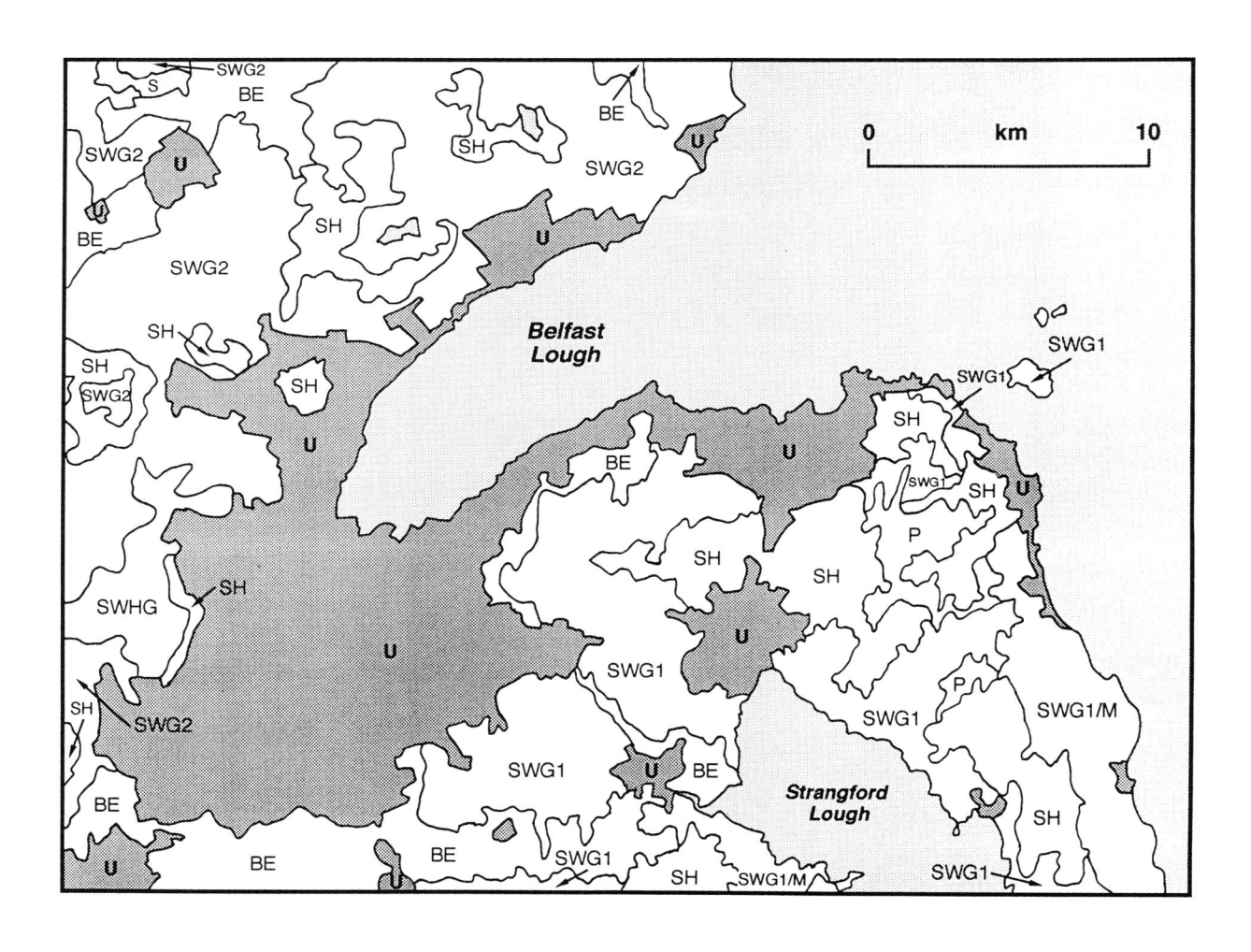

Soil Maps 15 and 21

The bedrock through 95 per cent of the land area on **Soil Maps 21 and 15** is Silurian shale, and it is remarkable that there is so much shale rock outcrop through the lowlands of this area of County Down. The soil map units, shown in a pale pink colour on the published map, are brown rankers on shale rock. In these extensive lowland rock-plains, on both sides of Strangford Lough, the shallow freely draining rankers are less than 40 cm in depth and shale rock outcrops frequently. These shale rankers burn quickly in a dry summer, as they did in 1995, and must be classified as poor quality agriculture land Class 4. They are of very limited agricultural production or land use potential in drought, and normally are constrained to medium to poor quality grass. Arable crops are found in adjacent till areas, which are of silt loam texture and do not dry out.

The most striking feature of Soil Maps 21 and 15 is the distribution of drumlins and related till soils. These areas are shown in pale grey-green colour, and represent slightly gleyed (SWG1), very stoney silt loam soils developing from shale till. The alignment of the drumlins on Map 21 is NNW to SSE, or between 290 and 330 degrees, with the stoss or blunt end of the drumlins facing NNW or towards the source of the oncoming, moving ice. Geomorphologists have dated this drumlin field to a time quite late in the last glaciation, probably about 20,000 to 25,000 years BP. The glaciation which formed these drumlins had its centre in the Lough Neagh area, moving outwards in a south-east direction over what is now counties Armagh and Down, and directly eastwards down the Lagan Valley over South Antrim. It is interesting that a Belfast-based geologist, called Close, introduced the term 'drumlin' into scientific literature from a study of this drumlin field in the 1860s, and that it was still being investigated in the 1960s! (see publications by Stephens, Hill and Vernon).

The density of the drumlins becomes greater and their form more sharply defined moving backwards from the south-east outer fringe of the drumlin field in the south of The Ards and around Downpatrick, towards the ice centre in the north-west. (Lough Neagh is a further 30 km further away). Perhaps most surprising thing about this drumlin field is that individual drumlins are usually isolated from each other, apparently moulded and deposited separately by sheet ice on the rock plain below. Most farms in the area would have a mix of land, including both till drumlins and rock plain. There are also a few small areas of drumlins developed on Triassic Red Sandstone around Comber, with brown earths being the commonly found soil profile type. Just off the area of Soil Map 21 to the north, the weakly developed drumlins of Soil Map 15 between Bangor and Holywood have a north to south alignment, and are thought to have had a two stage development, the earlier phase involving moulding by Scottish ice moving from the north and north-east.

One of the most unusual soil units found in Northern Ireland is the two-mile wide coastal till plain along the north and east coast of The Ards, as well as an outlier at its southern point. The parent material is red clay loam and clay, which does not appear to have any connection with the underlying shale bedrock, but rather is derived from Triassic Marl lying off-shore under the North Channel. The deposition of red clayey till starts at Groomsport-Bangor on the coast of North Down, and extends south to Donaghadee and Millisle. The material on the north and east coast of The Ards must come mainly from a Marl source, lying to the east under the North Channel, to explain the clay loam and even clay textures. Possibly because of the red colour, the profile does not seem to be gleyed further than SWG1. It is possible some soils are SWG2 but signs of gleying are masked by red colour. The east coastal plain is fairly level, being a dissected till plain ranging in altitude from 10 m and 30 m above present sea level. If there are any drumlins, and weakly developed drumlins have been identified around Ballyhalbert, the origin of this till plain would be in dispute. Geomorphologists have described this unit as being red clay, containing few stones, and being up to 5–6 m in thickness. In places, it contains marine shells and fossils, and is considered to be the product of a late-glacial marine transgression, extending up to 20 m a.s.l. at present, and being deposited about 12,000 BP - much more recent that the period of drumlin formation and from what is now a submarine area to the east. Soil Map 21 also shows other similar soil polygons of SWG1 on Mixed Shale and Red Triassic Sandstone till along both sides of Strangford Lough and again at its southern end in Lecale (J560502). This mixed till is usually sandy loam or sandy clay loam, and clearly has a large sandstone input.

The remaining areas of special soil interest are the line of gravel along the outer south-east coast of The Ards, and the south-facing slopes of Scrabo Hill, close to Comber, where the soils are slightly gleyed SWG1 on Red Triassic Sandstone till, inter-mixed with brown earths on sand and gravel. The latter join up with reclaimed marine sands to the east. The whole area, around Comber, is intensively cultivated with cereal and horticultural crops, and would be classified as Class 2 land, among the best in Northern Ireland. Generally, brown earths in this area would be very well drained and easily managed for a variety of crops.

Climatically, the whole area of Soil Maps 21 and 15 is well-favoured. Mean annual rainfall is less than 850 mm, and the area is one of the driest in the Province. The growing season is at least 280 days and can last up to 300 days. Agricultural production and land use potential are not usually constrained by temperature, but can be by a lack of rain. At one extreme, horticultural crops often require irrigation in the sandy soils around Comber. Lack of soil moisture is also a constraint on the shallow brown earths and rankers. However, on the till soils it is possible to grow a wide range of arable crops, and often experimental crops, such as linseed and maize at present, are successful in good seasons. Because of minimal climatic variation, agricultural land classification can be based on soils alone.

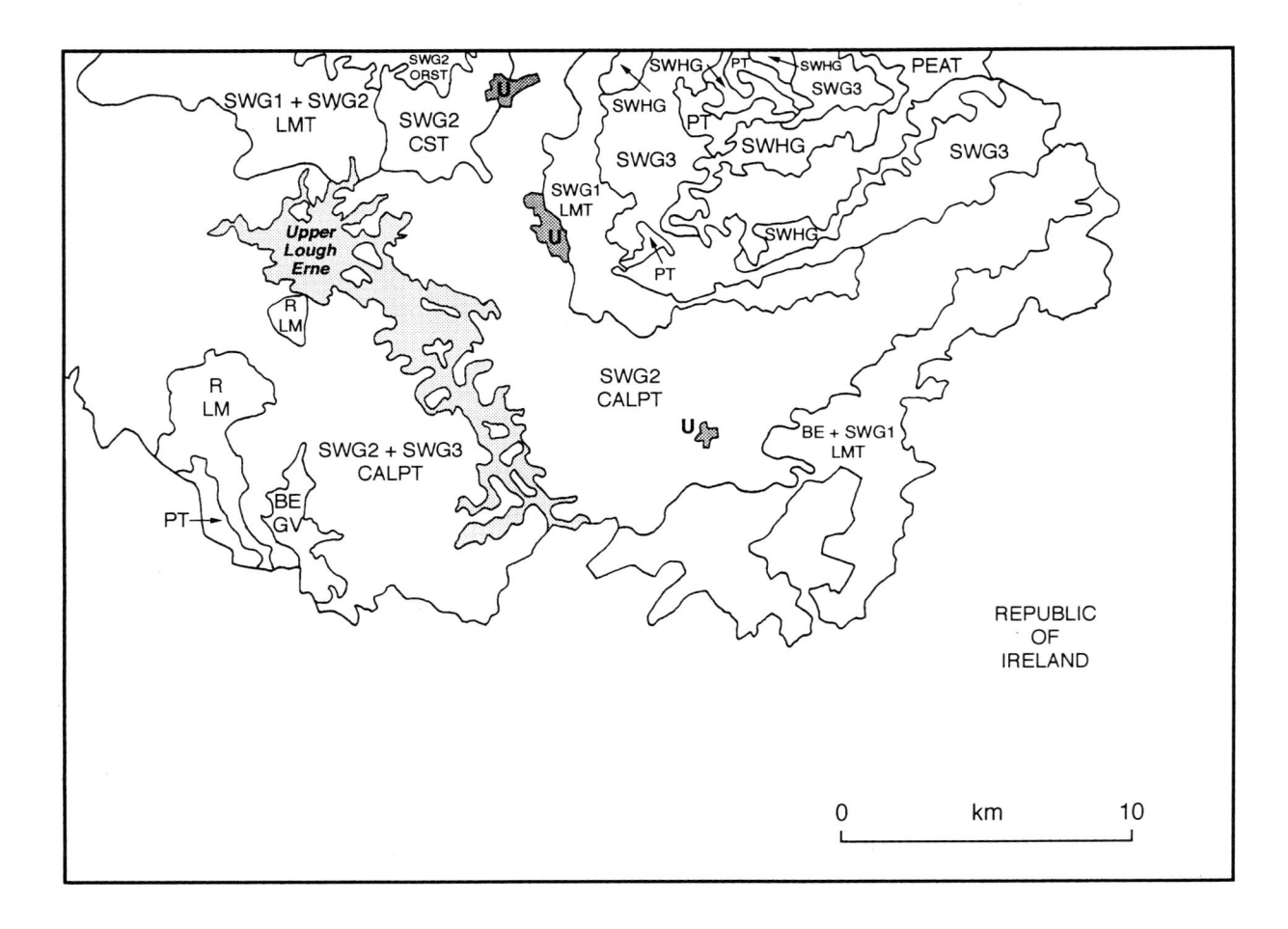

The distribution of the main soil series on Soil Map 27 (1:50.000)

Key for soil map sketch 27

In the area of **Soil Map 27**, the following rocks occur:

LM	=	Limestone	ORS	=	Old Red Sandstone
CALP	=	CALP rocks of the Middle Carboniferous	CS	=	Carboniferous

The soil profile codes without lower/pm codes belong to the YS = Yoredale Sandstone rocks.

T is added for till.

Soil Map 27

Very poorly drained SWG3 soils, developed on mixed till of shale, sandstone and limestone rocks from the 'Calp' or middle series of the Carboniferous, extend eastwards from Soil Map 26 into the area between the Arney and Swanlinbar rivers. This is a low-lying landscape with low ridges of clayey till rising slightly above the surrounding alluvium and basin, lowland peat. Farms are small in size, and the potential for agricultural improvement is limited, even with drainage. The 'Calp' till extends over most of the lowland south-west of Upper Lough Erne, around the isolated Knockninny limestone mountain (H275305) and the village of Derrygonnelly. There are some slightly better areas of SWG2, but the common situation is the continuing mixture of SWG3 Calp till, alluvium and peat, to the lough shore (Lough Erne Upper) and on to the political Border. There is one exception, and that is the stretch of 4-5km^2 of brown earth gravel (shown in bright yellow on the map) running from Derrygonnelly to the Border, near Ballyconnell. Much of this has been removed by quarrying for a local cement works. It has become a derelict landscape for agriculture.

The pattern of soils on the upland of Slieve Rushen (H240240) (on the Fermanagh-Leitrim Border) is of blanket peat above 300 m on the plateau summit, surrounded by humic rankers on Yoredale Sandstone and limestone rocks. To some extent, a similar pattern is found in the north-east on the upland of Slieve Beagh, to the east of Lisnaskea, where the blanket peat edge is between 200 m and 250 m, and is surrounded by humic gleys developed on Yoredale Sandstone till. Beyond the humic gleys, the SWG3 on Yoredale Sandstone till is very sandy clay overlying clay, and is extensively tree-planted. At the base of the slopes around Slieve Beagh coinciding quite well with the 100 m contour, much better soils, SWG1 on limestone till, extend to Lisnaskea and Moorlough Lake (H388300). Beyond these good limestone soils, lies the most extensive soil unit on Soil Map 27. This is at least 60 km^2 (the surface-exposed part only) of SWG2 on Calp till, running from Roslea to Newtownbutler, and seen as blue-grey in colour on the map. Near the shores of Upper Lough Erne, lowland depressions in the Calp till are infilled with basin peat, and in places covered with mineral alluvium, such as west of Lisnaskea. Here, there is about 10 km^2 of alluvium containing organic layers. It is very wet in winter, but can be used as summer grazing. In the same area, to the west of Maguiresbridge (H350385), there is a spread of moderate quality (Class 3B) land, SWG2 on Carboniferous Sandstone till (shown in pale blue). There is evidence of a southward movement by ice-push of about 3 km for this till, and also the adjacent Old Red Sandstone SWG2. The southern edge of the derivative ORS rock is 1.5 km north of the map edge of Soil Map 27 (McCabe, 1969).

The best quality agricultural land, possibly Class 3A, is the bank of SWG1 on Upper Limestone till, running around the marginal lowlands of Slieve Beagh upland, past the town of Lisnaskea (H365335), and mentioned in the previous paragraph. The relatively good quality of this limestone till was probably a reason for the location of the Plantation town of Lisnaskea. The limestone till is also ice-pushed southward about 1 km to Moorlough Lake (H388300). This landscape of the SWG1 limestone till is quite variable in soil depth to rock, and also in slope, and in quality it may be either Class 3A or 3B.

Adjacent on the southside is the great swath of SWG2 Calp till, which bridges across the Upper Lough Erne, and effectively crosses from Kinawley (H230310) to Roslea (H540325). This is a soil landscape which is the centre-piece of County Fermanagh soils, and one that has been greatly improved in the last 30 years by recent developments in drainage methods, especially of mole drains and gravel-filled tunnel drains. Its quality classification would have been raised from Class 4 to 3B.

In the north-west corner of Soil Map 27 (towards Enniskillen, just off this map), there is a soil map unit of poorly drained SWG2 on limestone till, around Carr Bridge (H296374). Although this is a landscape of low drumlins, at least half the area is infilled with basin peat. Around Mill Lough (H245385), the drumlin slopes are steeper and SWG1 soils (Class 3B) are found on limestone till.

The last remaining soil landscape lies along the political Border from Roslea to Clones, and Scothouse, which includes a portion of the Republic of Ireland, (County Cavan) almost encircled by County Fermanagh. It is a zone of good quality land of Class 3A in SWG1 soils on till of Lower Limestone, and brown earths on limestone till and limestone gravel (the latter being orange-yellow on the soil map). The brown earths might be Class 2 land in small units, but they lose value by isolated location next to the Border, the land-holding system and the extent of surrounding basin peat. It is land similar to that on Lower Limestone till in areas around Brookeborough, or Enniskillen to Ely Lodge in County Fermanagh.

Soil Map 26

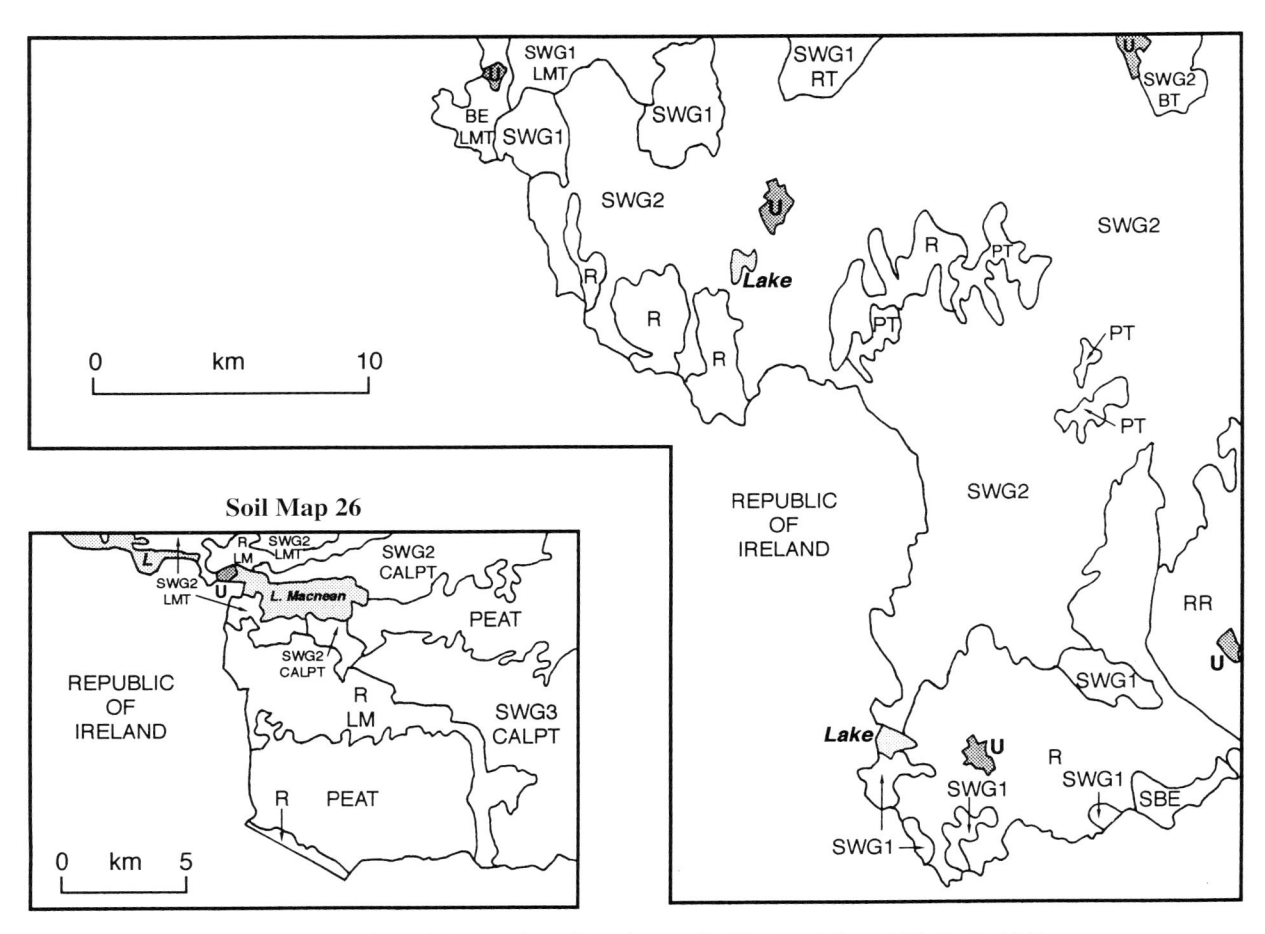

The distribution of the main soil series on Soil Maps 28 and 29 (1:50,000)

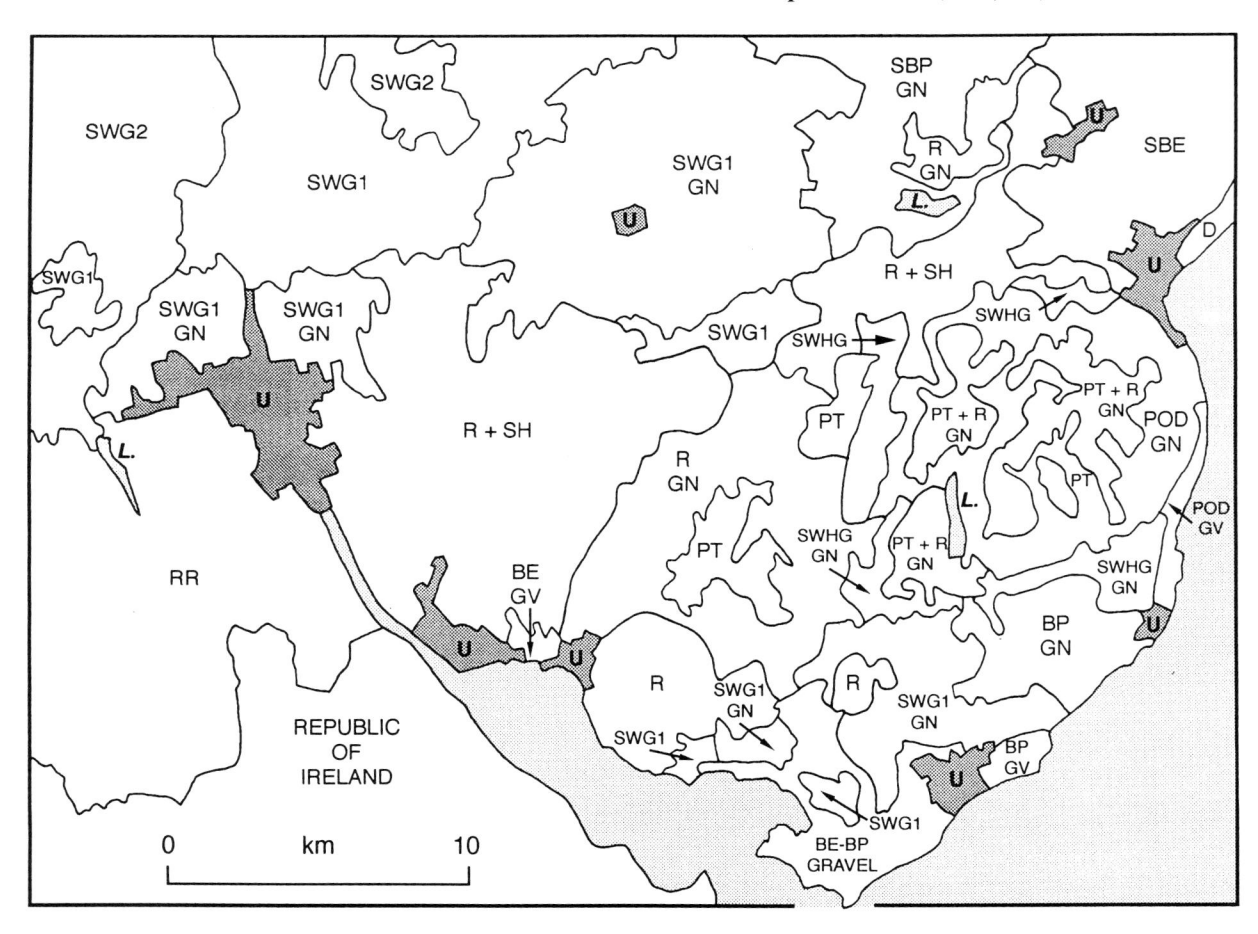

> **Key for soil map sketch (Soil Map 28 - top left)**
>
> Most of the area of **Soil Map 28** is underlain by shale, and those soils on shale rock are distinguished only by a soil profile code. Minority rocks present are:
>
> R = Red Trias Sandstone   B = Basalt
> LM = Limestone   T is added for till

On **Soil Map 28**, the majority area is SWG2 on shale till (shown in green colour). The till is moulded into drumlins, which stretch over 30 km to Armagh City and 50 km to the shores of Lough Neagh. The drumlins are found up to about 180 m above sea level, and above that, there are areas of shale rock rankers and peat. Just south of Keady (H845345), there are several high level platforms of blanket peat and surrounding humic rankers HR, at about 300 m above sea level. Deadman's Hill (H930315) rises to 356 m, and provides a good example. Annual precipitation is about 1100 mm at 300 m elevation. Southwards, the shale till drumlins are contiguous to near Cullyhanna (H925210), and end abruptly about 2 km south of the village. In the Border area, around Crossmaglen (H910150), there is an area of about 80 km² of shale rock outcrop and shallow brown rankers, with agricultural value no higher than Class 4. There are, however, about 22 drumlins, mostly isolated from each other and deposited on the rock plain, which provide some Class 3 land. There is a very small area of about 2 km² around Ballsmill (H985135) of shallow brown earths or Class 2 land, extending across the Border into Co Louth.

At the northern edge of Soil Map 28, there are extensions of SWG2 on the limestone till, mixed limestone and Red Triassic Sandstone till, and basalt till which extend from Soil Map 19 to the north. The descriptions of these soils can be found on Soil Map 19. Close to Middletown (H755387) on the political Border, there are several areas of better drained (SWG1) shale till, and even shallow brown earth on shale rock near Doogary Lough (H78S387), comprising about 16 km². Immediately adjacent to Middletown, there are very good quality (Class 2) brown earths and calcareous brown earths on limestone till, which extend into the Republic of Ireland towards Monaghan town. The extension northward on to Soil Map 19 includes about 1.5 km² of calcareous brown earth, which might qualify as grey-brown podzolic, CBE in Boneville House estate.

> **Key for soil map sketch 29 (lower left)**
>
> On **Soil Map 29**, most of the area is underlain by shale rock, and those soils are distinguished only by a soil profile code (see Chapter 2). The other rock type is GN = Granite.
>
> R = Rankers (Except HR = Humic Rankers)   GV = Gravel
> SH = Shallow soils

For varied soil landscapes as they both are, there is a continuity and natural soil connection from **Soil Map 28 into 29**. Most of the join line area is SWG2 on Silurian shale till, with the southern third being rankers on granite. The soil series distribution on Soil Map 29 closely reflects the underlying geology, with soil parent materials being co-incident with the related rock type. There are no examples of carry-over of till from one rock type to another, and there are only two major rocks (granite and shale) to identify on Soil Maps 28 and 29. There are minor occurrences of dolerite and felsite in the Ring Dyke Complex, west of Newry, but the soils developed are only very shallow rankers in both cases.

Colour is selected to give an indication of soil series groupings. On Soil Map 29, the granite soils are expressed in shades of red in the freely draining group of soils. Podzols developed on weathered granite rock, are seen most clearly in a dark purplish-red colour, which includes a peaty podzol complex on the steep slopes of the Mournes, and more commonly, a brown podzolic on the granite rock of the older Newry Granites of the Slieve Croob – Newry – Ring Dyke complex. These various podzols are intermixed with rankers on granite rock, and with blanket peat. There is only about 20 km² of blanket peat remaining in the Mournes, partly because of peat cutting and removal, and also because there are relatively few level surfaces to initiate peat formation.

Granite till is limited to two areas, though each is quite extensive. First is the Plain of Mourne, from Annalong to Attical and Kilkeel, divided almost equally between brown podzolics in the east, and SWG1 in the west between Attical and Kilkeel. The granite till does not have drumlinoid features, as it developed into a gently sloping till plain, smoothed-out by solifluction processes. South of this till unit, the coastal area around Kilkeel is a fringe of brown podzolics on fluvio-glacial gravels. The other (northern) unit of granite till is also SWG1, extending from Eastings 280 to west of Camlough (J040270) on the west edge of the map. The part of this unit, lying to the east of Rathfriland (J200335), which is the uppermost basin of the River Upper Bann is very strongly drumlinised. The drumlins, rising above a floor of river alluvium, are orientated almost east-west, which reflects the direction of ice movement (deflected around the mountain mass of the Mournes). Polygons of peaty gley SWHG on granite till stand-out clearly in strong blue colour.

Shallow brown earths on a thin cover of shale till occur in the north-east-corner of Soil Map 29, east of Castlewellan (J340360), and again, on south-facing slopes north of Warrenpoint (J150190). There are also a number of areas of SWG1 on shale till close to Warrenpoint, around Hilltown (J215290), and north-east of Newry. The last mentioned unit is mixed with SWG2 on shale till, and both units run off this soil map, northwards on to Soil Map 20 and westward on to Soil Map 28. Modal soil profiles of these SWG1 and SWG2 soils on shale till in this area are provided in Chapter 2.

CHAPTER 9
SOIL SURVEY APPLICATIONS

CHAPTER 9A
SOIL CARBON POOLS

MARGARET M. CRUICKSHANK, SCHOOL OF GEOSCIENCES,
THE QUEEN'S UNIVERSITY OF BELFAST

Introduction

The DANI Soil Survey has been used to quantify and map the distribution of organic carbon stored in Northern Ireland soils. This research is funded by the Department of the Environment (GB). It is part of a project to provide an inventory of carbon stored in soils and vegetation, and to quantify the fluxes of carbon between these reservoirs and the atmosphere. This is intended to help the UK meet its commitments under the Framework Convention on Climate Change. Countries party to the Convention are required (amongst other things) to adopt policies and take measures to protect and enhance reservoirs and sinks of greenhouse gases. Research on soil carbon pools and fluxes commenced in GB (Howard *et al.* 1995 and Milne *et al.* 1997) and was extended to Northern Ireland in 1994 to complete the national inventory (Cruickshank *et al.* 1996).

Methods used to estimate soil carbon pools in Northern Ireland

Compilation of a 1 x 1 km database from the soil maps. To assemble all the data required to estimate the total carbon pool and map its distribution, a database approach was needed. In GB, an existing 1 x 1 km database of dominant soil types was used for the first stage in estimating soil carbon stores. As a similar database was not available for Northern Ireland at the time, it was necessary to compile this from the 1:50,000 soil maps. Other sources were used to obtain the range of variables required. The seven variables were: Profile type, Parent material, Dominant cover, Peat depth, Carbon density (t/ha), Avery Soil Group and the Grid Reference of the SW corner of the grid square.

The database was in SIR (Version 2) on the mainframe computer at Queen's University of Belfast. Data on the different soil types were retrieved by SQL and to produce the map, the database was downloaded as a coverage in the ARC/INFO GIS (Geographical Information System).

Dominant soil types recorded from the soil maps. For each 1 x 1 km grid square, the dominant soil was read from the soil maps as the profile type – parent material combination; for example, a 'brown earth on shale till'. Use of this level of soil type was essential for comparability with GB. Comparison with map sheet statistics showed that this approach slightly overestimated soil types present in extensive units and underestimated those in small units. All grid squares on the sea coast and on the Border with the Republic of Ireland, which had any land in Northern Ireland, were included.

Checks were made on the accuracy of visually recording soils from the maps. Whether, given the complex soil patterns, the dominant soil in each grid square had been identified correctly, was checked when the eastern half of the Province had been completed. The DANI Soil Survey supplied a computer extraction from their mapping of all soil type areas for each grid square on Soil Sheet 20, which represented a 10 per cent sample of the eastern half of the Province. Under 3 per cent of the visual records were found to be incorrect and these were amended. The majority of errors occurred where two or more soils were nearly co-dominant. Subsequently, checks were made for grid squares where there was doubt as to the dominant soil; the error is likely to be less than 3 per cent. A second issue was whether grid co-ordinates and variables were entered correctly into the database. A printout was checked against the log of variables assembled to create the database and all errors were corrected.

The Soil Survey did not map in urban and disturbed areas, but soil maps of these areas were available from the HOST project (Chapter 9C). It was assumed that in completely built up urban areas (according to the CORINE image – Chapter 6) there was no vegetation cover, soil, or soil carbon. For soil types in suburban areas, which are only partly built over, allowance was made for the presence of soil carbon.

Soil property data and calculation of carbon density (tonnes/ha). For each soil type (except peat) recorded for the database, the Soil Survey supplied bulk density, organic carbon and depth data, these being required to calculate carbon density in tonnes per hectare. All property analyses were for soils under grass. The carbon density for soil types without property data, typically in small areas, was estimated in consultation with the Soil Survey and for peat, the carbon densities used in GB were applied.

Cover related differences in carbon density: dominant land cover from CORINE. As soil carbon density varies with vegetation cover (Howard *et al* 1995), each soil type can be divided into up to 5 variants depending on the dominant vegetation. Vegetation groups used in the GB study were: 'arable', 'permanent grass', 'semi-natural vegetation', 'trees' and 'no cover'. In Northern Ireland 'suburban' was added to this list. For work on fluxes it will be necessary to know the location and extent of forest on peat and peaty soils, so that for those soil types a distinction was made between planted forest (F) and other trees (T) – for example colonising cut-over raised bogs. Cover was interpreted on the 1:100,000 satellite images used for the CORINE project, and with the interpretation being made alongside the soil maps, the dominant cover type was obtained for the location of the dominant soil in each grid square.

Recognition of cover in relation to soil carbon density raised two problems in Northern Ireland: (1) the distinction between rotation grass and cropped land and (2) property data were available only for soils under improved grass. To ensure that the study was comparable with that of GB, it was accepted that soils under grass and arable cover would differ in carbon density, even though evidence was not available. The DANI Soil Survey was rarely able to dig large sample pits in arable fields. Neither were there property records for soils under semi-natural vegetation or trees, because Soil Survey sampling was concentrated on agricultural land. Therefore GB evidence was used to estimate the carbon density in soils with cover types other than grass.

Calculation of soil carbon density differences in relation to cover. Earlier work in GB showed contrasts in carbon density related to cover type at the Avery Major Soil Group level (Howard *et al.* 1995). The scale of these differences was used to estimate the likely carbon density under arable (A) and semi-natural vegetation (S) cover in Northern Ireland. In GB, the rarity of data for soils under tree cover (T) led to those soils being given the same carbon density as soils under semi-natural vegetation (Milne *et al.* 1997) and this practice was continued.

Three area-weighted means, for improved grass, arable, and semi-natural vegetation, across all the Avery Soil Groups for England and Wales were calculated (excluding peat, disturbed soils and urban areas):

      154 tonnes/ha under arable cover (A)
      183 tonnes/ha under improved grass (P)
      243 tonnes/ha under semi-natural vegetation (S).

On average, soils with A cover have 84 per cent of the carbon density of those under P, while soils with S cover have 133 per cent of the carbon density under P. These figures were applied to the known P carbon densities, to estimate the likely carbon density under A and S, T and F in Northern Ireland. Thus, the same percentage differences were applied to all soil types. This simple solution is in keeping with the recognition that to get a carbon density for all soil types under P, some 'best estimating' had been necessary. Suburban areas (Su) were allowed half the carbon density under grass for the same soil type (based on evidence of the proportion of the ground covered by different vegetation types). Built up urban areas and quarries were regarded as having a carbon density of zero.

Carbon in peat and peaty soils.

As these soils have high carbon densities and are extensive, it was important to estimate their carbon content as accurately as possible. This required consideration of: peat carbon density, peat depth and how to deal with 'peaty' soils. A major influence on peat carbon content is depth, which was estimated for each grid record where peat was the dominant soil. The estimates of peat carbon density used in GB were applied because the Soil Survey could not supply property data for peat. Limited property data (especially bulk density) were available for peaty soils. As with other soils, the calculated carbon densities should be amended according to cover, but it was important that carbon estimates for such soils would not exceed those for peat itself. The following section outlines the solutions devised for peat depth and carbon density, and for peaty soils.

Peat depth. The Soil Survey records peat where the organic layer exceeds 0.5 m and only grid squares on the soil maps where peat was dominant were taken as peat for the present study. A valuable source on peat depth was the survey by Double (1954) at Queen's University; for most of the 30 sites he examined, there were many depth soundings, which were used to calculate a mean depth for 1 x 1 km grid squares. For sites known from recent field surveys to have large scale peat extraction, a generalised estimate of consequent loss of depth was made. Elsewhere, field knowledge and consultation of the Peatland Survey maps (Cruickshank and Tomlinson, 1988) was used to estimate peat depth. In that survey, peatland was classified and mapped at 1:20,000 from aerial photographs dating mainly between 1975-83. Later surveys of peat extraction from blanket and lowland bogs (Cruickshank *et al.* 1991, and Cruickshank *et al.* 1995) also provided field knowledge of peatland, which helped in estimating mean depth to the nearest 0.5 m for 1 x 1 km squares. Since peat is defined as >0.5 m deep, the thinnest peat was taken as 1 m deep.

Peat carbon density. One of the properties required to calculate carbon density is bulk density, but there are

TABLE 9A.1
Peat carbon densities related to cover and depth

Total peat depth m	Blanket peat with S, F, T, N t/ha	Blanket peat with A, P t/ha	Basin Peat with S, F, T, N	Basin Peat with A, P
1	530		357	286
1.5	830		607	500
2	1130		857	714
2.5	1470		1179	*967*
3	*1860*	580 for all depths	1500	*1200*
3.5	2250		*1815*	
4	2630		*2124*	
4.5	*3030*		*2442*	
5	*3450*		*2784*	
5.5	*3933*		*3174*	
6.0			*3618*	
6.5			*4125*	
7.0			*4702*	
7.5			*5360*	

[Cover types: S - semi-natural, F - forest, T - trees, N - none; A - arable, P - pasture]

few measurements of peat bulk density in Northern Ireland. Previous estimates of bulk density for GB peat (Howard *et al.* 1995) were revised by Milne *et al.* (1997), which led to a reduction in peat carbon densities per ha. These revised densities were the best available (Table 9A.1) and the rate of increase of carbon with depth estimated in GB was used to extrapolate for extra densities required in Northen Ireland (shown in italics in Table 9A.1).

The carbon density of peat is also influenced by whether it is blanket or lowland peat and its cover. Although the soil maps do not distinguish between blanket and lowland peat, this could be found from the Peatland Survey and cover was interpreted from CORINE satellite images. After the grid records for blanket and lowland peat had been labelled for cover and peat depth, the carbon densities listed in Table 9.A.1 could be added to the database.

Peaty soils. These have a surface organic layer less than 0.5 m deep and include the following profile types: humic rankers, peaty podzols, surface water humic gleys, humic gleys and organic alluvium. When the carbon densities for peaty soils under grass cover were calculated, it was observed that they were relatively high and, as noted earlier, it had been decided on GB evidence to increase the carbon density under semi-natural vegetation, trees and forest to 133 per cent of the grass value. As such cover is common on peaty soils, it would have led some to have a carbon density higher than that being used for 1 m of peat. In order to keep the peaty soils in a sensible relationship to the thinnest peat, the following solutions were adopted.

The carbon density for podzols and peaty podzols under grass cover was around 300 t/ha for podzols and up to 450 t/ha for peaty podzols, so that an increase of 133 per cent could exceed the value for 1 m of blanket peat. As in both cases the grass would often be poor quality, permanent and almost semi-natural, the grass (P) carbon density was used also under S, T and F cover.

For surface water humic gleys, carbon densities under grass were 325–350 t/ha and such soils are usually around the lower edge of blanket peat. The common cover was S and on the satellite image these areas were often indistinguishable from adjacent peat with S cover, i.e. significantly different from poor grass on surface water humic gleys. 133 per cent of the P carbon density was used, yielding a carbon density between that and 1 m of blanket peat with S cover. For humic gleys, which are uncommon, the same solution was used, but these soils characterise wet valley bottoms.

Soils classified as organic alluvium are diverse, often in lowland valleys and, in some cases, are cut-away peat bogs. Such 'soils' can be fen peat, are highly organic and of variable depth. Commonly, the cover is wet, rushy, poor grassland. The carbon density used in GB for 1 m of lowland peat under P is only 286 t/ha – lower than recorded for peaty soils with P cover (as above). It was decided to use the carbon density for 1 m of lowland peat under S or T (357 t/ha) for organic alluvium, regardless of cover.

Size of the soil carbon pool and its distribution

The soil carbon pool is examined in relation to major soil groups and cover types in Table 9A.2. The total soil carbon pool for Northern Ireland of 386 Mt (385,683 kT) represents 4 per cent of the UK total. The area, as used for the carbon study, is about 6 per cent of the UK land area. This result places Northern Ireland just below the UK average in carbon density and between the mean carbon densities for England/Wales (low) and Scotland (high). Peat accounts for only 15 per cent of the area, but 43 per cent of the carbon pool (classes 10.1 and 10.2 on Table 9A.2); if peaty soils (as defined above) are included, together they account for 53 per cent of the soil carbon. In passing, it can be noted that the vegetation carbon pool at 4.4 Mt (Cruickshank et al. 1996) is only a small fraction of the soil pool. A difference is typical of temperate climates, but the scale of the contrast is explained by the presence of high carbon density peat which increases the soil pool and lack of forest reduces the vegetation pool.

The frequency of the cover types (Table 9A.2) contributes to contrasts in carbon density between the Avery Groups (last column). Semi-natural vegetation, trees or forest increased the carbon density and certain soil groups have a high proportion of these cover types. Arable cover, which reduced the carbon density, is uncommon and found mainly on rare brown earths and on widespread stagnogleys.

The distribution of the carbon pools is mapped in Colour Plate 17. The highest carbon densities are related to peat and peaty soils – as in the Sperrins, Antrim Plateau, Lough Navar and other upland regions. Although the key extends only to >1000 tonnes/ha, the highest values occur in deep lowland bogs which exceed 5000 tonnes/ha. Low values characterise the climatically drier lowlands of the south and east, and the Strule valley where better drained soils are found.

Table 9A.2 Soil groups, cover and carbon pools in Northern Ireland

Avery Soil Code	Area (km²)	Cover (km²) (A)	(P)	(S)	(T)	(F)	(N)	(Su)	Carbon pool (kT)	Carbon t/ha
1.2 Raw alluvial soils	534	74	390	26	9		24	11	9,651	180.7
1.3 Raw skeletal soils	30		1	29					0	0
3.1 Rankers	1,158	21	469	581	42	29	4	12	23,537	203.3
3.2 Sand-rankers	61	2	17	38	1		3		0	0
4.1 Calcareous pelosols	4		4						78	195
5.1 Brown calcareous soils	4	1	3						57	142.5
5.4 Brown earths	1,537	135	1,292	31	20		8	51	25,511	166.1
5.5 Brown sands	493	55	356	3	4		8	67	9,252	187.7
5.6 Brown alluvial soils	5	1	4						74	148
6.1 Brown podzolic soils	418	2	379	23	6		1	7	6,690	160
6.3 Podzols	202		108	84	8	2			6,826	337.9
7.1 Stagnogley soils	6,902	260	6,022	253	166		29	172	115,998	168.1
7.2 Stagnohumic gley soils	573		122	324		127			23,103	403.2
8.2 Sandy gley soils	18	12	2	1				3	213	118.3
8.4 Argillic gley soils	8		7				1		134	167.5
8.5 Humic-alluvial gley soils	9	1	7			1			275	305.6
8.7 Humic gley soils	5		3	1	1				211	420
10.1 Raw peat soils	1,510	1	15	1,193		297	4		121,152	801.9
10.2 Earthy peat soils	554	3	182	277	40	40	7	5	42,982	775.8
No soil (quarries)	16						16		0	0
Total	14,041	568	9,383	2,864	297	496	105	328	385,744	

Arable (A), Grass (P), Semi-natural (S), Trees (T), Forest (F), Suburban (Su), No cover (N)
Forest is used only on peat and peaty soils; on other soil types it is included under trees.
No cover is urban built over except for classes 10.1 and 10.2 where it is due to peat extraction.

TABLE 9A.2
Soil groups, cover and carbon pools

Conclusions

A first estimate of the soil carbon pools. The results presented here are a first estimate of soil carbon pools in a part of the UK where soil maps and property analyses were being completed during the work. Lack of an existing 1 x 1 km database at the time, necessitated extra work in creating one, but the database could be developed to suit the needs of carbon pools research. For example, dominant cover was estimated for the area of the dominant soil in each grid square. This was an advantage, but there were also problems which might be solved in future work. Also at the time of this study, soil property data analysis was incomplete, so that soil types occupying small areas, of which there are many, had their carbon density estimated through discussion with the Soil Survey. The soil carbon densities used should be regarded as the best achieveable at the date of this study. When soil property results are fully analysed, carbon densities may need to be revised, which could in turn lead to a revision of the total pool.

Carbon density and land cover. The general difference found in England and Wales in soil carbon density related to arable or grass cover was applied to Northen Ireland, but this may be inappropriate. Future work could commission soil property analyses for fields used regularly for arable crops, for grass >5 years old, and for grass which is part of a rotation including arable crops. Comparisons could be made also between soils under grass cut for silage or hay, and grass that is only grazed. In addition, the 4th Level subdivision of pasture in CORINE (Chapter 6) might have potential to locate the more productive grass, likely to be regularly re-seeded and/or cropped, but field survey would be required to establish the content of these CORINE classes. There is a need for further evidence of cover related differences in carbon density in a region where grass cover accounts for 77 per cent of the agricultural area.

Peat bulk density and depth. With awareness of the extent of past peat cutting in Northern Ireland, a lowland raised bog and a blanket peat site were sampled for bulk density and percentage carbon at different depths. The results were generally comparable with those from GB (Milne *et al.* 1997), but bulk densities were slightly higher, which might be caused by slight drying due to cutting. If this applies widely, then the carbon stores may have been underestimated. Another uncertainty arises from the rarity of peat depth measurements or depth profiling of peatbogs. The further work needed on bulk density and depth of peat is another reason to expect future revision of the total pool.

Soil carbon fluxes. Protection and enhancement of soil carbon requires better understanding of the fluxes into or out of the pools. To quantify the link between cover and soil carbon density is important towards knowing which cover reduces and which increases it. This is essential in understanding the effects of land use change on soil carbon and could inform land use policy changes to ameliorate global warming. For example, in GB, conifer forests have been planted mainly on drained peat or peaty soil and in Northern Ireland, 64 per cent of State Forests are on such soils. Cannell *et al.* (1995) indicate that the flux loss of carbon from draining the peat for forest planting may exceed the carbon sequestered by the trees. Therefore, protection of peat carbon may be helped by the 1992 change in forestry policy; planting on peat bogs will not be extended, but re-stocking forests on peaty soil and peat could continue.

Protection of soil carbon should be helped by ESA designations (since 1986), although farmer participation in these is voluntary. ESAs offer grants to reduce stocking levels, which could lead to changes in the grass cover and increased soil carbon. ESA policies also prevent reclamation of upland rough grazing and more widely, grants for reseeding (reclamation) of rough grazing ceased in 1988/89; this practice involves disturbance of the soil and reduces soil carbon. Maintenance of heather by controlled burning releases carbon stored in the heather and may also reduce carbon in organic matter in the topsoil. In these various ways, recent policies may be changing the fluxes and therefore the soil carbon pools.

Soil carbon and land classification. Recognition of the value of soil carbon pools challenges the traditional type of land classification developed out of soil survey, which ranks soil types by their potential agricultural productivity. To conserve and enhance soil carbon means protecting some of the soils poorest for production. These have traditionally been classified as of low value, but to assist in sequestering carbon from the atmosphere and thereby aid attempts to lessen global warming, such soils have higher values.

Acknowledgement. The research is funded by the Department of the Environment, Global Atmosphere Division (Contract EPG 1/1/3) through NERC to the Institute of Terrestrial Ecology. Work on Northern Ireland is sub-contracted to the School of Geosciences, Queen's University. This chapter is based on contract reports to the Department the Environment (UK), with their permission.

The author thanks the following for assistance, which was essential to completion of the project: DANI Soil Survey, Ronnie Milne, Institute of Terrestrial Ecology, Edinburgh Research Station for guidance and discussion in adapting the methodology used in GB to Northern Ireland. Paula Devine, Centre for Social Research, Queen's University is thanked for work on the database and map, and Roy Tomlinson, as co-leader of the carbon project in Northern Ireland.

REFERENCES

Cannell, M G R, Dewar, R C and Pyatt, D G 1995 Conifer plantations on drained peatland in Britain: a net gain or loss of carbon? *Forestry* 66, 353 - 369.

Cruickshank M M and Tomlinson, R W 1988 *Northern Ireland Peatland Survey*. Contract Report to Countryside and Wildlife Branch, Department of Environment (NI).

Cruickshank, M M, Tomlinson R W, Bond, D, Devine, P M, and Cooper, A 1991 *Survey of the scale, extent and rate of peat extraction from blanket bogs in Northern Ireland*. Contract report to Environment Service, Department of Environment (NI).

Cruickshank, M M, Tomlinson, R W, Bond, D, Devine, P M, and Edwards, C J W 1995 *Peat extraction, conservation and the rural economy in Northern Ireland*. Applied Geography, 15, 4, 365 383.

Cruickshank M M, Tomlinson R W, Devine, P M and Milne, R 1996 Carbon pools and fluxes in Northern Ireland. Chapter 4 in *Carbon sequestration in Vegetation and Soils*. DoE Contract EPG 1/1/3.

Double, K W W 1954 *A survey of the peat resources of Northern Ireland*. Unpublished MSc Thesis. Queen's University of Belfast.

Howard, P J A, Loveland, P J, Bradley, R I, Dry, F T, Howard, D M H, and Howard, D C 1995 The carbon content of soil and its geographical distribution. *Soil Use and Management,* 11, 9 - 15.

Milne, R, Brown, T A and Howard, D C 1997 Carbon in the vegetation and soils of Great Britain. *Journal of Environmental Management.* In press.

CHAPTER 9B

ACIDIFICATION OF SOILS

C. JORDAN, AGRICULTURAL AND ENVIRONMENTAL SCIENCE DIVISION, DEPARTMENT OF AGRICULTURE FOR NORTHERN IRELAND, NEWFORGE LANE, BELFAST BT9 5PX.
J.R. HALL, INSTITUTE OF TERRESTRIAL ECOLOGY, MONKS WOOD, HUNTINGDON, CAMBRIDGESHIRE, PE17 2LS.

Introduction

Acid deposition from rainfall is one of the major factors increasing the rate of soil acidification, particularly where organic-rich soils have developed on mountain slopes from granitic parent materials under coniferous and heath vegetation (Krug and Frink, 1983 and Reuss and Johnson, 1986). Susceptibility of these soils to acidification is attributed to their lack of basic substances capable of neutralising acid rain. Moreover, steep slopes act to minimise contact time between acid rain and neutralising materials in the soil. This is believed to be particularly important for the initial portions of heavy rains and rapid snowmelts that are often more acid than the later parts of the storm or melt. As a result, the soils and their runoff become acidified. The acidification of soil and water in mountainous regions heralds the future for the lowlands as acid rain eventually reduces the greater acid neutralising capacity of the deeper soils. If these effects are to be halted or, ultimately, reversed, action is needed to be taken now to reduce the impact of acid rain.

Rain is naturally acid due to the presence of small amounts of carbonic, sulphuric and nitric acids due to dissolution of oxides of carbon, sulphur and nitrogen, respectively, from the atmosphere. These oxides result from natural processes such as respiration, forest fires, volcanic eruptions and lightning discharges. Monitoring around the world has shown that, even in pristine conditions, these processes cause the pH of rain to fall as low as 5.0 (Charlson and Rodhe, 1982). On this basis, when the pH of rainfall is less than 5.0, we can conclude that the acidity in the rain has definite anthropogenic origins, resulting largely from burning fossil fuels. Thus, the generally accepted working definition of 'acid rain' is rain with a pH below 5.0. The three main pollutants contributing to acid rain are emissions of SO_2, NO_x and NH_3 whose major sources are power stations, vehicular traffic and intensive livestock farming, respectively.

Acid Rain in Northern Ireland

Rainfall chemistry across the Province has been monitored regularly over the last 10 years, as part of the UK acid rain monitoring network (Jordan, 1987, 1994 and 1996). Data gathered since 1986 show that the annual rainfall-weighted mean pH of rain in Northern Ireland, in common with Western Scotland and Wales, has always been in the lowest H^+ concentration band of the UK maps, 0-20 $\mu eqH^+ l^{-1}$ (RGAR, 1990). This equates to a pH greater than 4.7 and, by definition, slightly acid rain. One might conclude that Northern Ireland is unlikely to suffer greatly from the effects of acid deposition evident in more industrial regions of the UK and northern Europe where acid deposition is higher and acidification of soils and freshwaters has already taken place (Mason, 1990 and Patrick et al. 1995). This is far from the truth, partly because accurate pH measurements in poorly buffered samples of rain are difficult to achieve but also because a significant proportion of the free H^+ ions in rain are 'fixed' through interaction with ammonia emissions present in the atmosphere. The ammonium ions so formed release their H^+ ions again during subsequent assimilation by vegetation when the rain reaches the soil.

A better estimate of total acid deposition can be calculated using the concentrations of the various components contributing to rainfall acidity where:

Total acidity = nm ($SO_4S + NO_3N + NH_4N - Ca - Mg - Na - K$)

TABLE 9B.1
**Mean annual concentrations of ions in rainfall for the period 1992-94,
after removal of the marine contributions**

STATION	GRID REF [a]	RAIN mm	SO$_4$S µeql^{-1} [b]	NO$_3$N µeql^{-1}	NH$_4$N µeql^{-1}	Measured pH	Effective pH	K [c] µeql^{-1}	Ca µeql^{-1}	Mg µeql^{-1}
Altnaheglish	C698041	1425	26.8	17.4	20.9	5.27	4.37	3.8	18.1	0.6
Silent Valley	J306243	1624	35.4	22.8	26.6	5.01	4.25	3.6	18.9	6.0
Parkgate	J227897	1132	47.0	16.8	39.1	5.33	4.16	5.4	24.0	4.0
Lough Navar	H065545	1708	17.1	11.5	11.3	5.07	4.74	0	16.0	5.7
Hillsborough	J243577	667	42.1	19.8	44.4	4.85	4.06	1.1	13.9	5.0

a   Irish grid references for use with 1:50,000 scale maps.
b   1 ueq of SO$_4$S, NO$_3$N, NH$_4$N, H, K, Ca and Mg is equivalent to 16, 14, 14, 1, 39, 20 and 12 m g, respectively.
c   non-marine Na and Cl concentrations are zero.

The 'nm' stands for non-marine concentrations – that is, the concentration of each ion has been corrected for the amount of each ion naturally present in sea-spray. The sulphate (SO$_4$S), nitrate (NO$_3$N) and ammonium (NH$_4$N) concentrations generate H$^+$ ions while the calcium (Ca), magnesium (Mg), sodium (Na) and potassium (K) ions (base cations) will effectively neutralise some of the generated H$^+$ ions. The net concentration of H$^+$ ions deposited will then determine the effective pH of the rain. Annual mean pHs calculated in this way are significantly lower than those from measured pH values and provide a truer picture of the level of acidity in rainfall. Annual mean concentrations of ions in rainfall over the period 1992-94 for Northern Ireland stations, including measured and effective pH, are detailed in Table 9B.1 (Jordan, 1996). These data are based on wet-only deposition and do not include any allowance for dry or cloud droplet deposition or for the effects of altitude.

Mapping of rainfall acidity

The annual average concentrations of major ions can be used to construct an interpolated concentration field using Kriging methods (Webster et al. 1991). The interpolated concentration fields can then be used to estimate wet deposition using precipitation fields provided by the Meteorological Office.

wet deposition   = acid concentration x rainfall /100000
(keq H$^+$ ha^{-1} year^{-1})   (m eq l^{-1})   (mm)

However, because monitoring stations are few and mainly in lowlands below 500m above sea level, there is a systematic underestimate of wet deposition on high ground. The deposition data presented in this chapter for Northern Ireland are the sum of wet plus dry plus cloud deposition. Wet deposition data are provided by AEA Technology who are responsible for the UK monitoring network. At higher altitudes, where higher precipitation also occurs, the composition of the precipitation varies as a consequence of seeder-feeder scavenging processes (Fowler et al. 1988). The wet deposition values are corrected for these processes at the Institute of Terrestrial Ecology (ITE), Edinburgh using a simplified seeder-feeder model (CLAG, 1994). Estimates of cloud droplet deposition, i.e. the direct capture of cloud droplets by terrestrial surfaces, are also made and included in the total estimates of deposition. As there are no monitoring networks for dry deposition in the UK, a process-based model together with meteorology, land use and air chemistry data is used to estimate pollutant deposition velocities which, in turn, are used to estimate levels of dry deposition. The UK deposition data are available and mapped on a 20km grid. The mapping procedure used at 20km resolution has been shown to provide unbiased estimates of wet deposition (CLAG, 1994).

Maps showing mean total (wet + dry + cloud) deposition for the years 1989–92 are given in Figures 1a to 1d in Colour Plate 18. For non-marine sulphur (Figure 1a in Colour Plate 18), deposition increases from west to east across the Province. Oxidized nitrogen (Figure 1b in Colour Plate 18) and non-marine base cation (Figure 1d in Colour Plate 18) deposition vary little over most of the area. Reduced nitrogen (Figure 1c in Colour Plate 18) values are similar to oxidized nitrogen for most grid squares but with higher values in upland areas in the west, north-east and south-east of the Province, due largely to higher rainfall in these areas.

Total acid deposition is calculated in the same way as total acid concentrations, i.e. :

acid deposition = non-marine (sulphur + oxidized nitrogen + reduced nitrogen – base cations)

Due largely to higher sulphur and nitrogen deposition levels on the eastern side of the Province total acid deposition (Figure 3a in Colour Plate x) is also greatest in the eastern half of the Province where values up to 2.0 keq H$^+$ ha^{-1} year^{-1} are predicted.

If the soils on which this acid rain falls are unable to buffer or neutralise the acidity deposited, they will gradually become acidified with enhanced leaching of base cations (Jordan and Enlander, 1990). In order to prevent further acidification of soils, emissions of acidifying pollutants need to be controlled. The UK government and the United Nations Economic Commission for Europe (UNECE) have adopted the so-called 'Critical Loads' approach as an effects-based method for the development of emission control strategies (DOE, 1990).

The Critical Loads approach

A critical load is defined as 'a quantitative estimate of exposure to one or more pollutants below which significant harmful effects on sensitive elements of the environment do not occur according to present knowledge' (Nilsson and Grennfelt, 1988). To apply the critical loads concept, it is necessary to define a critical load in terms of the damage threshold for the response of a receptor, such as an ecosystem or individual plant species. Thus, a critical loads approach is consistent with the concept for sustainability since deposition below the critical load will not lead to harmful effects.

The critical load concept was developed to provide a receptor-oriented approach for use in developing emission control policies and setting emission targets. This approach determines the sensitivity of the environment to pollutant loadings and estimates the long-term emissions of the pollutants that can be sustained without causing damage to the environment (Hornung, 1993). The approach has been adopted and developed by a number of international co-operative programmes and activities under the UNECE Convention on Long Range Transboundary Air Pollution (LRTAP). Parties to the Convention used the critical loads approach in developing the Second Sulphur Protocol signed in Oslo in 1994 (UNECE, 1994).

Critical loads for soils

A number of methods for calculating critical loads of acidity for soils have been developed in recent years, including empirical approaches, equilibrium mass balance methods such as the PROFILE model, and dynamic models such as MAGIC and SAFE (CLAG, 1994). The empirical approach is the simplest and was first developed at a workshop held at Skokloster in Sweden in 1988 (Nilsson and Grennfelt, 1988). This method divides soil minerals into five classes depending on their dominant weatherable minerals. Critical loads were assigned to these classes according to the amount of acidity which would be neutralised by base cation production through weathering from the relevant minerals (Table 9B.2).

Soils with large stores of leachable base cations

TABLE 9B.2
Mineralogical and petrological classification of soil material and critical loads (after Nilsson and Grennfelt, 1988).

Skokloster class	Minerals controlling weathering	Parent Rock	Critical load * keq H$^+$ ha^{-1} year^{-1}
1	Quartz, K-feldspar	Granite	< 0.2
2	Muscovite Plagioclase Biotite (<5%)	Granite Gneiss	0.2 - 0.5
3	Biotite Amphibole (<5%) Schist Gabbro	Granodiorite Greywacke	0.5 - 1.0
4	Pyroxene Epidote Olivine (<5%)	Gabbro Basalt	1.0 - 2.0
5	Carbonates	Limestone Marl	> 2

* The critical load values can be converted from keq H$^+$ ha^{-1} year^{-1} to values in kg of sulphur by multiplying by 16 (1keq H$^+$ is equivalent to 16 kg S).

and/or weatherable soil minerals tend to be insensitive to acidification except in the very long-term. Therefore such soils are assigned high critical loads – for example, carbonates are assigned a value > 2.0 keq H^+ ha^{-1} $year^{-1}$. Conversely, soils with a lower buffering capacity such as those with a mineralogy dominated by quartz or potassium feldspar are assigned to the most sensitive critical loads class (Table 9B.2).

The critical loads may also be adjusted within the ranges shown in Table 9B.2 by applying modifying factors which may decrease or increase their sensitivity to acidification. For example, a freely draining soil in which leaching is greater, would be assigned a critical load towards the lower limit of the range.

Sverdrup and Warfvinge (1988) also developed a system of mineralogical classes of soil materials which incorporated additional minerals in the classification and suggested possible weathering rates for soils with different mineral contents. In Great Britain, the classifications proposed at Skokloster and by Sverdrup and Warfvinge were modified to enable soil map units to be allocated to a soils material class and assigned a critical load (Hornung et al. 1995a).

Critical loads mapping

The Critical Loads Advisory Group (CLAG) was set up by the DOE(GB) in 1990 to provide expert advice on the application of critical loads in the UK. The CLAG includes specialist sub-groups responsible for different areas of critical loads research eg soils, freshwaters, vegetation. In addition, the Critical Loads Mapping and Data Centre (MADC) was established at ITE, Monks Wood to coordinate the mapping activities of CLAG, provide a central database facility and to act as the UK National Focal Centre for the UNECE mapping programme.

Using the empirical approach, a map of critical loads of acidity for soils of Great Britain was collated at the MADC and is used as the national map of critical loads. These critical loads are set to prevent chemical change in the soil (Nilsson and Grennfelt, 1988) and, more specifically, to prevent an increase in soil acidity or a decline in base saturation. It is thought that such chemical changes may be prevented if acid inputs do not exceed the production of base cations by mineral weathering within the soil.

A map of critical loads at 1km resolution was developed jointly between the Institute of Terrestrial Ecology, the Soil Survey and Land Research Centre, the Macaulay Land Use Research Institute and Aberdeen University (Hornung et al. 1995a). Using 1:250,000 soil maps, each map unit was allocated to one of five critical loads classes based on the mineralogy of the dominant soil series present. Peat soils were originally assigned to a separate class as the empirical approach used for mineral soils is not appropriate for classifying peat soils. The response of peat soils to acid deposition is determined mainly by their hydrology, the chemistry of lateral drainage water flushing the peats and the chemistry of atmospheric inputs. As a result, models were developed at Aberdeen University for estimating critical loads for dystrophic, eutrophic and basin peats (Cresser et al. 1993, and Hornung et al. 1995a). Each 1km square of Great Britain was subsequently assigned to a critical loads class based on the dominant soil unit present.

Initially, Northern Ireland soils were excluded due to the unavailability of data. However, with the completion of the field mapping by the Northern Ireland Soil Survey team, information on soil types enabled the completion of the critical loads map of acidity for soils in the UK at 1km resolution.

The data provided by the Department of Agriculture for Northern Ireland to ITE consisted of the critical loads class assigned to each 1km square across the Province together with the Irish Grid coordinates. Two additional classes were defined to provide complete coverage of the Province when mapping i.e. class 0 = water, urban or disturbed areas and class 6 = peat soils. No critical loads were assigned to the class 0 squares – instead they appear on the critical loads map as white areas (Figure 2 in Colour Plate 18). As for Great Britain, the models developed for deriving critical loads for peat soils were used to allocate critical loads classes to the 1438 1km peat squares from the Northern Ireland data set.

A breakdown of the areas by critical load class is provided for Northern Ireland soils in Table 9B.3.

TABLE 9B.3
Areas of each critical load class for acidification of soils in Northern Ireland

Critical Load keq H$^+$ ha^{-1} year^{-1}	Area km^2	Percentage %
<= 0.2	0	0
0.2 - 0.5	3888	29.7
0.5 - 1.0	4335	33.1
1.0 - 2.0	4243	32.4
> 2.0	618	4.7

Exceedance maps *(see Colour Plates)*

To identify areas of the Province where the inputs from acid rain (i.e. acid deposition) exceed the critical load, exceedances are calculated, where :

exceedance = deposition - critical load

A map of exceedance (Figure 3b, Colour Plate 19) of empirical critical loads of acidity for soils by current (1989–92) acid deposition shows that 54.5 per cent of the 1km squares of the Province are exceeded. The highest exceedances are found in the extreme north-east, the north-west and south-west where low critical loads correspond with high deposition values. A breakdown of the areas in four exceedance classes is shown in Table 9B.4.

TABLE 9B.4
Areas of each exceedance class for acidification of soils in Northern Ireland by current (1989-92) acid deposition

Exceedance Class keq H$^+$ ha^{-1} year^{-1}	Area km^2	Percentage %
not exceeded	5953	45.5
0.0 - 0.2	1301	9.9
0.2 - 0.5	3107	23.8
0.5 - 1.0	2488	19.0
> 1.0	235	1.8

Future

Abatement strategies to reduce emissions of acidifying pollutants have been developed under the UNECE LRTAP Convention. The aim is to reduce emissions, and thereby deposition, below the critical load. However, in practice, this would not be attainable in many areas of Europe. Therefore strategies are aimed at achieving "target" deposition loads. For the 1994 Second Sulphur Protocol, a '60% gap closure' was used – this aimed at closing the gap between 1980 levels of deposition and a 5-percentile critical load (i.e. a critical load value set to protect 95 per cent of the ecosystem area within a grid square) by 60 per cent (Bull, 1995). For the UK, this translated to a requirement to reduce emissions of sulphur by 80 per cent from the 1980 level. The UK agreed to achieve these reductions by the year 2010.

In the 1980s, acid emissions were dominated by oxides of sulphur and, until recently, abatement strategies concentrated on reducing these emissions. As a result, the modelling of deposition has been focused on sulphur, though with the need to develop a revised Protocol to control emissions of nitrogen pollutants, the current emphasis is on models to predict nitrogen pollutant deposition. However, the effects of nitrogen pollutants are more complicated, as they can contribute to both acidification and eutrophication. Methods for deriving critical loads of nitrogen for terrestrial and freshwater ecosystems have been developed at a number of international workshops (Grennfelt and Thornelof, 1992 and Hornung *et al.* 1995b). These have to take account of nitrogen uptake and removal processes by ecosystems.

The effects of future emissions and depositions of sulphur have been explored using the Hull Acid Rain Model, HARM (CLAG, 1994). A number of emission scenarios have been considered in recent years during the development of the Second Sulphur Protocol. HARM version 10.4 has been used to predict sulphur deposition for the UK for 2010, following implementation of the Protocol. This gives values of <= 0.5 keq ha^{-1} year^{-1} across the whole Province and, as there are no areas with very low critical loads, no grid squares are exceeded by 2010 sulphur deposition alone. However, in Great Britain, the critical loads for 8 per cent of 1km squares remain exceeded, the majority of which are exceeded by <= 0.2 keq ha^{-1} year^{-1}. These are mainly scattered across upland areas of north and west Britain or in the south-east on poorly buffered soils where critical loads are very low.

Some areas which currently receive high levels of oxidized and reduced nitrogen deposition may remain exceeded in the future. At present, no official scenarios for future nitrogen emission reductions are available. Exceedance of current acid deposition can, however, be compared with a possible future scenario for acid deposition by using modelled non-marine sulphur for 2010 and assuming current levels of nitrogen and base cation deposition (Figure 4a, Colour Plate 19). Exceedance of the empirical critical loads of acidity by this deposition field (Figure 4b, Colour Plate 19) results in 30.9 per cent of 1 km squares being exceeded in the Province, with highest values in the Sperrins, Mournes and north-east Antrim. Table 9B.5 summarises the areas in the exceedance classes.

TABLE 9B.5
Predicted areas of exceedance for acidification of soils in Northern Ireland using modelled non-marine sulphur deposition for the year 2010 with current levels of nitrogen and base cation deposition

Exceedance Class keq H$^+$ ha^{-1} year^{-1}	Area km^2	Percentage %
not exceeded	9036	69.1
0.0 - 0.2	1515	11.6
0.2 - 0.5	2202	16.8
0.5 - 1.0	331	2.5
> 1.0	0	0

By comparison, using this scenario for Great Britain, 56.3 per cent of 1km squares exceed the critical load, particularly in the upland regions of the north and west. It is clear, therefore, that emission reductions of nitrogen as well as sulphur will be necessary to offer a high level of protection to both soils and freshwaters throughout the UK.

References in Appendices

CHAPTER 9C

HYDROLOGY OF SOIL TYPES (HOST)

A. HIGGINS, AGRICULTURAL AND ENVIRONMENTAL SCIENCE DIVISION,
DEPARTMENT OF AGRICULTURE NORTHERN IRELAND,
NEWFORGE LANE, BELFAST BT9 5PX

Soils play a crucial role in the storage and transmission of water within river catchments. Their physical properties largely govern the rate at which precipitation enters the river system and in conjunction with this, their chemical and biological properties affect water quality in the rivers. The determination of this influence from soil distribution maps alone is extremely difficult, and therefore some interpretation is required. The Hydrology Of Soil Types (HOST) map of Northern Ireland (Colour Plate 21) represents a regional classification of soils by their hydrological response, providing a tool which is of significant benefit in the prediction of catchment flow parameters.

Development

The development of HOST was initiated by the completion of soil mapping at 1:250,000 by the Soil Survey and Land Research Centre (SSLRC) in 1983 for England and Wales, and by the Macaulay Land Use Research Institute (MLURI) for Scotland in 1984. The definition of hydrological classes was further assisted by detailed hydrological databases held at the Institute of Hydrology (IoH). Northern Ireland was included in the HOST project in 1988 through the availability of detailed soil distribution and attribute data from the DANI Soil Survey and the funding of research work by the Rivers Agency of the Department of Agriculture for Northern Ireland.

HOST can be viewed as a logical extension of the Winter Rainfall Acceptance Potential (WRAP) classification developed for the Flood Studies Report (FSR)(NERC, 1975). In this system soils were divided into 5 hydrological classes by the following 4 site properties:
1. Soil-water regime
2. Depth to an impermeable horizon
3. Permeability above an impermeable horizon
4. Slope.

The WRAP classification was applied to the soil map and the percentage of each class in the relevant WRAP catchment was calculated and incorporated into the FSR predictive equation through the SOIL term. (A full description of the development of WRAP can be found in Farquharson *et al.* 1978).

The FSR predictive equation worked well in the rest of the United Kingdom, but when applied to Northern Ireland it was found to significantly underpredict the size of the mean annual flood at all sites where flow data were available (Hanna and Wilcock, 1984). Although a number of other possible reasons were identified to account for this underprediction Wilcock and Hanna (1987) indicate that the SOIL term used in the FSR, as derived from the WRAP maps, does not accurately reflect soil conditions in Northern Ireland, (see Chapter 5). This situation reflects the generalised nature of the Northern Ireland soil map (McAllister and McConaghy, 1968) on which the WRAP map was based, and the lack of physical data on the soil types identified. This earlier poor predictive ability was a major factor in the development of a HOST map for Northern Ireland.

HOST Classification

The HOST classification was defined using three sources of information: (a) catchment-scale hydrologic parameters; (b) soil properties; (c) substrate hydrogeology. There follows a brief examination of the role of these data in the development of HOST.

(a) Catchment-scale hydrological parameters
A significant advantage of HOST over WRAP was the use of catchment variables during the development of the classification. With WRAP, the classification was produced and then calibrated for the required parameter, such as the mean annual flood. Two variables were chosen to verify and calibrate the

HOST classification, calculated from the National Water Archive, held at the IoH. The base flow index (BFI) can be regarded as a measurement of the percentage of river flow which originates from stored sources, i.e. groundwater rather than direct runoff. The index varies between 0.1, (catchments with a very flashy response to rainfall and consequently low proportion of groundwater in the riverflow), and 1 (high groundwater component in the riverflow). Data from 575 catchments throughout England, Wales and Scotland were available for the use of the HOST project. Standard percentage runoff (SPR) is the percentage of a rainfall event which causes a short term increase in the riverflow. Some 170 catchment values were used in the project (Boorman *et al.* 1995).

(b) Soil properties.
Four soil properties were used to indicate soil influence on the hydrological response of a catchment.
(i) Depth to a slowly permeable layer. Such soil layers have a lateral hydraulic conductivity of less than 10 cm day^{-1}. This value can either be measured directly by such instruments as the Guelph permeameter or estimated from soil texture and soil structure. This soil layer has the effect of reducing the infiltration capacity (see Chapter 5), causing seasonal saturated conditions with a consequent decrease in storage potential and increasing the response to heavy rainfall. In terms of HOST, the identification of such a layer is only required if it occurs within one metre of the surface.
(ii) Depth to a gleyed layer. Gleying is caused by intermittent waterlogging in the soil. The definition in HOST is used to identify soils which are saturated for at least 30 days of the year. The presence of such a layer reduces the storage capacity of the soil and increases the response to heavy rainfall. Again the presence of a gleyed layer is only included if it exists within the top 1m of soil.
(iii) Integrated air capacity (IAC). IAC measures the volume of pores in the soil greater than 60 microns in diameter, integrated over the top 1m of soil. IAC is of primary use as an indicator of storage capacity in some slowly permeable and impermeable soils and as a substitute for hydraulic conductivity in permeable soils.
(iv) The presence of a peaty surface layer. A peaty surface layer is defined as having more than 20 percent organic matter and being 7.5 to 40 cm thick. The presence of such a layer indicates soils that are (or were before drainage) saturated to the surface over long periods. Peaty surface layers have the capacity to store considerable amounts of water and are usually slowly permeable which encourages surface runoff.

(c) Substrate hydrogeology
The HOST working group developed a classification of soil substrates based on a number of hydrological criteria. The classification is summarised in Table 9C.1. Initially all soil substrates were allocated to one of the 32 hydrological classes, shown in Table 9C.2. These hydrological classes were then grouped on the basis of permeability into three general classes based on the definitions of Bell (1985).
(i) Permeable substrate, where the saturated hydraulic conductivity is greater than 10cm day^{-1}.
(ii) Slowly permeable substrate, where the saturated hydraulic conductivity is between 0.1 and 10cm day^{-1}.
(iii) Impermeable substrate, where the saturated hydraulic conductivity is less than 0.1 cm day^{-1}.

The permeable classes were further subdivided into 6 classes on the basis of general pathways of vertical water movement. Finally, each hydrological class was allocated to one of three categories based on the presence or absence of an aquifer or significant groundwater and the depth to that.

The soil and substrate properties outlined above could define a very large number of categories. However, conceptual models, used by the HOST working group (Boorman *et al.* 1995) to describe dominant water movement through soil, were used to group properties which would give a similar hydrological response. Various classification schemes were constructed and tested using multiple linear regression analysis, where the BFI of gauged catchments (575) were regressed against the percentage cover of each HOST class within the catchments. A similar procedure was carried out using the 170 catchments in the SPR data-set. Utilising this approach, a 29 class HOST classification was constructed (Table 9C.1), which explained 79 percent and 60 percent of the observed variation in BFI and SPR respectively. Comparable figures from WRAP are 52 percent and 47 percent.

Since a number of important catchments in Northern Ireland extend into the Republic of Ireland (Blackwater, Erne and Finn), HOST mapping was continued across the Border. The lack of soil survey information in the Border counties of the Republic necessitated a more general approach to HOST classification, based on available soil information and topographic maps.

Northern Ireland HOST Map

The complex distribution pattern of the 308 soil series identified by the DANI Soil Survey, is simplified by the HOST classification system. Of the 29 HOST classes, 23 are defined in Northern Ireland. Table 9C.3

SUBSTRATE HYDROGEOLOGY		MINERAL SOILS				PEAT SOILS	
	Groundwater or aquifer	No impermeable or gleyed layer within 100cm	Impermeable layer within 100cm or gleyed layer at 40-100cm		Gleyed layer within 40cm		
Weakly consolidated, microporous, by-pass flow uncommon (Chalk)	Normally present and at >2m	1 4.31	13 0.87		14 0.66	15 9.93	
Weakly consolidated, microporous, by-pass flow uncommon (Limestone)		2 2.12					
Weakly consolidated, macroporous, by-pass flow uncommon		3 1.58					
Strongly consolidated, non or slightly porous, by-pass flow common		4 3.33					
Unconsolidated, macroporous, by-pass flow very uncommon		5 5.07					
Unconsolidated, microporous, by-pass flow common		6 2.61					
Unconsolidated, macroporous, by-pass flow very uncommon	Normally present and at <2m	7 1.01			IAC <12.5 [<1m day⁻¹]	IAC >=12.5 [>1m day⁻¹]	Drained	Undrained
Unconsolidated, microporous, by-pass flow common		8 1.62			9 3.68	10 2.21	11 0.55	12 2.94
			IAC>7.5	IAC<=7.5			
Slowly permeable	No significant groundwater or aquifer	16 0.43	18 5.4	21 4.02	24 13.85	26 2.49	
Impermeable (Hard)		17 9.28	19 2.16	22 1.10		27 0.83	
Impermeable (Soft)			20 0.69	23 1.31	25 3.64		
Eroded peat						28 0.58	
Raw peat						29 5.73	
Unclassified	Urban 5.15 Lakes 0.74						

TABLE 9C.1
The HOST Classification. Large numbers are the HOST class, small numbers are the percentage of England, Wales and Scotland in that class.

indicates the areal cover of each of these classes together with BFI and SPR values associated with that HOST class. The distribution of these classes is discussed in terms of soil series and physical setting below and is illustrated in Colour Plate 21.

HOST class 1 is freely draining mineral soils developed on chalk. The areal coverage of only 0.178 percent reflects the limited exposure of chalk in Northern Ireland, and is confined mainly to areas of the Antrim coast (e.g. Soil Map 9). HOST class 4 is similar in that the soils are defined by predominantly downward water movement, indicated by BFI and SPR values of 0.791 and 2 percent. This class consists of freely draining mineral soils such as brown earths and brown podzolics developed on rock types like basalt, Old Red Sandstone and limestone (e.g. Soil Maps 18 and 20). The combination of freely drained soils and permeable substrate gives rise to high water storage capacity which in turn means that river flow will be governed by a high base flow component and low runoff component. Associated with the distribution of class 4 are HOST classes 13 and 14, which contain mainly shallow mineral soils, gleyed to varying degrees, overlaying permeable substrates. The two classes vary markedly in their hydrological response, with class 13 dominated by downward water movement and class 14 with a strong lateral component as indicated by the SPR value of 25 percent. The presence of a gleyed layer in class 14

Class Number	Class description
1	Soft sandstone, weakly consolidated sand
2	* Weathered/fissured intrusive/metamorphic rock
3	* Chalk, chalk rubble
4	Soft Magnesian, brashy or Oolitic limestone and ironstone
5	* Hard fissured limestone
6	* Hard coherent rocks
7	* Hard but deeply shattered rocks
8	Soft shales with subordinate mudstones and siltstones
9	Very soft reddish blocky mudstones
10	Very soft massive clays
11	Very soft bedded loams, clays and sands
12	Very soft bedded loam/clay/sand with subordinate sandstone
13	* Hard (fissured) sandstones
14	* Earthy peat
15	* River alluvium
16	* Marine alluvium
17	Lake marl or tufa
18	Colluvium
19	* Blown sand
20	Coverloam
21	* Glaciolacustrine clays and silts
22	* Till, compact head
23	Clays with flints or plateau drift
24	* Gravel
25	* Loamy drift
26	Chalky drift
27	* Disturbed ground
34	* Sand
35	Cryogenic
36	Scree
43	* Eroded blanket peat
44	* Raw peat
50	Unsurveyed
51	* Lake

TABLE 9C.2
Hydrogeological classes identified within HOST. Classes denoted with * have significant areal distribution within Northern Ireland

restricts the amount of vertical water movement, reduces the potential for water storage and increases runoff. Typical soils of class 14 would be gleyed rankers developed on Carboniferous Sandstone and basalt (Soil Map 13).

The distribution of class 15 is most commonly associated with upland regions. Here, a peaty surface layer, formed as a result of lower temperatures and higher rainfall, is developed on a wide range of permeable substrates. The BFI and SPR values (0.38 and 48.4 percent respectively) indicate a soil group dominated by lateral movement, giving quick runoff and low storage. Soils in this group vary from peaty podzols on sands and gravels to peat rankers on Yoredale Sandstone (e.g. Soil Maps 13 and 27).

HOST class 7 and 8 are of limited occurrence and are comprised mainly of soils developed on freely draining lake sands and brown earth alluvium respectively (e.g. Soil Map 8). Although freely draining, the BFI and SPR values reflect the reduced storage potential caused by the watertable being within 2m of the surface. HOST classes 9 and 10 have similar distribution patterns along rivers and areas of coastal and lacustrine alluvium. The two classes are distinguished on the basis of IAC, used as an indicator of hydraulic conductivity. Class 9 consists of soils with a saturated hydraulic conductivity of less than 1m per day and mainly defines G1 and G2 alluvium. The BFI value of 0.734 indicates a fairly high base flow component with an intermediate 25 percent SPR value. HOST class 10 contains soils with a higher hydraulic conductivity which is reflected in the lower BFI value of 0.52. This class comprises mainly groundwater gleys developed on sand and gravel. Again HOST class 11 shares a similar distribution pattern with classes 9 and 10 as it is comprised mainly of organic alluvium. The BFI value of 0.927 and SPR of 2 percent indicate the large storage potential liberated by drainage of the surface peaty layer.

HOST classes 16, 18 and 21 are all well drained mineral soils developed on slowly permeable glacial tills, but differentiated by the depth to gleying and IAC values (here used to indicate storage capacity). Class 18 is the most common and is characterised by intermediate values of BFI and relatively high values of SPR . This class is typified by such soils as brown earths on basalt till (e.g. Soil Map 8) and brown podzolics on mica-schist till (e.g. Soil Map 7). HOST class 24 is the most extensive of the defined classes, consisting of gleyed soils developed on fine textured glacial till, reflecting the extensive glaciation of Northern Ireland. Typical soils of this class would be SWG1 and SWG2 basalt tills in the east of the Province (e.g. Soil Map 20). As would be expected, these soils are dominated by low storage capacity and quite high rates of runoff, resulting from a high water table overlying a slowly permeable substrate. This situation is amplified in HOST class 26, which consists of gleyed till soils with a peaty surface layer which increases the SPR value to 58.7 percent and reduces the BFI to 0.244. This class is most extensive in upland regions and is typified by soil series such as SWHG on Old Red Sandstone (e.g. Soil Map 18).

Classes 17, 19 and 22 comprise mainly well drained mineral soils developed on an impermeable substrate. Again, a division is made on the basis of depth to an impermeable layer and IAC value. The most common class of this group, 17, is associated with brown earths developed on impermeable rock types typified by soil series such as brown earths on shale (Soil Map 21) and brown podzolics on schist (Soil Map 7). This class is characterised by quite high BFI values (0.609) reflecting the high storage capacity in the soil. The BFI decreases for 19 and 22 as the depth to the impermeable region decreases, lowering the amount of potential storage available. This situation is also evident in class 27 which consists of similar soil types with a peaty surface horizon.

Virtually all peat areas in Northern Ireland have some peripheral cutting and consequent alterations to drainage characteristics, however, areas included in class 29 are predominantly undrained bog including both raised and blanket bog. HOST class 28 is distinguished on the basis of a higher BFI (0.581) caused by increased water storage potential in eroded blanket bogs. Both classes are characterised by lateral water movement, with 60 percent of rainfall appearing as quick response.

The HOST classification of Northern Ireland is currently undergoing testing at IoH, where BFI and

HOST CLASS	Percentage cover	BFI*	SPR(%)*
1	0.178	1.000	2.0
2		1.000	2.0
3		0.900	14.5
4	3.540	0.791	2.0
5	3.587	0.900	14.5
6		0.645	33.8
7	0.009	0.792	44.3
8	0.075	0.560	44.3
9	4.640	0.734	25.3
10	0.210	0.520	25.3
11	6.535	0.927	2.0
12		0.170	60.0
13	0.003	1.000	2.0
14	0.183	0.380	25.3
15	2.775	0.380	48.4
16	0.196	0.778	29.2
17	2.950	0.609	29.2
18	5.798	0.518	47.2
19	1.199	0.469	60.0
20		0.524	60.0
21	0.156	0.340	47.2
22	0.147	0.315	60.0
23		0.218	60.0
24	46.179	0.312	39.7
25	0.004	0.170	49.6
26	4.857	0.244	58.7
27	1.178	0.259	60.0
28	0.027	0.581	60.0
29	8.179	0.226	60.0
Lakes	4.151		
Urban	3.246		

TABLE 9C.3
Percentage land cover for the 23 HOST classes occurring in Northern Ireland, with associated BFI and SPR values. (* Boorman *et al.* 1995)

SPR values from catchments with long duration flow records will be used to assess its predictive ability. Although HOST has been applied to the 1:50,000 soil maps of Northern Ireland, its development from soil and substrate properties means that it can be applied at any scale. Future work will concentrate on the refinement of HOST classes based on IoH results and its use in the prediction of other flow parameters such as mean annual flood. HOST has already been used in the United Kingdom to estimate low flow indices (Gustard *et al.* 1992) and assess the vulnerability of aquifers to pesticide pollution (Hollis, 1991).

Soil Moisture Properties

Data regarding soil moisture retention are among the most important physical factors recorded on any soil profile as this property affects the soil's agricultural potential, (e.g. susceptibility to droughtiness), and mechanical strength, (e.g. susceptibility to poaching). In the DANI Soil Survey, 441 soil horizons, representing 170 soil profiles, were sampled for moisture retention properties. The measured properties, sampling procedure and results are discussed below.

Measured Properties

Soil is a porous medium in which there is a dramatic variation in pore space size, shape and interconnectivity. Pores can vary in size from diameters of a few centimetres (macropores) to very fine micropores with diameters of less than 1 micron. Porosity and pore size distribution are influenced by a number of factors: particle size distribution; type and development of soil structure; bulk density; organic content; the presence of swelling clays. The total porosity of a soil, and the pore size distribution within it, control the volume, retention characteristics and movement of water within any particular soil. Water is held in these pores by a combination of three different forces. (i) Adsorption. Here water molecules are attracted to the surface layers of colloidal material. This is most significant in soils with high clay or organic contents. (ii) Capillarity. In this case water is held in pores by a combination of absorptive forces between the water and the pore surface and surface tension at the water surface. (iii) Osmosis. Here water is held by the differential in ionic concentrations of soil solutions, water moving from a low to a high concentration.

In the Northern Ireland Soil Attribute Database (see Appendix 3) porosity is defined as the percentage of total pore space (T_t) in the sampled volume of soil. This is calculated as follows:

$T_t = (1 - (D_{bt}/D_p))/100$

Where D_p = The particle density of the soil
        D_{bt} = The soil bulk density.

D_{bt} is reported here as dry field bulk density (stones included).

The standard procedure adopted for moisture analysis (Hall *et al.* 1977) involves the determination of volumetric moisture content of the soil sample at five different suction levels, each of which has its own significance. The sample cores are initially saturated until all available pore space is filled with water. Moisture content is then determined at the three low suctions, 5, 10, and 40 Kilopascals (kPa) (0.05, 0.1 and 0.4 Bar respectively) on sand and kaolin suction tables (Stakman *et al.* 1972). Determination at the higher values of 200 and 1500 kPa (2 and 15 Bar) was carried out on pressure membrane apparatus, where air pressure is used to simulate the high suctions found in drying soils (Richards, 1947).

The volume of water retained in a sample at 5 kPa ($Q_{vt}(5)$), also known as retained water capacity, has been equated to the volume of water retained in most soils at field capacity (Reeve *et al.* 1973). Field capacity is the volume of water left in a soil when excess water has drained away under the influence of gravity. In sandy textured soils, field capacity has been equated to the volume of water retained at the slightly higher suction of 10 kPa ($Q_{vt}(10)$) (e.g. Bennett and Entz, 1989). The retained water volume at 40 kPa ($Q_{vt}(40)$) is used as an intermediate suction point. The volume of water retained at suctions of less than 200kPa ($Q_{vt}(200)$) is termed the volume of easily available water, that is, the amount of water a plant can remove from the soil without any stress. As such, Reeve *et al.* (1973) consider it a potentially useful gauge of a soil's tendency to droughtiness. At suctions of 1500 kPa, the retained water volume ($Q_{vt}(1500)$) is used to indicate the upper limit of water which a plant can remove from the soil. Above suctions of 1500 kPa, termed wilting point, water can only be removed from the soil by evaporation. Two values derived from the retained moisture data are also included in the database to describe the soil. The total available water (A_{vt}) is the volumetric water content between $Q_{vt}(5)$ and $Q_{vt}(1500)$, giving the total volume of water available for plant use. Air capacity (C_{at}) is the difference in volume between T_t and $Q_{vt}(5)$. This value equals the volume of largest pores (greater than 60 microns), also termed drainable porosity, which are air filled at field capacity and can be used to infer relative values of permeability. Thus, in two horizons with high and low C_{at} values, permeability would be expected to be higher in the horizon with the high C_{at} value.

Again, soil moisture sampling followed the standard procedure initiated by the Soil Survey of England and Wales (Hall *et al.* 1977), and will be outlined only briefly here.

Certain 5 kilometre inspection pits, sampled as part of the DANI Soil Survey sampling strategy (see Chapter 2), were selected as being representative of a soil series. Soil moisture samples were taken, thus giving these pits a full range of descriptive, chemical and physical data. In each case triplicate, 222 cm³ undisturbed cores were taken from each of the sampled horizons using an impact driven volumetric soil sampler. The stoniness of certain soil series (e.g. Brown earth shale till), did prevent some horizons from being sampled. Although it is possible to abstract a disturbed sample from these sites it was felt that such a sample would not accurately reflect the prevailing field conditions. The samples were then stored in moisture-proof conditions in a cold room to await processing.

Northern Ireland Soils

Averaged bulk density and moisture retention characteristics are presented in Table 9C.6 for 24 of the 26 major soil series of Northern Ireland. Subsets of this group will be used to illustrate the various relationships found.

In Table 9C.4 it will be noted that D_{bt} decreases down the profile for all three soil series. This recognised general trend is a result of decreasing carbon content, structural development and bioturbation with depth, leading to a reduction in total pore space and thus resulting in greater compaction. In general terms, for any individual soil type, bulk density is always lower under permanent pasture, which tends to encourage good structural development and high levels of organic carbon content. In arable situations bulk density undergoes a cyclic pattern of change with low bulk density immediately after ploughing, increasing in the months following as a result of compaction from secondary cultivation.

The basalt soils illustrate clearly the tenacity with which clay rich soils (in this case greater than 30 percent clay) retain water. Large volumes of water are retained at $Q_{vt}(5)$ which are only slowly released across the spectrum of soil types. This has the effect of giving the soil low shear strength at or near field capacity, with the consequence of low load bearing capability and susceptibility to poaching. Even at $Q_{vt}(1500)$ considerable volumes are retained (not less than 26.88 percent), indicating the large percentage of small pores (smaller than 0.2 microns) and large quantities bound on colloidal material and in mineral structures. This large volume of unavailable water means that despite the retained water capacity the amount available to the plant is relatively small. Thus, in terms of rainfall in a brown earth basalt till Ap horizon of 35 cm depth, the total amount of water stored would be 18.07 cm with only 8.12 cm available for plant use. The C_{at} values in Table 9C.4 illustrates the presence of more large pores in the brown earth profile compared to the two gleyed profiles, and a decrease with increasing gleying and depth.

Table 9C.5 shows results from two soil series which are in contrast to each other and to the basalt soils. The soil developed on diatomite has a silty texture (greater than 50 percent silt) and is relatively unique in its properties. The variation of bulk density with depth does not conform to the expected pattern. The uniformly low D_{bt} values seem to be a result of the structure of the silicaceous material of which the diatomite is composed (see Chapter 2). The very low bulk densities give rise to high T_t values (the highest recorded in Northern Ireland), which in turn give rise to very high values of $Q_{vt}(5)$. In contrast to the basalt soils, this is released more evenly across the range of applied suctions, resulting in a lower $Q_{vt}(1500)$ value in the Ap and Bg horizons. In practice this means that large volumes of water are available for plant use, thus in an Ap horizon of 35 cm depth, 24.59 cm of water in total is stored with 17.16 cm available for plant use. It is obvious that these diatomite soils well deserve their local reputation for drought tolerance.

In contrast to the above examples, the brown podzolic soil developed on granite rock, Table 9C.5, exhibits the characteristics common to sand and sandy textured soils. Lower volumes of water are retained at field capacity (indicated by $Q_{vt}(10)$ because of the sandy texture) as a result of the higher proportion of large pores, indicated by the high C_{at} values. In this example, the relatively low value of $Q_{vt}(1500)$ in both horizons means that there are still quite high volumes of water available for plant exploitation (7.68 cm in a 35 cm Ap profile, which is similar to the amount in the more retentive basalt soil).

Acknowledgement

This research of the HOST project was funded by the Rivers Agency of the Department of Agriculture of Northern Ireland, over the period 1988 to 1995, when Alex Higgins, Mervyn Henry and Mary Mallon held the post of Research Assistant. The author and DANI Soil Survey (AESD) wish to record their appreciation of this financial support

Series	Horizon	D_{bt} g/cc	T_t %	5 kPa %	10 kPa %	40 kPa %	200 kPa %	1500 kPa %	A_{vt} %	C_{at} %	C %
Brown Earth Basalt Till	Ap	0.88	64.73	51.63	49.76	43.43	39.10	28.44	23.19	13.10	7.75
	Bw	1.06	60.62	46.99	45.31	40.43	37.98	27.78	19.21	13.63	0.92
	C	1.20	56.62	46.35	45.30	42.15	38.39	26.88	19.47	10.27	0.41
SWG1 Basalt Till	Ap	0.95	61.66	51.98	50.18	44.42	41.64	28.39	23.59	9.69	6.23
	Bg	1.21	55.87	46.39	45.22	42.26	39.57	30.82	15.57	9.49	0.53
	C	1.36	51.08	45.20	44.31	42.43	40.20	29.66	15.54	5.88	0.17
SWG2 Basalt Till	Ap	1.00	59.46	52.06	50.16	44.51	37.45	29.95	22.11	7.41	5.07
	Bg	1.17	57.28	50.42	49.58	46.02	41.50	31.82	18.60	6.86	0.64
	C	1.35	51.11	46.95	46.40	43.63	40.55	30.99	15.96	4.15	0.19

TABLE 9C.4

Bulk density, organic carbon and soil moisture data for the Brown Earth, SWG1 and SWG2 basalt soil series.

Series	Horizon	D_{bt} g/cc	T_t %	5 kPa %	10 kPa %	40 kPa %	200 kPa %	1500 kPa %	A_{vt} %	C_{at} %	C %
Diatomite	Ap	0.65	73.64	70.27	66.15	61.16	50.77	21.23	49.04	3.37	6.74
	Bg	0.62	74.55	71.70	65.62	62.88	51.55	17.82	53.88	2.85	1.53
	Bg/C	0.50	79.02	75.12	69.17	65.56	47.01	26.74	48.39	3.89	2.02
Brown Podzolic Granite Rock	Ap	1.21	52.81	35.25	33.05	27.84	20.51	11.12	21.93	17.56	2.86
	Bw	1.36	48.06	28.59	26.11	20.26	15.91	7.26	18.85	19.46	1.29

TABLE 9C.5

Bulk density, organic carbon and soil moisture data for diatomite and brown podzolic soil developed on granite.

Series	Horizon	D_{bt} g/cc	T_t %	5 kPa %	10 kPa %	40 kPa %	200 kPa %	1500 kPa %	A_{vt} %	C_{at} %	C %
Brown Earth Basalt Till	Ap	0.88	64.73	51.63	49.76	43.43	39.10	28.44	23.19	13.10	7.75
	Bw	1.06	60.62	46.99	45.31	40.43	37.98	27.78	19.21	13.63	0.92
	C	1.20	56.62	46.35	45.30	42.15	38.39	26.88	19.47	10.27	0.41
SWG1 Basalt Till	Ap	0.95	61.66	51.98	50.18	44.42	41.64	28.39	23.59	9.69	6.23
	Bg	1.21	55.87	46.39	45.22	42.26	39.57	30.82	15.57	9.49	0.53
	C	1.36	51.08	45.20	44.31	42.43	40.20	29.66	15.54	5.88	0.17
SWG2 Basalt Till	Ap	1.00	59.46	52.06	50.16	44.51	37.45	29.95	22.11	7.41	5.07
	Bg	1.17	57.28	50.42	49.58	46.02	41.50	31.82	18.60	6.86	0.64
	C	1.35	51.11	46.95	46.40	43.63	40.55	30.99	15.96	4.15	0.19
Brown Earth Shale Till	Ap	1.21	52.62	43.80	41.38	37.06	32.54	19.72	24.08	8.82	4.25
	Bw	1.17	54.24	39.08	37.07	31.97	25.22	16.33	22.75	15.16	1.79
SWG1 Shale Till	Ap	1.11	56.21	42.08	40.54	35.92	28.33	20.25	21.83	14.13	3.14
	Bg	1.54	42.59	32.45	31.43	28.18	24.28	17.40	15.05	10.14	0.44
	C	1.69	36.88	31.92	31.45	29.83	26.38	19.15	12.76	4.97	0.16
SWG2 Shale Till	Ap	1.08	56.84	44.84	42.69	37.48	32.34	21.77	23.07	12.00	4.55
	Bg	1.50	43.17	30.57	28.77	26.05	24.88	17.16	13.41	12.59	0.53
	C	1.54	42.49	37.12	36.16	32.53	23.65	15.21	21.90	5.37	0.17
Brown Earth Mica-Schist Till	Ap	1.23	52.61	47.27	44.69	38.43	20.71	13.58	31.11	5.33	3.03
	Bw	1.37	49.03	33.72	31.35	22.73	17.68	9.12	22.23	15.32	1.29
	C	1.63	40.34	33.12	29.82	23.77	14.98	13.25	16.57	7.22	0.48
SWG1 Mica-Schist Till	Ap	1.22	50.06	45.10	43.29	39.47	28.33	18.33	24.96	4.96	4.05
	Bg	1.52	43.46	33.42	31.95	28.00	24.49	12.72	19.23	10.04	0.51
	C	1.71	36.75	27.39	26.03	23.50	17.96	12.93	13.10	9.36	0.25
SWG2 Mica-Schist Till	Ap	1.12	57.17	47.87	45.96	39.65	28.85	17.63	28.23	9.30	3.97
	Bg	1.50	44.12	33.89	31.54	26.27	20.15	9.77	21.77	10.23	0.34
	C	1.75	36.10	26.42	25.37	22.37	17.47	12.58	12.79	9.68	0.31

TABLE 9C.6 Averaged values for 24 of the 26 major soil series identified by the DANI Soil Survey.

The values are defined as;
- D_{bt} = Dry field weight bulk density.
- T_t = Total sample porosity.
- 5 -1500 = Volume of water retained at each suction value.
- A_{vt} = Volume of water available for plant use.
- C_{at} = Volume of air filled pores at a suction of 5 kPa.
- C = Organic carbon content of soil horizon

Series	Horizon	D_{bt} g/cc	T_t %	5 kPa %	10 kPa %	40 kPa %	200 kPa %	1500 kPa %	A_{vt} %	C_{at} %	C %
Brown Podzolic Granite Rock	Ap	1.21	52.81	35.25	33.05	27.84	20.51	11.12	21.93	17.56	2.86
	Bw	1.36	48.06	28.59	26.11	20.26	15.91	7.26	18.85	19.46	1.29
SWG1 Granite Till	Ap	1.11	55.50	47.58	46.20	38.36	27.27	21.80	24.40	7.92	4.43
	C	1.60	38.86	27.90	26.84	23.82	22.01	15.09	11.75	10.96	0.97
SWG2 Lough Neagh Clay Till	Ap	1.21	51.57	45.27	43.40	38.68	31.39	21.40	23.87	6.30	3.57
	Bg	1.51	43.51	36.05	34.75	31.57	27.94	21.25	14.80	7.46	0.58
	C	1.65	37.44	31.36	30.01	27.55	26.48	18.33	13.03	6.07	0.12
Brown Earth Old Red Sandstone Till	Ap	1.10	56.13	46.46	44.68	39.19	33.49	18.17	28.29	9.67	4.35
	Bw	1.33	50.79	35.50	33.16	28.31	24.07	15.69	19.81	15.29	1.29
SWG1 Old Red Sandstone Till	Ap	0.97	59.89	51.53	50.22	42.30	38.15	26.68	24.85	8.36	5.66
	Bg	1.47	45.27	35.27	33.32	28.26	28.21	21.06	14.21	10.00	0.74
	C	1.78	33.45	28.13	27.58	22.26	19.81	13.56	14.57	5.32	0.41
SWG2 Old Red Sandstone Till	Ap	0.80	67.57	59.55	56.82	47.81	35.42	21.98	37.57	8.02	4.31
	Bg	1.59	41.82	35.99	35.52	33.16	30.28	24.07	11.92	5.83	0.40
	C	1.73	36.52	33.29	33.00	30.78	28.92	21.10	12.20	3.22	0.24
Brown Earth Carb. Limestone Till	Ap	1.20	53.43	44.26	42.43	36.36	29.25	21.22	23.04	9.17	3.54
	Bw	1.42	47.07	35.64	33.13	28.74	23.60	15.15	20.49	11.43	0.97
	C	1.50	45.10	41.30	40.34	37.19	30.56	20.37	20.93	3.81	0.85
SWG1 Carb. Limestone Till	Ap	1.00	59.96	55.92	54.86	49.03	35.99	29.31	26.61	4.04	5.55
	Bg	1.33	51.17	50.97	50.30	48.17	43.25	38.85	12.12	0.20	0.78
	C	1.35	50.58	43.60	42.63	37.83	32.99	27.48	16.12	6.98	0.61
SWG2 Carb. Limestone Till	Ap	0.92	63.77	60.53	57.70	51.78	28.99	17.38	43.14	3.25	3.24
	Bg	1.42	45.85	28.38	26.34	21.19	20.03	17.44	10.94	17.47	0.39
SWG1 Calp Till	Ap	0.99	60.88	53.99	51.55	43.05	29.55	22.07	31.92	6.89	2.86
	Bg	1.42	46.04	37.36	35.83	30.55	28.31	20.83	16.53	8.67	1.00
	C	1.38	48.38	43.19	42.33	37.90	38.82	25.81	17.39	5.19	0.46
SWG2 Calp Till	Ap	0.89	64.08	56.56	53.41	45.39	35.49	24.18	32.39	7.52	4.97
	Bg	1.41	47.72	42.61	41.73	38.27	33.14	24.70	17.91	5.11	0.75
	C	1.51	45.17	42.16	42.18	39.81	45.74	27.97	14.19	3.01	2.84

TABLE 9C.6 (Continued)

Series	Horizon	D_{bt} g/cc	T_t %	kPa 5 %	10 %	40 %	200 %	1500 %	A_{vt} %	C_{at} %	C %
SWG2 Carb. Sandstone Till	Ap	1.07	58.23	52.41	52.26	48.16	42.09	32.55	19.71	5.83	3.63
	Bg	1.09	59.05	55.00	54.51	56.46	46.67	35.07	19.44	4.04	0.71
	C	1.32	51.60	50.14	49.56	47.08	36.86	30.00	19.56	1.46	0.3
SWG3 Yoredale Sandstone Till	Ap	0.78	68.63	57.05	54.18	46.74	35.40	24.39	32.66	11.58	14.17
	Bg	1.42	46.67	41.39	40.26	36.97	31.64	21.34	20.05	5.28	0.77
	C	1.46	46.48	44.33	43.82	40.17	32.28	25.38	18.95	2.15	0.48
Brown Earth Trias Sandstone Till	Ap	1.28	50.93	42.64	40.63	34.53	26.12	14.94	27.69	8.29	2.93
	Bw	1.47	44.81	33.77	31.76	26.66	23.05	16.61	17.16	11.04	0.65
	C	1.67	37.80	32.21	31.50	27.58	24.62	17.31	14.90	5.59	0.18
SWG1 Trias Sandstone Till	A(2)	1.24	50.93	43.73	41.71	34.33	27.91	19.09	24.65	7.19	4.66
	B(2)	1.58	41.38	33.24	31.76	27.93	21.79	17.63	15.61	8.14	0.68
	C(2)	1.54	42.51	40.27	40.14	37.99	33.41	26.49	13.78	2.24	0.22

TABLE 9C.6 (Continued)

REFERENCES

Avery, B.W., 1980. *Soil classification for England and Wales (Higher Categories)*. Soil Survey Technical Monograph No.14. Lawes Agricultural Trust, Harpenden.

Bell, F.G., 1985. *Engineering properties of soils and rocks*. Butterworths, London.

Bennett, D.R. and Entz, T., 1989. Moisture-retention parameters for coarse-textured soils in southern Alberta. *Canadian Journal of Soil Science*, 69, 263-272.

Boorman, D.B., Hollis, J.M., Lilly, A., 1995. *Hydrology of soil types: a hydrologically-based classification of the soils of the United Kingdom*. Report No. 126, Institute of Hydrology.

Farquharson, F.A.K., Mackney, D., Newson, M.D. and Thomasson, A.J., 1978. *Estimation of run-off potential of river catchments from soil surveys*. Special Survey No.11. Soil Survey of England and Wales. Lawes Agricultural Trust, Harpenden.

Gustard, A., Bullock, A. and Dixon, J.M., 1992. *Low flow estimation in the United Kingdom*. Report No. 108, Institute of Hydrology.

Hall, D.G.M, Reeve, M.J., Thomasson, A.J. and Wright, V.F., 1977. *Water retention, porosity and density of field soils*. Soil Survey Technical Monograph No.9. Lawes Agricultural Trust, Harpenden.

Hanna, J.E. and Wilcock, D.N., 1984. The prediction of mean annual flood in Northern Ireland. *Proceedings Institution of Civil Engineers*, Part 2, 77, 429-444.

Hollis, J.M., 1991. *Mapping the vulnerability of aquifers and surface waters to pesticide contamination at the national/regional scale*. BCPC Mono. No. 47 Pesticides in soils and water.

Natural Environment Research Council, 1975. *Flood Studies Report*.

McAllister, J.S.V. and McConaghy. S., 1968. Soils of Northern Ireland and their influence upon agriculture. *Record of the Ministry of Agriculture*, 101-105.

Reeve, M.J., Smith, P.D. and Thomasson, A.J., 1973. The effect of density on water retention properties of field soils. *Journal Soil Science*, 24, 355-367.

Richards, L.A., 1947. Pressure-membrane apparatus, construction and use. *Agricultural Engineering*, 28, 451-454.

Stakman, W.P., Valk, G.A. and Harst, G.G. van der, 1972. *Determination of soil moisture retention curves. I. Sand box apparatus (range pF0-2.7)*. Institute for Land and Water Management. Wageningen.

Wilcock, D.N. and Hanna, J.E., 1987. Derivation of flow duration curves in Northern Ireland. The prediction of mean annual flood in Northern Ireland. *Proceedings Institution of Civil Engineers*, Part 2, 83, 381-396.

CHAPTER 9D

SOIL SURVEY AND LAND CLASSIFICATION

Soil maps provide a statement of the natural environment as it is expressed through soil. It is an almost permanent statement, as soil profiles, and related soil series, are not likely to change in the lifetime of use of a soil map. But, it should be noted that soil maps are not an end in themselves, and to be useful as an aid to any form of land management, they have to be interpreted, often with supplementary information, for the potential user. Such a person or group may find it difficult even to read a soil map, despite the kind of guidance given at the opening of the previous chapter (Chapter 8 on Soil Maps), and may not realise the potential value of soil maps to their kind of land management interest. It may be necessary for a professional expert to produce 'interpretative' maps, based on soil maps, but adding other information relevant to the specific application of the map.

A very wide range of land management problems can be addressed, from those concerned with environmental problems far below the ground, perhaps in rock aquifers, and where geological information is required, to land classification just below the surface for pipelines, to those on the land surface concerned with recreation, amenity, nature and wild life conservation, planning issues, forestry and agricultural development. The latter will require additional information about the nature of the ground surface, as well as the climate in the atmosphere above the ground. If any of the additional information is considered to be 'limiting' on the success of a particular crop, or restricting on the range of possible land uses, then the soil polygons of the soil map will be downgraded for the agricultural application of the soil map.

Agricultural Land Classification

The valuation of land for agriculture is an ancient practice. It was used by those in authority, government or church, as a basis for land taxation, from ancient China, to classical Rome, and post-Napoleonic Europe. Ireland had two land valuation surveys, pre- and post-Famine (in 1830s and 1850s), directed by a geologist Sir Richard Griffith. Systems are still in use in Australia and New Zealand for land taxation, but they were abandoned for that purpose in the UK in 1920, and in the Republic of Ireland in 1980. A new view of land classification emerged from work in the USA and in Canada in the 1960s, which is still in use at present, and is seen as a management policy for sustainable land use. These are land classification systems which define properties and delimit areas of different classes of land for potential crop yield or versatility of land uses. They began as Land Use Capability (LUC) systems of the USA (Klingebiel and Montgomery, 1961), in Canada (Canada Land Inventory, 1965) and in Britain (Bibby and Mackney, 1969). These systems, as well as others employed in the Republic of Ireland and Australia were reviewed in the Geographical Magazine in the 1970s, (Cruickshank, 1974).

A development of the Land Use Capability scheme has been adopted by the Soil Survey of Scotland, and described in a monograph (Bibby *et al.* 1982). In the scheme, which classes land by its range or versatility of possible agricultural land uses, the best and most versatile is Class 1, and the worst or most restricted is Class 7. The first 4 classes include possible use for exacting, specialised arable crops and vegetables, as well as grass. The lowest 3 classes are suited only for improved grassland and rough grazing in marginal land, Class 7 being of very limited agricultural value. The classification is intended as a scheme to assess the value of land for agriculture, and in Scotland, has been based on existing soil maps, but recognising the importance of other limiting physical factors such as climate, gradient, wetness, erosion, and pattern or variability of the surface. Land use capability is a land classification which can be mapped, and is appropriately based on soil maps. It is usually a generalised, 'broad-brush' classification which can be used even at 1:250,000 scale. It has not been employed in Northern Ireland so far, but may be in the future.

Prior to 1997, the only agricultural land classification map available for Northern Ireland was published in 1963 in 'Land Use in Northern Ireland'

and modified in 1982 (Cruickshank and Wilcock, 1982) as the generalised map shown in Figure 9D.1. The land classification scheme being introduced now in Northern Ireland is the Agricultural Land Classification of England and Wales, which is fully described with revised guidelines and criteria for grading in a MAFF publication (MAFF, 1988). At the end of 1996, the Agricultural Land Classification (ALC) had been introduced at the level of soil only, and all 17 soil maps (1:50,000) have been converted to ALC maps, using the criteria of Table 9D.1. These criteria shown on Table 9D.1 are a simplification of the exercise of agricultural land classification, which must give greater weighting where necessary, to limiting factors of climate, site, and interactive or compensating factors. At the time of writing, DANI holds climate data for average annual rainfall (AAR), potential evapotranspiration (PT), and length of growing season, but does not hold data for field capacity days (FCD), accumulated temperature January to June (ATO), soil moisture deficit (SMD), or any micro-climate data. All or any of these groups of climate can be limiting on land quality for agriculture, so the presentation in Table 9D.1 is a temporary simplification of ALC. There is one qualification, however, which does improve the scheme, and that is the use of soil profile types which in themselves reflect climate.

The Agricultural Land Classification of England and Wales was first published in the 1960s (MAFF, 1966), as a 5 class scheme, and modified in the 1970s (MAFF, 1976) by dividing Class (or Grade) 3 into two classes. A much more specific definition of criteria for grading was provided for the new 6 classes in 1988 (MAFF, 1988), as is shown in the following extract from that publication.

Grade 1 – excellent quality agricultural land
Land with no or very minor limitations to agricultural use. A very wide range of agricultural and horticultural crops can be grown and commonly includes top fruit, soft fruit, salad crops and winter harvested vegetables. Yields are high and less variable than on land of lower quality.

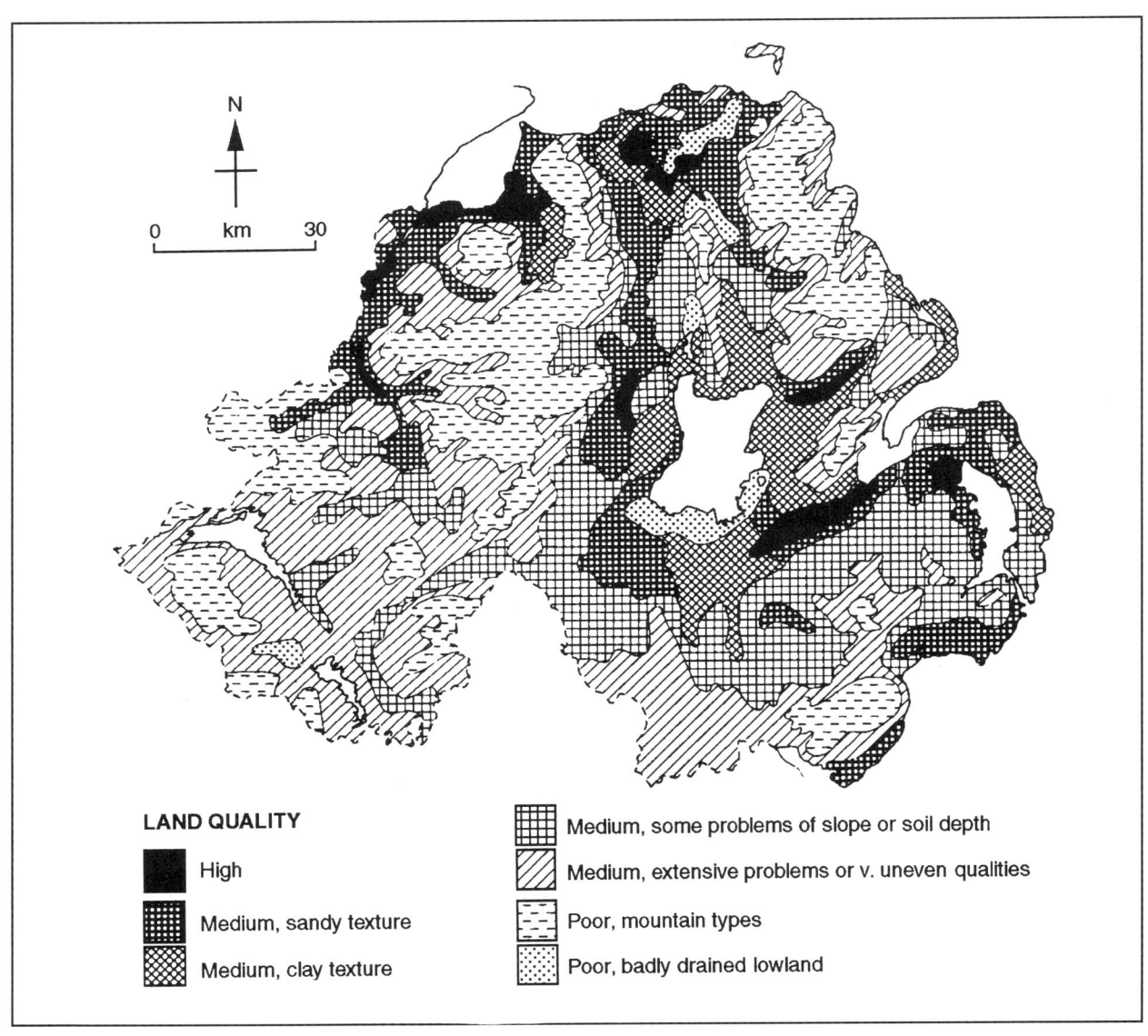

Figure 9D.1 Generalised agricultural land classification in Northern Ireland.

TABLE 9D.1
AGRICULTURAL LAND CLASSIFICATION IN NORTHERN IRELAND

Class or Grade	Soil profiles and soil parent materials	Annual Rainfall mm	Growing Season days	Altitude m (asl)	Slope degrees	Soil cm
1	Brown Earths on Sand, Gravel, or Sandy Till	<850	250-300	<100	<7	>60
2	Brown Earth and some SWG1 on Sand or Sandy Till	<1050	250-280	<150	<7	45-60
3A	SWG1/G1 on other parent materials Shallow Brown Earth Brown Podzolic	<1150	250-280	<200	<7	30-45
3B	SWG2/G2, Lake Clays G2, Alluvium and Organic Alluvium	<1400	250-280	<200	7-11	20-30
4	Most Rankers, Podzols & Peat Podzols, Humic Gleys, Peat, and Organic Alluvium	1100-1500 or <800 lowland	230-250	<300	11-18	15-20
5	Urban, Disturbed, Blanket Peat, Humic and Rock Rankers	800-2000	<230	0-1000	>7	<15

Grade 2 – very good quality agricultural land
Land with minor limitations which affect crop yield, cultivations or harvesting. A wide range of agricultural and horticultural crops can usually be grown but on some land in the grade there may be reduced flexibility due to difficulties with the production of the more demanding crops such as winter harvested vegetables and arable root crops. The level of yield is generally high but may be lower or more variable than Grade 1.

Grade 3 – good to moderate quality agricultural land
Land with moderate limitations which affect the choice of crops, timing and type of cultivation, harvesting or the level of yield. Where more demanding crops are grown yields are generally lower or more variable than on land in Grades 1 and 2.

Subgrade 3a – good quality agricultural land
Land capable of consistently producing moderate to high yields of a narrow range of arable crops, especially cereals, or moderate yields of a wide range of crops including cereals, grass, oilseed rape, potatoes, sugar beet and the less demanding horticultural crops.

Subgrade 3b – moderate quality agricultural land
Land capable of producing moderate yields of a narrow range of crops, principally cereals and grass or lower yields of a wider range of crops or high yields of grass which can be grazed or harvested over most of the year.

Grade 4 – poor quality agricultural land
Land with severe limitations which significantly restrict the range of crops and/or level of yields. It is mainly suited to grass with occasional arable crops (eg cereals and forage crops) the yields of which are variable. In moist climates, yields of grass may be moderate to high but there may be difficulties in utilisation. The grade also includes very droughty arable land.

Grade 5 – very poor quality agricultural land
Land with very severe limitations which restrict use to permanent pasture or rough grazing, except for occasional pioneer forage crops.

It should be noted that the soil limitations are not necessarily soil profile types, although the division of gleys into three degrees of gleying is relevant for agricultural land classification, but rather are soil properties such as texture, structure, depth and stoniness of any soil horizon. The reason being that any one soil horizon may be limiting in itself. In addition, there are important possible limitations of soil wetness and soil drought, which are called interactive limitations (produced by interaction among soil, climate and site properties). Such detailed data on the specific limiting factor, and its quantification, in defining a class or grade of land quality, are more likely to be used in classifying a site or a field. The creation of an agricultural land classification map on a scale between 1:50,000 and 1:250,000, has to be a generalisation, masking differences that exist in reality. The scale used and the approach adopted will depend also on the purpose of the land classification map. The Agricultural Land Classification (ALC) is given for most soil series discussed in Chapter 8 on Soil Maps.

Groundwater Vulnerability Map of Northern Ireland

A groundwater vulnerability map on 1:250,000 scale has been prepared by the British Geological Survey, the Soil Survey in England (SSLRC) and the DANI Soil Survey, partly using Northern Ireland soil maps, and published by the Department of the Environment for Northern Ireland (available for purchase from the latter). The map (Figure 5.8 in Chapter 5) is discussed by Wilcock, and is based on the principles that assessment of vulnerability of groundwater depends on 1) the distribution of aquifers, 2) the nature of the geological strata above, and 3) the physico-chemical properties of surface soils. A combination of these conditions is used to produce a 7 class groundwater vulnerability classification, divided first by permeability of rock types, and secondly by the leaching potential of the soils.

The 7 classes fall into 3 main geological groups. Type A includes aquifers which are highly permeable, and are able to store large amounts of water. These include Cretaceous Chalk and Sandstone, Triassic Sandstones and Magnesian Limestone, and Carboniferous Pure Limestone. The most vulnerable aquifers belong to these rocks where they underlie soils of high leaching potential, such as rankers, shallow brown earths, and brown earths, particularly if developed on sand, gravel or sandy texture parent material. The next class in the sequence belongs to the same geology, but has soils of intermediate leaching potential such as most brown earths and brown podzolic soils, as well as slightly gleyed SWG1, and drained peat, on glacial fill parent material. The third class are soils of low permeability, including all other gleys (SWG2 and SWG3) with a high clay content, and humic gleys.

Type B includes mainly superficial deposits, such as brown sand, beach deposits, glacial sands and gravel, and river alluvium. These function as very local aquifers, never able to store much water, but often supplying base flow to rivers. These Type B aquifers may overlie the Type A aquifers, and thus facilitate flow into the lower aquifers. The Type B group is again subdivided by the three degrees of soil leaching potential.

Type C is the geological group with negligible permeability and not containing exploitable groundwater. This Type C includes the Precambrian Mica-schists and Quartzites, Silurian and Ordovician shales and greywackes of Down and Armagh, and also Triassic Mudstones and Marl.

The groundwater vulnerability map has been produced partly as a management guide for the DoE (NI) Environment Service for the protection of groundwater. It will allow for better assessment of new developments above aquifers, where these could have an impact on groundwater quality. In Northern Ireland, there are only 3 large regions where groundwater quality might be at risk from contamination by pollutants, and the risk assessment is different in all 3 cases. The first lies in lowland Fermanagh, on either side of both the Lough Ernes, and extending east into the Clogher Valley. There is little risk from intensification of agriculture in this area, and in addition, the land close to the Lough Ernes, in the Sillees and Arney rivers in west Fermanagh, is impermeable and strongly gleyed clay till. The second area lies over the Triassic Red Sandstone rock from Armagh to Cookstown and Moneymore, with an outlier of Carboniferous Limestone beyond, at Maghera. There is some risk here from intensive agriculture, and associated fertiliser pollution and excessive slurry and silage effluent. However, the third area of the Triassic Red Sandstone in the Lagan Valley and through to Comber at Strangford Lough is an area of very high risk from intensification of agricultural practices, offset to a large degree by urbanisation, and hence the sealing of the ground surface by concrete and tarmacadam. The soils in this third area have a high leaching potential, so pollutants would be rapidly transferred to aquifers.

Land Evaluation For Pipelines – Assessment of Corrosion Risk

Land classification is required before the laying of underground pipelines. The zone of interest is within 2 metres of the ground surface, and more-or-less, co-incides with maximum soil profile depth into the parent material. The soil properties used as criteria for defining classes will be those that affect the risk of corrosion on the pipeline surface, and will vary according to the metal used in the pipe. The Soil Survey and Land Research Centre (SSLRC) in England has reclassified mapped soil series by their potential risk of corrosion on buried ferrous metal (Jarvis, M.G. and M.R. Hedges, 1994). The soil properties used were moisture content, acidity, aeration, soluble salt concentration, and electrical resistivity. Under direction from SSLRC, the DANI Soil Survey has undertaken a similar corrosion risk soil classification for coated pipes, which was commissioned by a commercial contractor. In this case, corrosion risk was increased by acidity at low pH values, by wetness, by high clay content, and by soluble sulphates and chlorides. The resulting soil corrosivity map for Northern Ireland is not reproduced here because of the commercial nature of the contract, but it is mentioned as an example of below-ground

TABLE 9D.2
Comparison of mean values in mg/kg for a selection of elements between England and Wales and Antrim, NI

Element		England and Wales Mean values mg/kg	Basalt Soil of Co. Antrim Mean values mg/kg
CHROMIUM	Cr	41.2	92.59
COBALT	Co	10.6	34.44
COPPER	Cu	23.1	68.96
LEAD	Pb	74.0	26.68
MAGNESIUM	Mg	3,736	9,387
NICKEL	Ni	24.5	107.39
PHOSPHORUS	P	848	1188.05

land classification which is based on a conversion of the soil polygons on soil maps.

Soil Geochemical Atlas of Northern Ireland

A valuable database exists for total concentrations of 15 chemical elements in the top soil (Ap) of every km² in the lowland area of the Province. 7,000 top soil samples were analysed. Total concentration values are available for Ca, Cd, Co, Cr, Cu, Fe, K, Mg, Mn, Mo, Na, Ni, P, Pb and Zn, and these total values can be used to delimit areas where maximum permitted values are close, or have been reached, and hence prohibit any further addition of that element to the soil surface by sewage sludge spreading, or by any other application.

The Geochemical Atlas NI Project is a joint venture between The Land Research Group of AESD (in DANI) and DoE (NI). Field work and field sampling was undertaken by AESD and financial support for soil analyses and publication of results was received from EU Interreg through DoE (NI). The project was modelled on 'The Soil Geochemical Atlas of England and Wales', (McGrath and Loveland, 1992), and methodology adopted from that source.

Soil sampling was done in every km square in the lowland area of Northern Ireland. The upper altitudinal limit of the lowlands was flexible, but was usually the upper limit of enclosure. In each km², the most common or dominant soil was selected for sampling, so sampling was biased to that extent. The soil samples were taken in the top soil (Ap) between 0 and 15 cm. Large bulk samples of about a kilogram weight were taken in the field, but after drying and preparation in the laboratory, only 25g subsamples were used for soil analyses.

The soil samples were dried and milled prior to digestion with aqua regia (4:1 hydrochloric: nitric acids by volume). The total element analysis was carried-out using ICP-OES, at the Macaulay Institute for Land Use Research (MLURI) in Aberdeen. All analyses were subject to formal quality procedures, using an in-house quality control soil, and all results were expressed as mg/kg dry matter. Further detail on soil analyses are provided in the publication mentioned earlier (McGrath and Loveland, 1992).

Although the Soil Geochemical Atlas project for Northern Ireland is well advanced, it is not complete, and only preliminary and provisional results are given in 4 maps in Colour Plates for 4 total elements (Phosphorus, Lead, Copper and Nickel in the basalt region of County Antrim). Province-wide data has not been examined at the time of writing, and it is planned to publish the complete Soil Geochemical Atlas later in 1997.

At this early stage, and considering only the results for basalt soils, the mean values for total concentrations compared with results for England and Wales, are lower or similar for Cadmium, Calcium, Iron, Potassium, Sodium, Lead, Zinc and Manganese. NI results (in the basalt region only) are significantly higher for Chromium, Cobalt, Copper, Magnesium, Nickel and Phosphorus in mg/kg dry soil. It is among this second group of high local values that Copper and Nickel occur and these are significant for possible sewage sludge application limits. A comparison of values in mg/kg dry soil between England and Wales (overall) and the basalt soils in Antrim is given for selected elements in Table 9D.2.

The geographical distribution of actual values in mg/kg dry soil is provided for 4 elements (Cu, Ni, P and Pb) in the basalt area of Antrim in Colour Plate 23. A number of km² have no values because of urban or inclusive development, or because sampling was incomplete. This report is given at an early stage in the project, and all results must be provisional.

Phosphorus P

In common with most soils in England and Wales, top soil P levels of the basalt soils in Antrim are generally in the range 500 – 1500 mg/kg. Mean values in both areas are around 1000 mg/kg. In Antrim, values exceed 1000 mg/kg in the lowland north and east of Ballymena, as well as around Antrim and Ballyclare, in the south of the county. Some extremely high total values of P (over 2000 mg/kg) are to be found in the same areas. However, it must be remembered that these are total values, and in basalt soil known for its capacity to fix P, the available P may be much less. This is an issue of major interest, particularly in the mid and western part of Antrim in the catchment of the River Bann and Lough Neagh where levels of available P are rising annually. The accumulative effect of this increase is making Lough Neagh, and other lakes, more eutrophic.

Lead Pb

In most lowland km squares in Antrim, total values of lead are less than 28 mg/kg, similar to most of lowland Eastern England. The highest values (over 123 mg/kg) of the Pennines, Cumbria and part of upland Wales, are reached in only 3 or 4 isolated km squares in Antrim. Early indications from the NI Soil Geochemical Atlas project show much higher values in the shale soils of Co Down, but there are no indications of very high Pb in any agricultural soil in the Province. Mean values in Antrim are only about one third of those mean values in England and Wales.

Copper Cu

Copper levels are low in chalk soils, chalky drift, sandy soils and peaty soils in England and Wales, and high only in mining and industrial areas (such as Cornwall, South Wales and Lancashire). In the Antrim basalt soils, almost all soils sampled are showing these high values of over 43.3 mg/kg, and mean values are three times higher than in England and Wales. There are many km squares in Antrim measuring over 100 mg/kg and these have reached the EU maximum permissible levels of potentially toxic elements (PTE). No further application of copper-rich materials should be made to such soils.

TABLE 9D.3
Upper limits (mg/kg dry soil) for total concentrations of heavy metals

	CEC Directive 1986		UK Limits
	Lower	Upper	
Copper	50	140	135
Nickel	30	75	75

These values are for soils with pH 6 to 7, and are reproduced in a full table (Table 1.7) in 'The Soil Geochemical Atlas for England and Wales (McGrath and Loveland, 1992).

Nickel Ni

The situation with Nickel is similar to Copper in Antrim soils. In England and Wales, very few areas (sites of artificial increase around smelters, and on some few areas of rock mineralisation) record values of over 60.0 mg/kg for Nickel concentration. Again, almost all the basalt soils have Ni levels in this highest category, which is at the maximum permissible concentration for PTE by EU directive. No further additions of Nickel could be tolerated in basalt soils in Antrim. Even mean values for basalt soils are more than four times greater than mean values in England and Wales (24.5 compared to 107.39 mg/kg).

Although work is on-going in 1996 regarding the Soil Geochemical Atlas project for 15 elements in Northern Ireland, and even field sampling and data analysis are not yet complete, early and provisional results are given here to signal that there does not appear to be a potential problem with Lead Pb, or most of the elements measured, but there may be a problem developing with total Phosphorus P and dangerous levels of total Copper Cu and total Nickel Ni have been reached at least in basalt-derived top soils (0-15 cm). These early results form the Soil Geochemical Project are supported by those in the Provisional Geochemical Atlas of Northern Ireland (1973), from Imperial College London, which published highest values for totals of Chromium, Copper, Magnesium and Nickel in basalt soil river deposits. Dickson and Stevens (1983) showed that the highest levels of extractable Copper in the Province were found in west and south-west Antrim on basalt soils.

Current and Future Work in Land Classification

The completion of soil mapping in the Province, the publication of 17 soil maps at 1:50,000, the provision of soil map data in digital form, the forming of a Soil Attribute Database and a 1km^2 Environmental Database (for dominant soil series, solid and drift geology, HOST class and various climate data) means that a considerable amount of basic information is now available for environmental and land classification projects. The unknown element of future demands makes it impossible to describe or predict what these projects might be. However, it is the intention of the newly-formed DANI Land Research Group to attempt various types of land classification, such as the potential for nitrate loss through leaching, the potential of land to accept applied slurry (or slurry acceptance potential), a risk assessment of heavy metal pollution through sewage sludge spreading, and possibly various less urgent projects such as assessing land suitability for particular arable crops and tree crops in forestry, and estimating optimal grazing capacity levels to ensure soil protection from damage and future sustantial yield. River catchment studies, using GIS techniques, will be included also in future research.

APPENDICES

APPENDIX 1
FIELD METHODS OF SOIL MAPPING

Soil surveys may be based either on regularly spaced inspection pits, or on a subjective spacing and free survey inspection pits. The former approach is necessary for soil mapping of small, areal units such as fields or farms up to about 100 ha, but the latter approach of free survey is appropriate for fast, reconnaissance mapping of large areas on scales from 1:25,000 – 1:100,000. The DANI Soil Survey never had more than 3 teams (a total of 6 soil surveyors) in the field at any one time, and was limited to a period of 7 years (1988-1994) for the mapping of 13,550 km^2 of the Province. Soil survey had to progress at a consistently fast rate for completion within the time available. The publication scale was to be 1:50,000 for the soil maps.

Soil surveyors prepared for field work by making themselves familiar, from existing publications, with the underlying geology, geomorphology and topography of the field area. A wide range of previously prepared and published maps was consulted in advance of field work. Land owners were consulted for permission for access to land, and also about detail of their soils known to them. Fieldwork was done using Ordnance Survey 1:10,000 maps, and free survey methods. Small diameter inspection pits were spade dug to about 80-90 cm, and full morphological details recorded on the field maps. Dividing lines or boundaries between soil profile types were established by digging pits on opposite sides of possible boundaries. In this way, soil polygons were delimited, within which soil profiles and parent materials were the same. Almost 25,000 soil polygons were delimited. These were the basic mapping units or polygons of each soil series, and were the components of the field maps used later for the production of the published 1:50,000 soil maps. This second stage process is described in the chapter on the soil maps (Chapter 8).

APPENDICES

APPENDIX 2
SUMMARY OF METHODS USED FOR ANALYSIS OF SOILS: SOIL SURVEY PROJECT

1. Stone Content: Determined as the weight of particles > 2 mm obtained by sieving the air dried soil.

2. pH: Determined on a 1:2½ soil/water ratio using a combination glass pH electrode.

3. Phosphorus: Extracted with 0.5M sodium bicarbonate solution a pH 8.5 and determined colorimetrically (Olsen's Test).

4. Potassium: Extracted with M ammonium acetate solution and determined by flame photometry.

5. Magnesium: Extracted with M ammonium acetate solution and determined by atomic absorption spectrophotometry.

6. Exchangeable Cations (Na, K, Ca, Mg): Extracted with M ammonium acetate solution, Na and K determined by flame photometry, Ca and Mg determined by atomic absorption spectrophotometry.

7. Cation Exchange Capacity (CEC): Determined by the barium chloride/triethanolamine method.

8. Iron and Alumium: Extracted with potassium pyrophosphate solution, Fe determined by atomic absorption spectrophotometry, Al determined by ICP emission spectrophotometry.

9. Iron: Extracted with sodium dithionite solution, determined by atomic absorption spectrophotometry.

10. Total Phosphorus: Digested with a nitric/perchloric acid mixture and determined colorimetrically.

11. Total Nitrogen: Determined by dry combustion using an elemental analyser.

12. Total Carbon: Determined by dry combustion using an elemental analyser.

13. Loss on Ignition (LOI): Determined as loss in weight on ignition at 850°C.

14. Particle Size Analysis: Determined by pipette method on < 2mm peroxide treated soil.

REFERENCES

Avery, B.W. and Bascomb, C.L. (ed) 1982, *Soil Survey Laboratory Methods*, Soil Survey Technical Monograph No. 6.

MAFF 1986. *The Analysis of Agricultural Materials*, third edition, Reference Book 427, HMSO, London.

APPENDICES

APPENDIX 3
SOIL ATTRIBUTE DATABASE

The soils attribute database is a comprehensive compilation of information on the observed, physical and chemical nature of all soil horizons at 5 kilometre inspection pits sampled as part of the DANI Soil Survey. The database is currently held on the Department of Agriculture's DEC ALPHA computer at Newforge Lane as a number of ORACLE tables linked by the grid reference of the actual site. The name of each field and a description of the property recorded is given below.

FLDDATA (Site environment data)

GRIDREF	Twelve figure 5k site reference
OBSDATE	Sampling data
OBSERVER	Initials of Soil Survey staff member
PROJNO	Project number (internal use)
ALTITUDE	Site altitude in metres
SUBGRP	Secondary site identifier
SERIES	Identified soil series name
ANGLE	Angle of slope on which 5k pit is sited
SHAPE	Slope shape
LANDTYPE	The Landform on which the 5k pit is sited.
VEG	The vegetation type of the surrounding area
LANDUSE	To what agricultural or other use the surrounding area is put
GRAZING	The grazing value of the site area
ROCKTYPE	The underlying rock type
GRID	Six figure grid 5k site reference

FLDDATA2 (Profile description)

GRIDREF	Twelve figure 5k site reference
HORIZON	Horizon name
DEPTH	Depth to lower boundary of horizon
TEXTCLASS	Hand textured particle size class
MUNSELL	Soil matrix colour
MOTTLE	The percentage distribution of mottles
COLOUR	The identified mottle colour
MOTTLE2	The percentage distribution of secondary mottles
COLOUR2	The identified mottle colour
STONEABUND	The assessed stone abundance
STONESIZ	The average size of stone
STONEHARD	Stone hardness
STONESHAPE	Stone angularity
OTHERSTONE	The presence of stones other than from underlying rock type
ROCK	Presence or absence of rock in the profile
ROCKTYPE2	The underlying rock type
HUMOSE	The estimated percentage of humose
MINERAL	The estimated percentage of ferri-manganiferous minerals
MINERAL2	The estimated percentage of jarosite minerals
CARBONATE	The estimated percentage of carbonate
STRUCTSHAPE	Identified soil structure shape
STRUCTSIZE	Dimensions of identified soil structure
DEVELOPMENT	Degree of development of structure
CONSISTENCY	Assessed soil consistency in horizon
PAN	Presence or absence of soil pan in the horizon
CEMENTING	Degree of cementing within the pan
ROOTS	Presence or absence of roots
ROOTFREQ	Abundance of roots within the horizon
NATURE	Type of roots
WATER	Degree of saturation of horizon
SEEPAGE	Presence or absence of seepage in the horizon
SAMPLES	Number of samples taken in the horizon
GRID	Six figure grid 5k site reference

LABDATA (Laboratory description)

GRIDREF	Twelve figure 5k site reference
HORIZON	Horizon designation
SAMPLE	Number of samples in the profile
LABNO	DANI laboratory processing number
PH	Honzon pH
P	Horizon phosphorus levels (sodium bicarbonate-extractable)
K	Horizon potassium levels (ammonium acetate - extractable)
MG	Horizon magnesium levels (ammonium acetate - extractable)
STONES	The percentage of stones in the horizon
MG2	Exchangeable magnesium levels in the horizon
CA	Exchangeable calcium levels in the horizon
K2	Exchangeable potassium levels in the horizon
CEC	Cation exchange capacity
FE	Pyrophosphate-extractable iron
AL	Pyrophosphate-extractable aluminium
FE2	Dithionite-extractable iron
P2	Total phosphorus
N	Percentage of nitrogen
C	Percentage weight of organic carbon
LOI	Percentage weight loss on ignition
FINE	Percentage weight of fine sand in horizon
MED	Percentage weight of medium sand in horizon
CO	Percentage weight of coarse sand in horizon
SILT	Percentage weight of silt in horizon
CLAY	Percentage weight of clay in horizon
GRID	Six figure grid 5k site reference

HYDDATA	(Soil moisture retention data)	**GEODATA**	(Geochemical data for A horizons and selected B horizons)
GRIDREF	Twelve figure 5k site reference	GRIDREF	Twelve figure 5k site reference
HORIZON	Horizon designation	CU	Total copper content
DBT	Horizon bulk density	CR	Total chromium content
TT	Horizon total porosity	CD	Total cadmium content
QVT5	Volume of retained water at 5 kPa suction	CO	Total cobalt content
		CA	Total calcium content
QVT10	Volume of retained water at 10 kPa suction	FE	Total iron content
		K	Total potassium content
QVT40	Volume of retained water at 40 kPa suction	MG	Total magnesium content
		MN	Total manganese content
QVT200	Volume of retained water at 200 kPa pressure	MO	Total molybdenum content
		NA	Total sodium content
QVT1500	Volume of retained water at 1500 kPa pressure	NI	Total nickel content
		PD	Total lead content
AVT	Volume of water available for plant use	P	Total phosphorus content
CAT	Volume of air fi~led pores at 5 kPa suction	ZN	Total Zinc content
		GRID	Six figure grid 5k site reference
GRID	Six figure grid 5k site reference		

APPENDICES

APPENDIX 4
TABLES OF FERTILISER RECOMMENDATIONS

The following tables of fertiliser recommendations are taken from - Fertiliser Recommendations for Agricultural and Horticultural Crops, MAFF, 1993, Reference Book 209, HMSO London.

TABLE 1
Classification of 'available' soil phosphorus levels

Phosphorus Index	Phosphorus concentration (mg/l)	Interpretation
0	0-9	- Deficient for all crops
1	10-15	- Low for all crops
2	16-25	) Adequate for grassland,
3	26-45	) cereals and outdoor fruit crops
4	46-70	- Adequate for most outdoor vegetable crops
5	71 - 100	- Adequate for potatoes
6	101 -140	- Adequate for most glasshouse crops
7	141-200	)
8	201-280	) Excessive
9	>280	)

TABLE 2
Classification of 'available' soil potassium levels

Potassium Index	Potassium concentration (mg/l)	Interpretation
0	0-6	- Deficient for all crops
1	61-120	- Low for all crops
2	121-240	- Adequate for grazing
3	241-400	- Adequate for silage
4	401-600	- Adequate for vegetables, fruit and most glasshouse crops
5	601-900	- Adequate for potatoes, celery and rhubarb
6	901-1500	- Adequate for tomatoes
7	1501-2400	)
8	2401-3600	) Excessive
	>3600	)

TABLE 3
Classification of 'available' soil magnesium levels

Magnesium Index	Magnesium concentration (mg/l)		Interpretation
0	0-25		- Deficient for all crops
1	26-50		- Low for potatoes, swedes and fodder beet
2	51-100		- Adequate for grassland, cereals and most outdoor vegetable crops
3	101-175		- Adequate for potatoes, swedes, fodder beet and most outdoor crops
4	176-250	)	Adequate for most glasshouse crops
5	251-350	)	
6	351-600		- Adequate for tomatoes
7	601-1000		
8	1001-1500	)	Excessive
	> 1500		

CEC and exchangeable cation ratings

The cation-exchange capacities of soils vary considerably. This is mainly due to variations in the humus content and the amounts and kinds of clay present. The Table below gives typical ratings of cation-exchange for British soils.

Cation exchange capacity ratings

Rating	Cation exchange capacity (m.e./100 g)	Total exchangeable cations (m.e./100 g)	Percentage base saturation
Very high	40	25	80 - 100
High	25 - 40	15 - 25	60 - 80
Moderate	12 - 25	7 - 15	40 - 60
Low	6 - 12	3 - 7	20 - 40
Very low	6	3	0 - 20

As a result of the exchange properties of clay and humus, greater or lesser amounts of bases can be held within the soil. In British soils the elements of Ca++, Mg++, K+ and Na+ have the exchangeable ratings shown below.

Exchangeable cation ratings (m.e./100 g) Ca++, Mg++, K+, Na+

Rating	Calcium	Magnesium	Potassium	Sodium
Very high	20	6	1.2	2
High	10 - 20	3 - 6	0.8 - 1.2	0.7 - 2
Moderate	5 - 10	1 - 3	0.5 - 0.8	0.3 - 0.7
Low	2 - 5	0.3 - 1	0.3 - 0.5	0.1 - 0.3
Very low	2	0.3	0.3	0.1

SOIL CARBON DENSITY

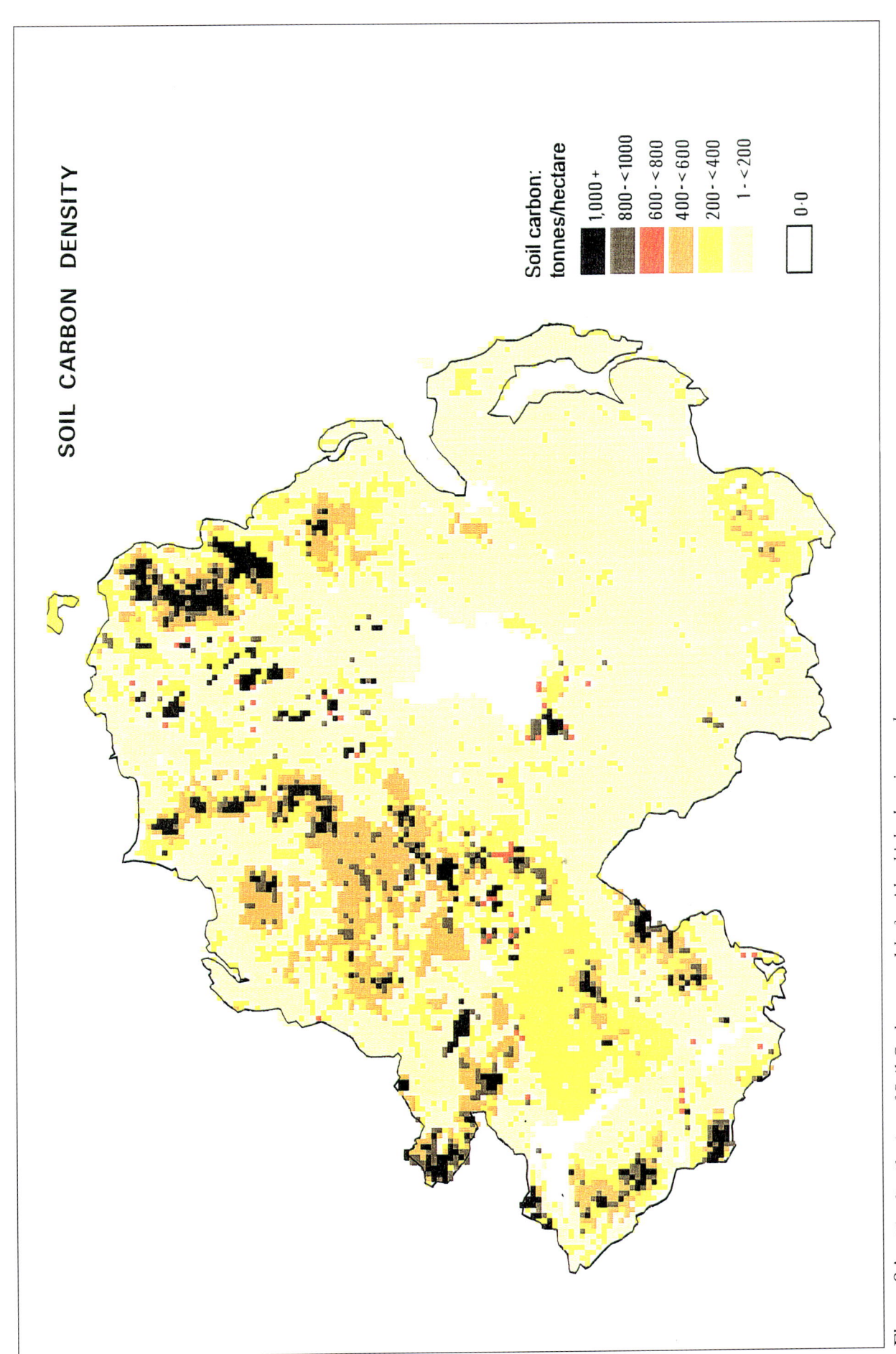

Soil carbon: tonnes/hectare
- 1,000 +
- 800 - <1000
- 600 - <800
- 400 - <600
- 200 - <400
- 1 - <200
- 0·0

Figure 9A *The distribution of Soil Carbon on a 1 km² grid, which also gives scale.*

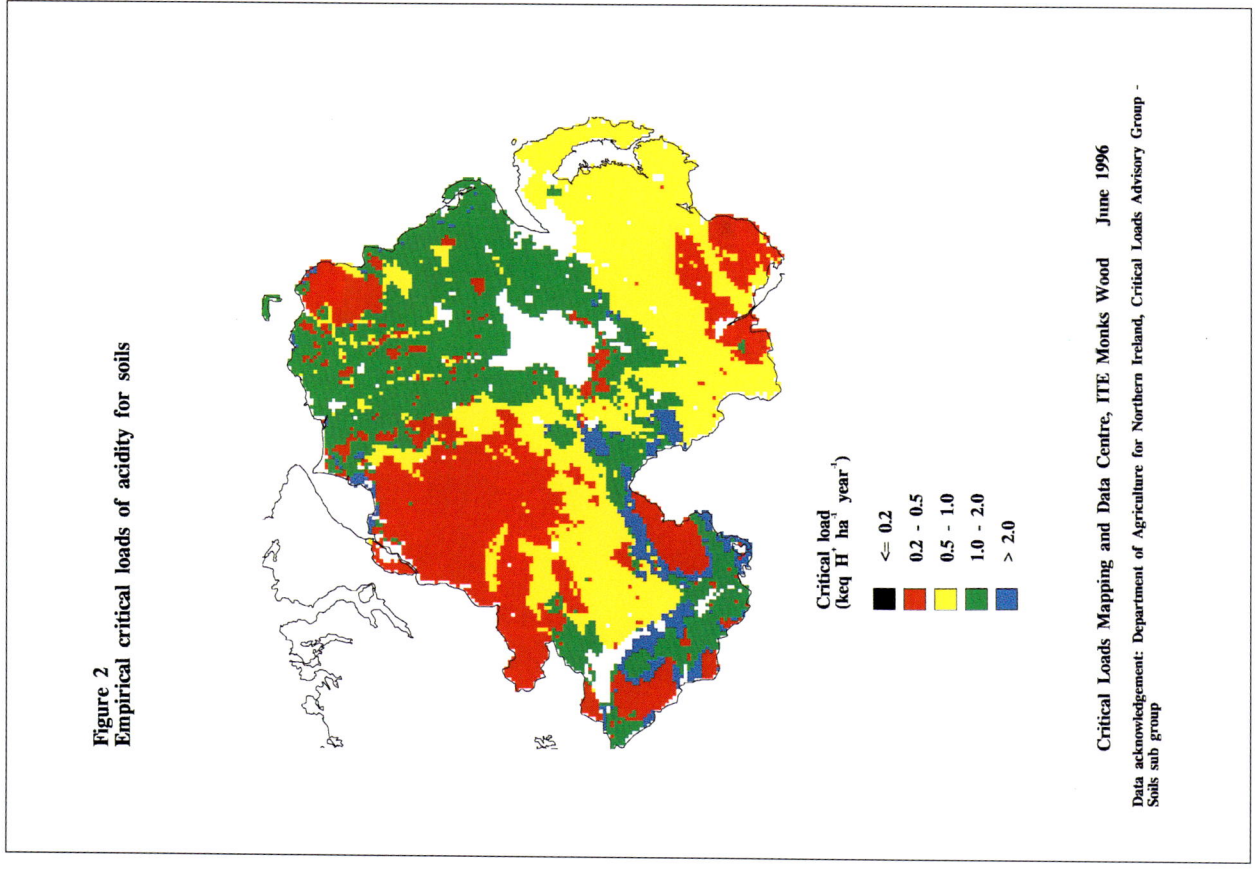

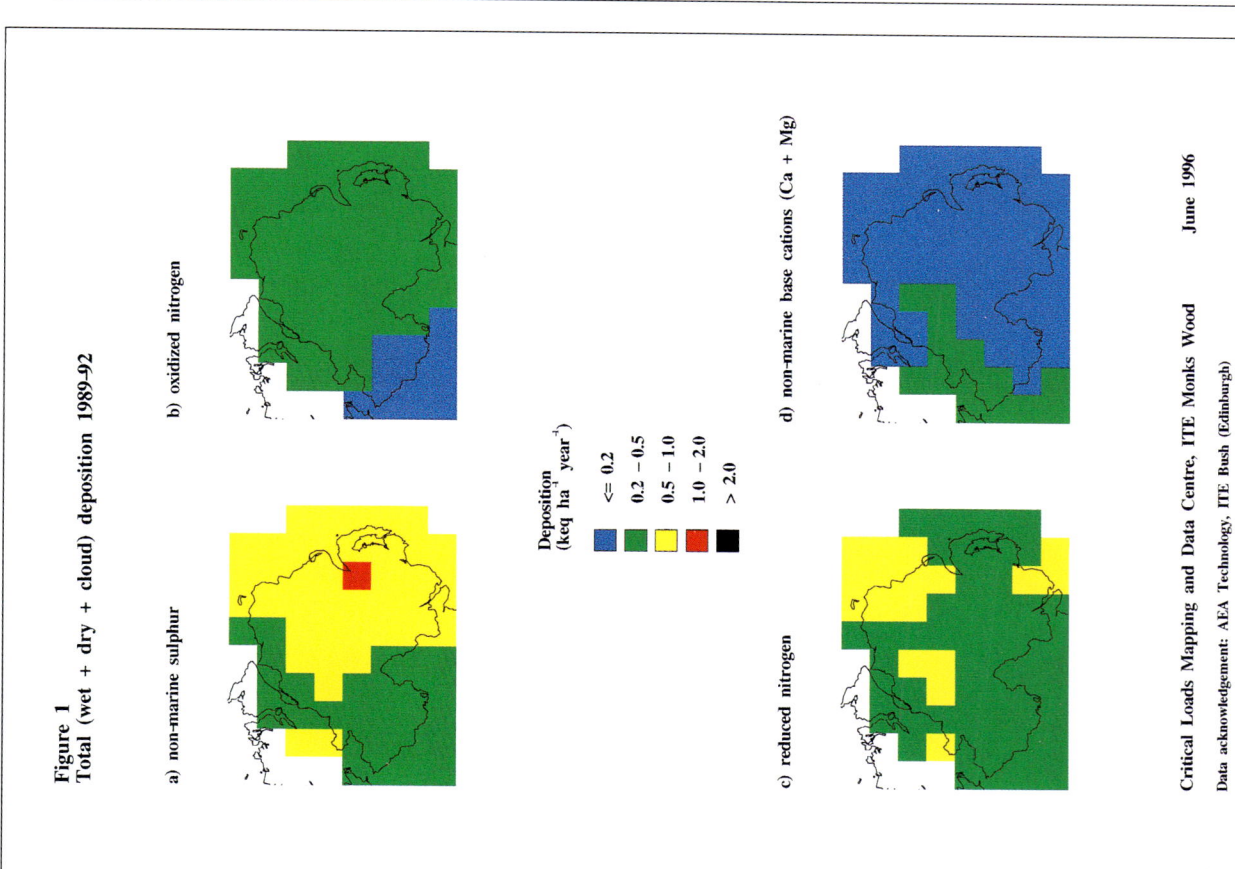

Figure 2
Empirical critical loads of acidity for soils

Critical load (keq H⁺ ha⁻¹ year⁻¹)
- ■ ≤ 0.2
- ■ 0.2 - 0.5
- ■ 0.5 - 1.0
- ■ 1.0 - 2.0
- ■ > 2.0

Critical Loads Mapping and Data Centre, ITE Monks Wood   June 1996

Data acknowledgement: Department of Agriculture for Northern Ireland, Critical Loads Advisory Group - Soils sub group

Figure 1
Total (wet + dry + cloud) deposition 1989-92

a) non-marine sulphur

b) oxidized nitrogen

c) reduced nitrogen

d) non-marine base cations (Ca + Mg)

Deposition (keq ha⁻¹ year⁻¹)
- ■ ≤ 0.2
- ■ 0.2 − 0.5
- ■ 0.5 − 1.0
- ■ 1.0 − 2.0
- ■ > 2.0

Critical Loads Mapping and Data Centre, ITE Monks Wood   June 1996

Data acknowledgement: AEA Technology, ITE Bush (Edinburgh)

Figures for Chapter 9B

Figure 4a
Modelled (HARM 10.4) non-marine sulphur deposition for 2010 plus measured oxidised and reduced nitrogen less non-marine base cation deposition 1989-92

Deposition
(keq ha^{-1} year^{-1})
- ■ <= 0.2
- ■ 0.2 - 0.5
- ■ 0.5 - 1.0
- ■ 1.0 - 2.0
- ■ > 2.0

Figure 4b
Exceedance of empirical critical loads of acidity for soils by modelled (HARM 10.4) non-marine sulphur deposition 2010 plus measured oxidised and reduced nitrogen less non-marine base cation deposition 1989-92

Exceedance
(keq ha^{-1} year^{-1})
- ■ Not exceeded
- ■ 0.0 - 0.2
- ■ 0.2 - 0.5
- ■ 0.5 - 1.0
- ■ > 1.0

Critical Loads Mapping and Data Centre, ITE Monks Wood   June 1996

Data acknowledgement: Department of Agriculture for Northern Ireland, Critical Loads Advisory Group - Soils sub-group, AEA Technology, ITE Bush (Edinburgh) and Hull Universities

---

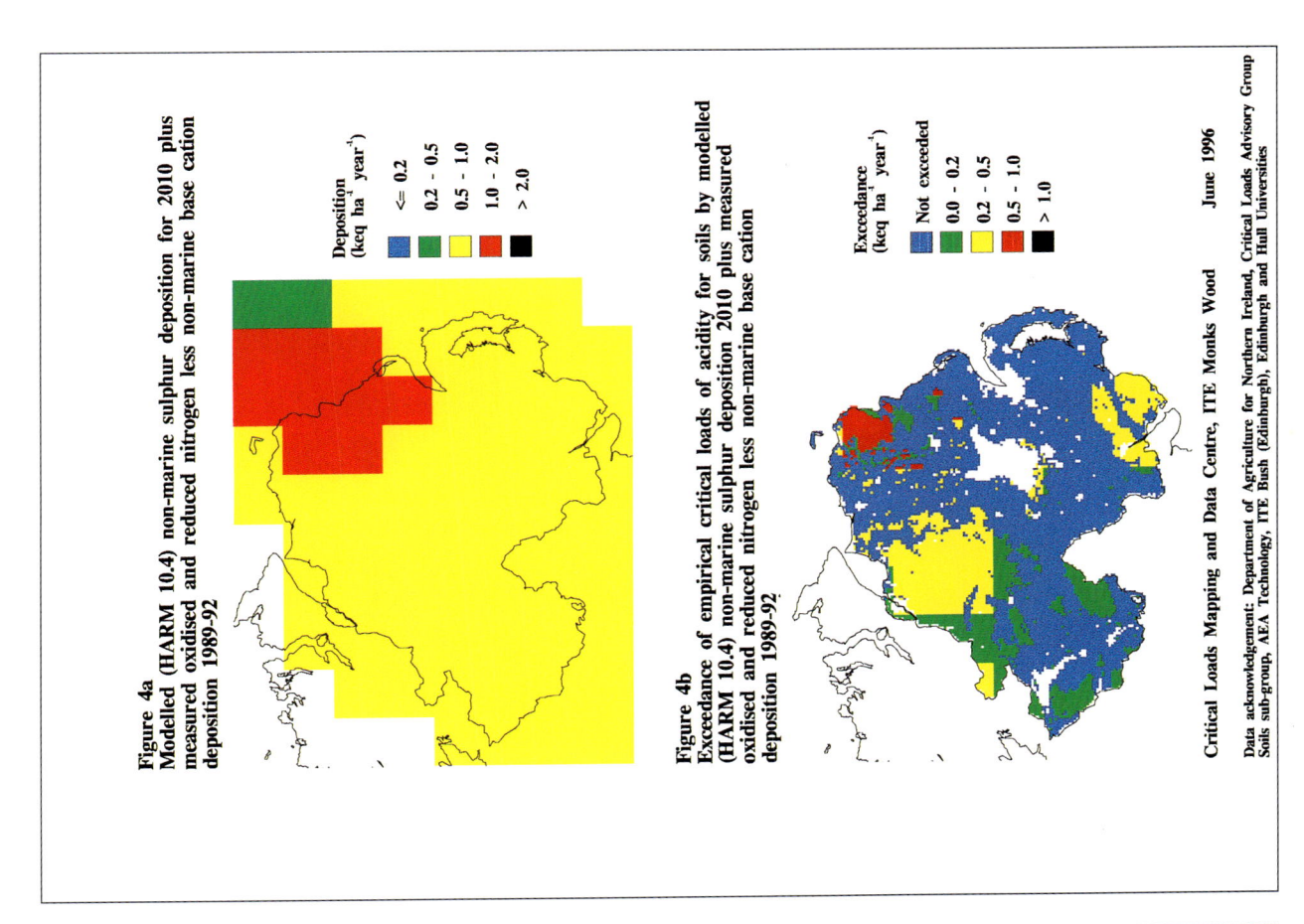

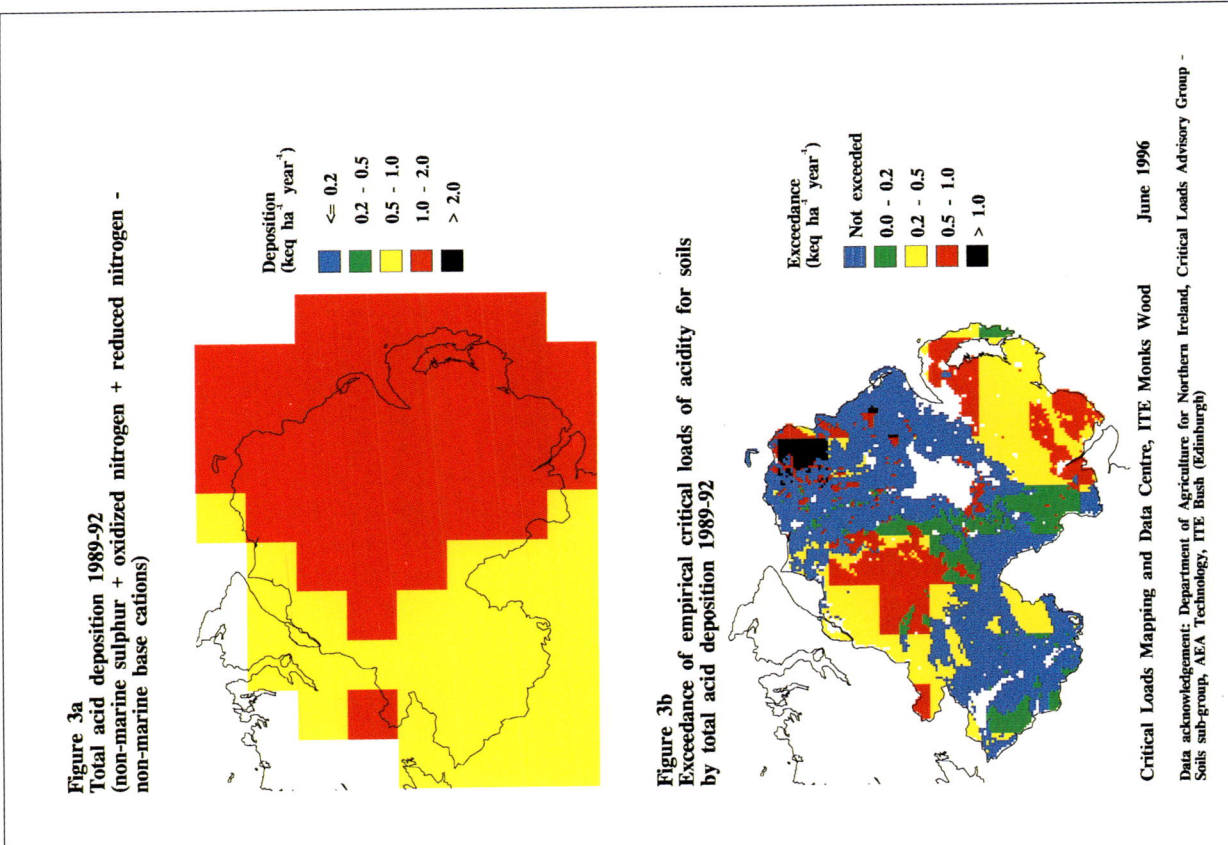

Figure 3a
Total acid deposition 1989-92
(non-marine sulphur + oxidized nitrogen + reduced nitrogen - non-marine base cations)

Deposition
(keq ha^{-1} year^{-1})
- ■ <= 0.2
- ■ 0.2 - 0.5
- ■ 0.5 - 1.0
- ■ 1.0 - 2.0
- ■ > 2.0

Figure 3b
Exceedance of empirical critical loads of acidity for soils by total acid deposition 1989-92

Exceedance
(keq ha^{-1} year^{-1})
- ■ Not exceeded
- ■ 0.0 - 0.2
- ■ 0.2 - 0.5
- ■ 0.5 - 1.0
- ■ > 1.0

Critical Loads Mapping and Data Centre, ITE Monks Wood   June 1996

Data acknowledgement: Department of Agriculture for Northern Ireland, Critical Loads Advisory Group - Soils sub-group, AEA Technology, ITE Bush (Edinburgh)

Figures for Chapter 9B

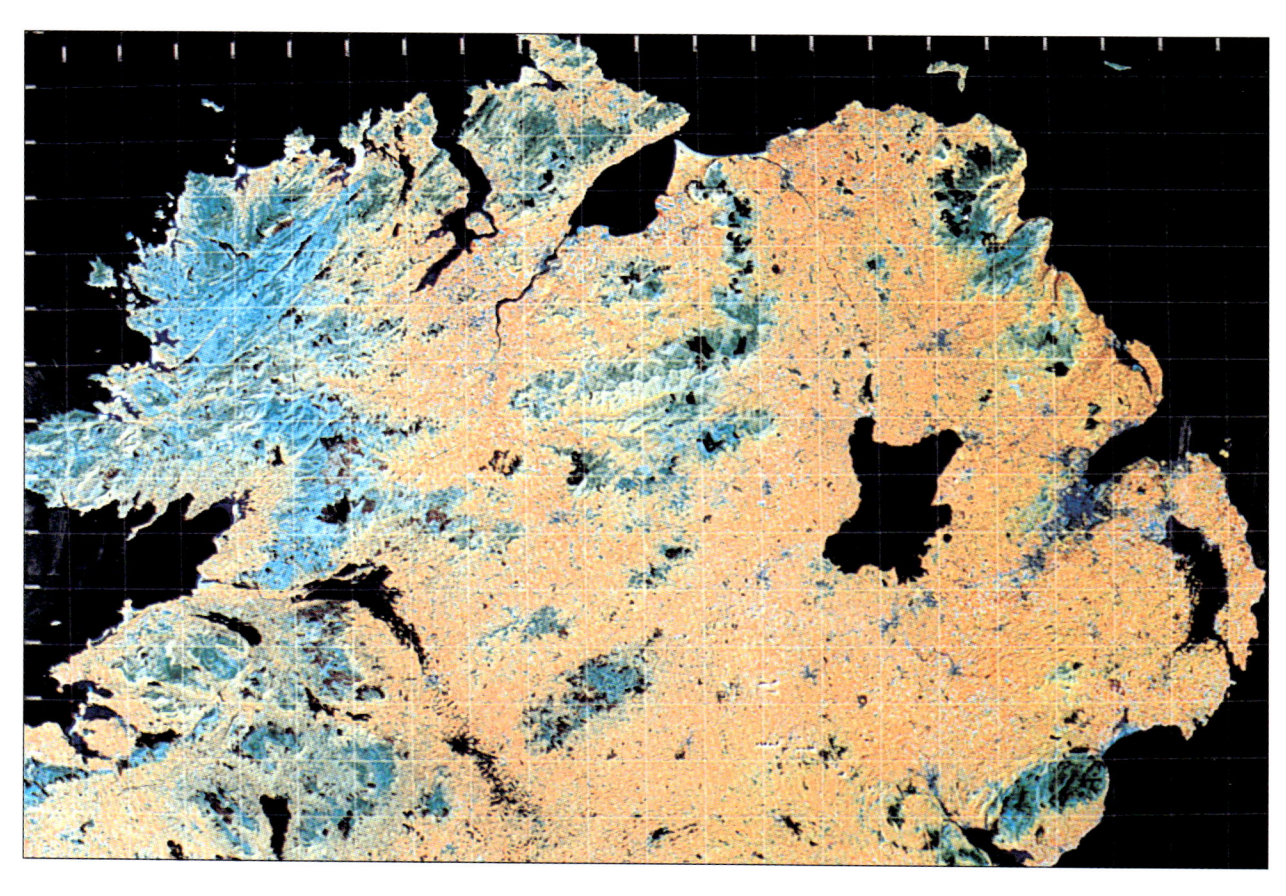

Figure 6.1 *This image of **Ireland North** is reproduced from reflectance values recorded by the LANDSAT TM satellite in three energy wavelengths (bands) – red, near infra-red (NIR) and middle infra-red (MIR). The human eye cannot see NIR nor MIR reflectance, therefore objects which reflect these bands strongly must be colour coded if they are to be seen on an image. This produces a false-colour image in which objects primarily reflecting red light are shown as blue, those primarily reflecting NIR are shown as red and those primarily reflecting MIR are shown as green. Objects reflecting strongly in more than one band produce other false colours (e.g. pinks, browns). An object which is red on the image indicates high NIR reflectance, which is a property of healthy vegetation. Objects with high MIR, but low NIR reflectance and are light green on the image, may indicate dry vegetation of poor growth. Blue objects on the image are generally bare soils, rocks or man-made structures.*

On this image, arable land (2.1.1) can be recognized prinicipally by the blue colour of bare (soil) fields, as for example to the east of Comber, or on the polderlands north of Limavady. Oil seed rape is pink and winter sown cereals are a brownish-red (both of these are difficult to see at this scale of reproduction). It should be noted that some 'blue' fields may be for re-seeded pastures rather than for arable agriculture. Pasture (2.3.1) is seen predominantly in shades of red – the redder the image the greater amount of grass in the field. The dominance of pasture in the agricultural land is apparent. Forests (3.1) are identified readily, as for example the rich red-brown blocks of conifers in the Sperrins and the northern Antrim plateau. Deciduous and mixed woodlands, generally of small area in NI, are difficult to distinguish at this scale. In this early May image, natural grasslands (3.2.1) tend to be a bluey-green colour and dominate on the lower margins of the hills; they may be seen particularly well on the south facing slopes above the Glenelly river. These natural grasslands merge into me darker green shades of peat bogs on the mountains, the darkness of colour depending on the amount of heather cover or of water in the peat. Lowland peat bogs of similar colours may be identified to the east and west of Omagh.

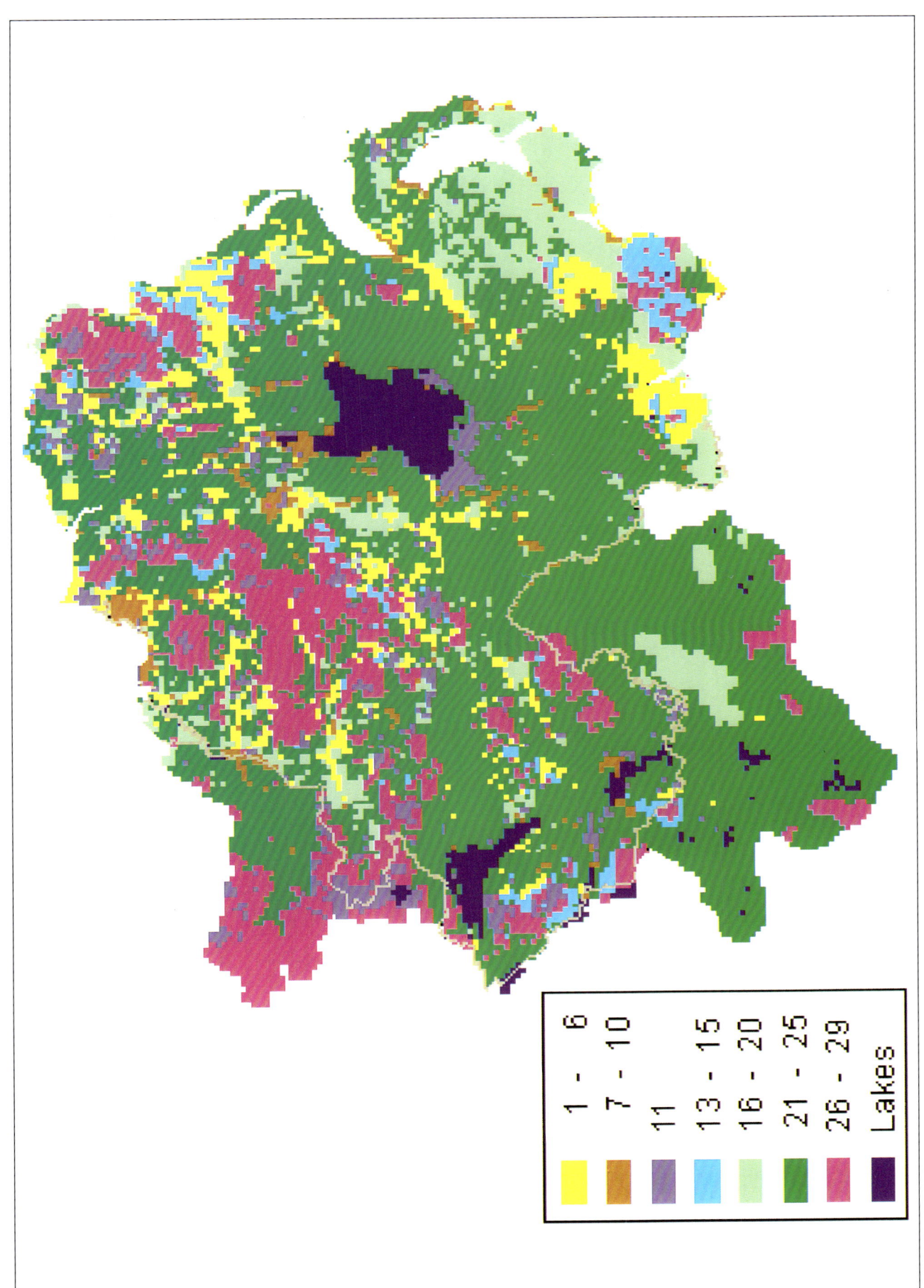

Figure 9C The HOST map of Northern Ireland and catchments extending into the Republic of Ireland expressed as the dominant class within each 1 kilometre square.

A

Horticulture in peat overlying gleyed Lough Neagh Clay till, in north Co. Armagh.

B

Old oak woodland and woodrush ground flora at Breen Wood, Glenshesk, north Co. Antrim. Soils are humus-iron podsols on mica-schist (see Profile 8 on CP2).

C

Demesne mixed woodland in the Lagan Valley, near Belfast. Soils are brown earths on hummocky glacial sands.

All three photographs illustrate topics from Chapter 6 on Land Cover.

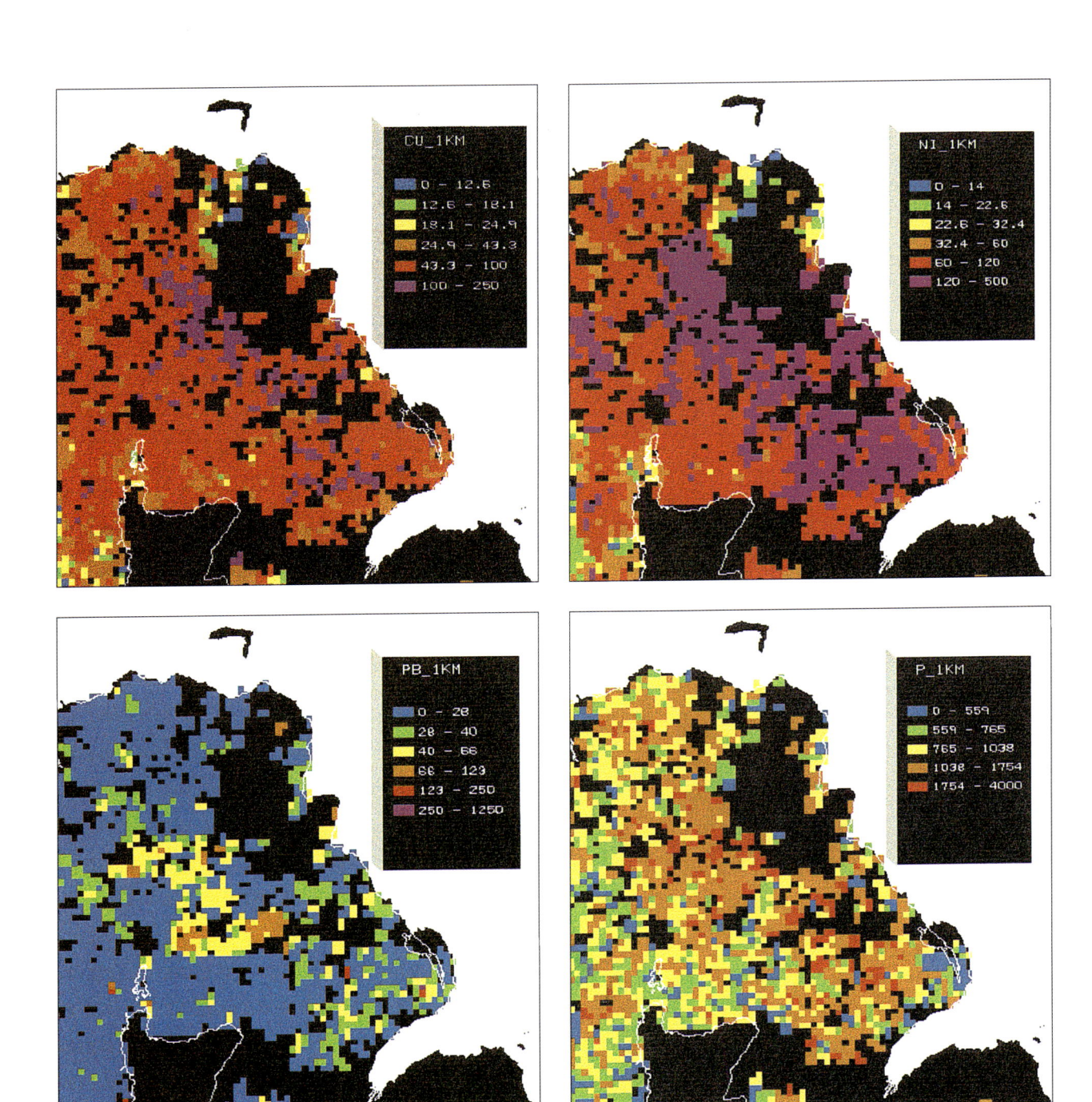

Four Figures from Chapter 9D Map extracts from a forthcoming NI Geochemical Atlas, showing total values of Copper (Cu), Nickel (Ni), Lead (Pb) and Phosphorus (P) in Co. Antrim. Copper and Nickel values are very high, Lead is low and Phosphorus is moderate to high.

Various types of field drains, assembled by John Courtney, then DANI Drainage Field Officer (WCMD), in SWG2 soil on Calp till in Co. Fermanagh.

SOIL TEXTURE

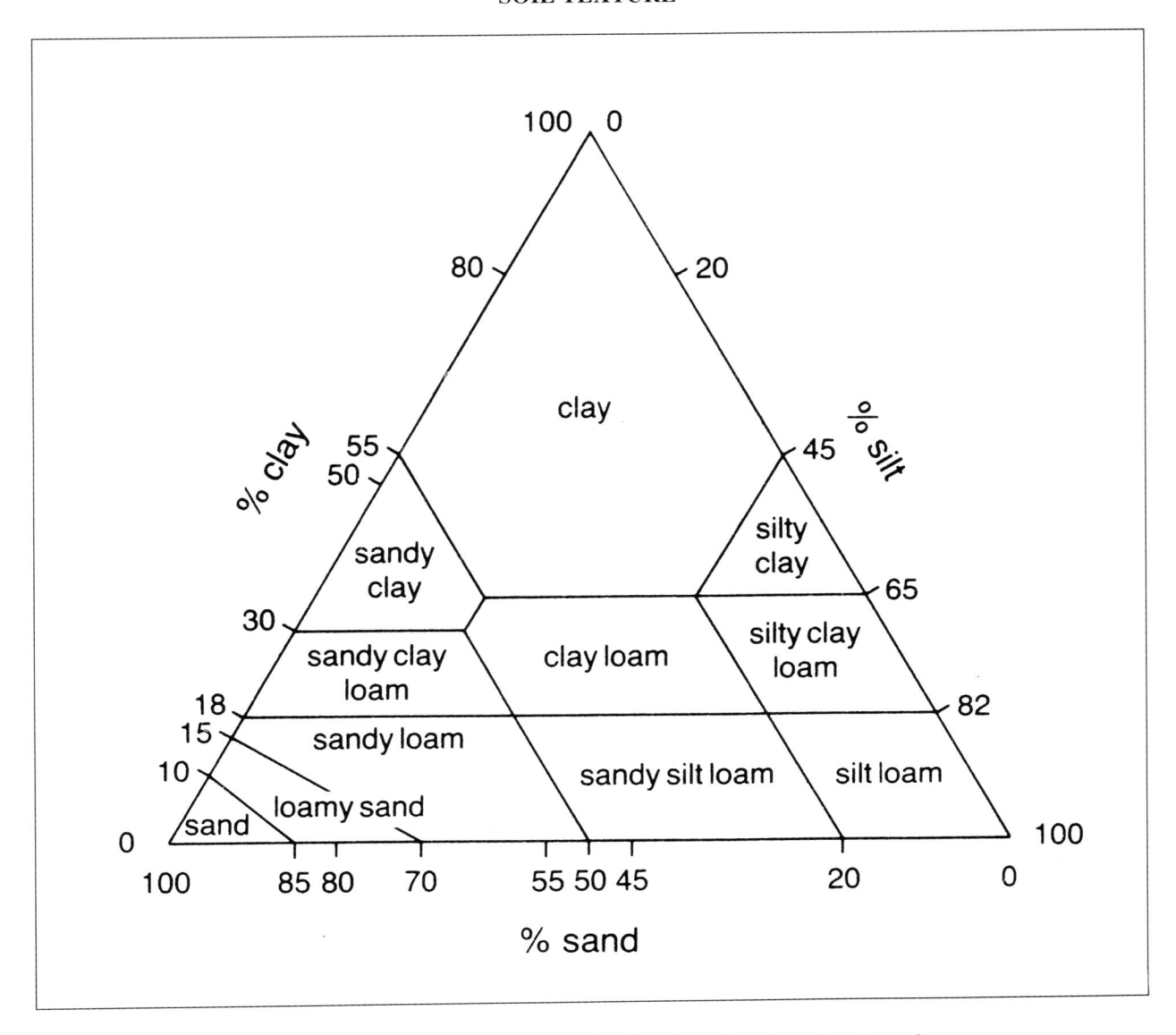

Limiting percentages of sand, silt and clay fractions for mineral texture classes

APPENDICES

APPENDIX 4
REFERENCES

CHAPTERS 1, 2 AND 8 – THE MAIN SOIL CHAPTERS

Adams, S.N., Jack, W.H. and Dickson, D.A., 1970. The growth of Sitka spruce on poorly drained soils in Northern Ireland. *Forestry*, 43 (2), 125-133.

Adams, S.N., 1986. Some relationships between soil chemical and physical properties, geology and pasture quality in Northern Ireland, *Record of Agricultural Research* (DANI), 34, 1-4.

Avery, B.W., 1980. *Soil Classification for England and Wales*, (Higher Categories), Soil Survey Technical Monograph No. 14, Harpenden.

Avery, B.W., 1990. *Soils of the British Isles*, C.A.B. International, Wallingford, UK, pp463.

Bailey, J.S., 1992. Effects of gypsum on the uptake, assimilation, and cycling of 15N-labelled ammonium and nitrate-N by perennial rye grass, *Plant and Soil*, 143, 19-31.

Bailey, J.S., 1995. Liming and nitrogen efficiency. *Commun. Soil Sc, Plant Anal.*, 26, 1233-1246.

Brown, W.O., 1954. Some soil formations of the basaltic region of north-east Ireland, *Irish Nat. Jour.*, 11, 120-132.

Colhoun, E.A., 1970. On the nature of the glaciations and final deglaciation of the Sperrin Mountains and adjacent areas in the north of Ireland. *Irish Geography*. 6(2), 162-85.

Colhoun, E.A., 1972. The deglaciation of the Sperrin Mountains and adjacent areas in counties Tyrone, Londonderry and Donegal, Northern Ireland. *Proceedings of the Royal Irish Academy*. 72B (8), 91-137.

Cruickshank, J.G., 1970. Soils and pedogenesis in the north of Ireland. N. Stephens and R.E. Glasscock (eds), *Irish Geographical Studies*, 89-104, Belfast.

Cruickshank, J.G., 1972. Soils and changing agricultural land values in part of County Londonderry, *Irish Geography*, 6 (4), 462-79.

Cruickshank, J.G., 1972. *Soil geography*, Newton Abbot, pp256.

Cruickshank, J.G., 1975. Soils of the northern and central parts of County Armagh, *Irish Geography*, 8, 63-71.

Cruickshank, J.G., 1978. Soil properties and management levels of marginal hill land in the Sperrin Mountains, Co. Tyrone and Co. Londonderry. *Irish Journal of Agricultural Research*, 17, 303-14.

Cruickshank, J.G., and Armstrong, W.J., 1971. Soil and agricultural land classification in County Londonderry, *Transactions of the Institute of British Geographers*, 53, 79-94.

Cruickshank, J.G. and Cruickshank, M.M., 1977. A survey of neglected agricultural land in the Sperrin Mountains, Northern Ireland, *Irish Geography*, 10, 36-44.

Cruickshank, J.G. and Cruickshank, M.M., 1981. The development of humus-iron podzol profiles, linked by radiocarbon dating and pollen analysis to vegetation history, OIKOS 36, 238-253.

Cruickshank, J.G., 1982. Soil, Chapter 8 in J.G. Cruickshank and D.N. Wilcock (eds), *Northern Ireland: Environment and Natural Resources*, The Queen's University of Belfast and New University of Ulster, Belfast, 165-84.

Cruickshank, M.M. and Tomlinson, R.W., 1990. Peatland in Northern Ireland: inventory and prospect, *Irish Geography*, 23, 17-30.

Cruickshank, M.M. and Tomlinson, R.W., 1990. An evaluation of LANDSAT TM and SPOT Imagery for monitoring changes in peatland. *Proceedings of the Royal Irish Academy*, 90, B, 7, 109-25.

Cruickshank, M.M. Tomlinson, R.W., Bond, D., Devine, P.M. and Edwards, C.J.W., 1995. Peat extraction, conservation and the rural economy in Northern Ireland. *Applied Geography*, 15,4, 365-83.

Curtin, D. and Smillie, G.W., 1981. Composition and origin of smectite in soils derived from basalt in Northern Ireland. *Clays and Clay Minerals*. 29, 277-284.

Dardis, G.F., 1985. Till facies associations of drumlins and some implications for their mode of formation. *Geografisker Annaler*, 67A, 13-22.

Dickson, E.L. and Stevens, R.J., 1983. Extractable Copper, Lead, Zinc and Cadmium in Northern Ireland soils. *Jour. Sci. Food Agric.*, 34, 1197-1205.

Edwards, K.J. and Warren, W.P., (Edit), 1985. *The Quaternary History of Ireland*, Academic Press, London.

Gardiner, M.J. and O'Callaghan, J.P., 1967. Studies on the clay minerals of Irish soils, *The Scientific Proceedings of the Royal Dublin Society*, B, 2, 87-97.

Garrett, M.K., Watson, C.J. Jordan, C., Steen, R.W.J. and Smith, R.V. 1992. The Nitrogen Economy of Grazed Grassland, *The Fertiliser Society Proceedings*, No.326: 1-32.

Green, F.H.W., 1979. *Field drainage in Europe: a quantitative survey*. (Institute of Hydrology, Report No. 57), Oxon.

Griffith Valuation. See valuation books and valuation maps (e.g. for 1834 and 1858), Public Record Office of Northern Ireland, Belfast.

Hammond, R.F., 1979. *The peatlands of Ireland*. An Foras Taluntais, Dublin.

Herries Davies, G.L. and Stephens, N., 1978. *Ireland*. (Geomorphology of the British Isles), London.

Hill, A.R., 1971. The formation and spatial distribution of drurnlins in a portion of north-east Ireland, in relation to hypotheses of drumlin origin in County Down, *Annals of the Association of American Geographers*, 63, 226-240.

Hill, A.R., 1973. The distribution of drumlins in County Down, *Annals of the Association of American Geographers*, 63, 226-240.

Jones, G.W., 1974. *A study of the cation supplying power of Northern Ireland soils*. PhD. Thesis. Faculty of Agriculture, The Queen's University of Belfast.

Kilroe, J.R., 1907. *Soil Geology of Ireland*.

McAleese, D.M. and McConaghy, S., 1957-1958. Studies on the basaltic soils of Northern Ireland, *Journal of Soil Science*, 8, 127-34, 135-40; 9, 66-75, 176-80, 81-8, 289-97.

McAllister, J.S.V. and McConaghy, S., 1968. Soils of Northern Ireland and their influence upon agriculture, *Record of Agricultural Research*, Ministry of Agriculture, Northern Ireland, 17, 101-8.

McCabe, A.M., 1969. The glacial deposits of the Maguiresbridge area, Co. Fermanagh, Northern Ireland, *Irish Geography*, 6, 63-77.

McCabe, A.M. 1993. The 1992 Farrington Lecture: Drumlin Bedforms and Related Ice-Marginal Deposition Systems in Ireland, *Irish Geography*, 26, 1, 22-44.

McConaghy, S. and J.V.S. McAllister, 1963. Soils, in *Land Use in Northern Ireland*. Edit Symons, L.J., ULP., 93-108.

McEntee, M.A. and Smith, B.J., 1993. The use of magnetic susceptibility measurements to interpret soil history: and example from mid-County Down, *Proceedings of the Royal Irish Academy*, 93B, 3, 175-80.

O'Neill, D.G., 1980. An investigation of the gravel tunnel drainage system. *Annual Report on Research and Technical Work for 1980: Enniskillen Agricultural College*, 10-13. Enniskillen, and in subsequent annual reports to 1989.

Proudfoot, V.B., 1958. Problems of Soil History. Podzol development at Goodland and Torr Townlands, Co Antrim, Northern Ireland, *Journal of Soil Science*, 9, 187-97

Savill, P.S. and Dickson, D.A., 1975. Early growth of Sitka spruce on gleyed soils in Northern Ireland, *Irish Forestry*, 32 (1), 34-A9.

Smith, A.G., 1970. Late-and post-glacial vegetation and climatic history of Ireland: a review. N. Stephens and R.E. Glasscock (eds.), *Irish Geographical Studies*. Belfast, 65-88.

Smith, B.J. and McAllister, J.J., 1986. Tertiary weathering environments and products in north-east Ireland, *International Geomorphology, Part II*, Edit. V. Gardiner, Wiley.

Smith, J., 1957. A mineralogical study of weathering and soil formation from olivene basalt in Northern Ireland, *Journal of Soil Science*, 8, 225-39.

Stevens, R.J. and Adams, T. McM., 1983. The sulphur status of some soils from counties Armagh, Londonderry and Tyrone, *Record of Agricultural Research (DANI)*, 31, 83-88.

Stevens, R.J. and McAllister, J.V.S., 1985. Evaluation of dolomite from County Fermanagh as a source of Magnesium and as a liming material, *Record of Agricultural Research (DANI)*, 33, 71-4.

Stevens, R.J. and Jones, G.W., 1985. Non-exchangeable Potassium in soils derived from different parent materials in Northern Ireland, *Record of Agricultural Research (DANI)*, 33, 81-6.

Stephens, N., Creighton J.R. and Hannon, M.A., 1975. The late-Pleistocene period in north-eastern Ireland: an assessment 1975, *Irish Geography*, 8, 1-23.

Symons, L.J., ed., 1963. *Land use in Northern Ireland*, London.

Tomlinson, R.W. and Gardiner, T., 1982. Seven bog slides in the Slieve-an-Orra hills, Country Antrim, *Journal of Earth Sciences, Royal Dublin Society*, 5, 1-9.

Wilcock, D.N., 1979. Post-war land drainage, fertiliser use and environmental impact in Northern Ireland, *Journal of Environmental Management*, 8, 137-49.

Wilcock, D.N., 1982. Chapter 2 Rivers in J.G. Cruickshank and D.N. Wilcock(eds.), *Northern Ireland: Environment and natural resources*. Belfast, 43-71.

Wilson, P., Griffiths, D. and Carter, C. 1996. Characteristics, impacts and causes of the Carntogher bog-flow, Sperrin Mountains, Northern Ireland. *Scottish Geographical Magazine*, 112, 1, 39-46.

Vernon, P., 1966. Drumlins and Pleistocene ice flow over the Ards peninsular Strangford Lough area, County Down, Ireland, *Journal of Glaciology*, 6, (45), 401-9.

CHAPTER 9D, SOIL SURVEY APPLICATIONS

Applied Chemistry Research Group, Imperial College London, 1973. *Provisional Geochemical Atlas of Northern Ireland*, Tech. Comm. No 60.

Bibby, J.S., Douglas, H.A., Thomasson, A,J and Robertson, J.S., 1982. *Land Capability Classificaion for Agriculture*. Macaulay Institute for Soil Research, Aberdeen.

Bibby, J.S., and Mackney, D., 1969. *Land Use Capability Classifcation*, Soil Survey Technical Monograph No. 1.

Cruickshank, J.G., 1974. The value of land, *The Geographical Magazine*, XLVl, 12, 684-91.

Dickson, E.L., and Stevens, R.J., 1983. Extractable Copper, Lead, Zinc, and Cadmium in Northern Ireland soils, *Jour. Sci. Food Agric.*, 34, 1197-1205.

Jarvis, M.G., and Hedges, M.R., 1994. Use of soil maps to predict the incidence of corrosion and the need for mains renewal. *Jour. I. W.E.M.*, 8, 68-75.

Klingebiel, A.A. and Montgomery, P.H., 1961. *Land capability classificafion*, United States Department of Agriculture Handbook No 210.

MAFF, 1966. *Agricultural Land Classification*, Agricultural Land Service Technical Report No. 11.

MAFF, 1976. Agricultural Land Classification of England and Wales. The Definition and Identification of Sub-Grades within Grade 3, *Technical Report No. 11/1*.

MAFF, 1988. Agricultural Land Classification of England and Wales: Revised guidelines.

McGrath, S.P., and Loveland, P.J., 1992. *The Soil Geochemical Atlas of England and Wales*, Blackie, London.

Symons, L.J., 1963. *Land Use in Northern Ireland*, LUP, London.

CHAPTER 4 – CLIMATE

Atkinson, B. W. and Smithson, P. A., 1976. Precipitation. T. J. Chandler and S. Gregory (eds.), *The Climate of the British Isles*. Longman, London, 129-182.

Bailie, S., 1980. *About the growing season in Northern Ireland 1941–70*. Unpublished BSc dissertation, Department of Geography, Queen's University of Belfast.

Barrett, E. C., 1976. Cloud and Thunder. T. J. Chandler and S. Gregory (eds.), *The Climate of the British Isles*. Longman, London, 199–210.

Bergeron, T., 1965. On the low-level redistribution of atmospheric water caused by orography. *Proceedings of the International Conference on Cloud Physics* May 24-June 1, 1965. Tokyo and Sapporo (Supplement: 96-100).

Betts, N. L., 1978a. The problem of water supply in Northern Ireland. *Water Services*, 82, 10–16.

Betts, N. L., 1978b. Water supply in Northern Ireland. *Irish Geography*, 11, 161–166.

Betts, N. L., 1982. Climate. J. G. Cruickshank and D. N. Wilcock (*eds.*), *Northern Ireland: Environment and Natural Resources*. Queen's University of Belfast and New University of Ulster, Belfast, 9–42.

Betts, N. L., 1984. The summer drought of 1983 and the re-emergence of Northern Ireland's water supply problem. *Journal of Meteorology*, 9, 37–40.

Betts, N. L., 1987. Extreme cold in Northern Ireland, 11–14 January 1987. *Journal of Meteorology*, 12, 154–157.

Betts, N. L., 1989. *A synoptic climatology of precipitation in Northern Ireland*. Unpublished PhD thesis, Department of Geography, School of Geosciences, Queen's University of Belfast.

Betts, N. L., 1990. Meso-scale precipitation distributions within a frontal depression. *Transactions of the Institute of British Geographers*, N.S., 15, 277–293.

Betts, N. L., 1992. The North Antrim flood of October 1990. *Irish Geography*, 25, 138–148.

Betts, N. L., 1994. Storm force winds of February 1994 black out Ulster. *Irish Geography*, 27, 61–67.

Bleasdale, A., 1963. The distribution of exceptionally heavy daily falls of rain in the United Kingdom, 1863 to 1960. *Journal of the Institution of Water Engineers*, 17, 45–55.

Bleasdale, A., 1970. The rainfall of 14th and 15th September 1968 in comparison with previous exceptional rainfalls in the United Kingdom, *Journal of the Institution of Water Engineers*, 24, 181–189.

Boucher, K., 1975. *Global Climate*. English Universities Press.

Burt, S., 1987. A new North Atlantic low pressure record. *Weather*, 42, 53–56.

Burt, S., 1993. Another new North Atlantic low pressure record. *Weather*, 48, 98–102.

Carter, T. R., Parry, M. L. and Porter, J. H., 1991. Climatic change and future agroclimatic potential in Europe. *International Journal of Climatology*, 11, 251–269.

Clayton, K. M., 1989. Implications of climatic change. *Coastal Management*. Proceedings of the Conference held by the Institute of Civil Engineers at Bournemouth, May 1989. Thomas Telford, London.

Collins, J. F. and Cummins, T. (*eds.*), 1996. *Agroclimatic Atlas of Ireland*. AGMET, Dublin.

Collins, J. F., Larney, F. J. and Morgan, M. A., 1986. Climate and Soil Management. T. Keane (*ed.*), *Climate, Weather and Agriculture*. AGMET, Meteorological Service, Dublin, 101-133.

Connaughton, M. J., 1969. *Air Frosts in Late Spring and Early Summer*. Agrometeorological Memorandum No. 2. Meteorological Service, Dublin.

Connaughton, M. J., 1975. *First and last occurrences of ground frost*. Agrometeorological Memorandum No. 6. Meteorological Service, Dublin.

Conrad, V., 1946. Usual formulas of continentality and their limits of validity. *Transactions of the American Geophysical Union*, 27, 663-664.

De la Mothe, P. D., 1968. Middle latitude wavelength variation at 500 mb. *Meteorological Magazine*, 97, 333–339.

Dixon, F. E., 1959. An Irish weather diary of 1711-25. *Quarterly Journal of the Royal Meteorological Society*, 85, 371–385.

Doornkamp, J. C., Gregory, K. J. and Burn, A. S. (eds.), 1980. *Atlas of Drought in Britain 1975–76*. Institute of British Geographers.

Fitzgerald, D., 1992. Climatological Data Archive of the Irish Meteorological Service – an Agricultural Resource. J. F. Collins (ed.). *Role of Climate*. Proceedings of 1992 AGMET Conference. AGMET, Dublin.

Glassey, S. D., 1967. *Average and extreme dates of first and last screen frosts in Northern Ireland*. Climatological Memorandum No. 61. Meteorological Office, Belfast.

Glassey, S. D. and Durbin, W. G., 1971. *Wind at Ballykelly*. Climatological Memorandum No. 68. Meteorological Office, Bracknell.

Grindley, J., 1972. Estimation and mapping of evaporation. *Symposium on World Water Balance*. International Association of Scientific Hydrology Publ. No. 92, 200–213.

Hardman, C. E. *et al*, 1973. *Extreme Wind Speeds Over the United Kingdom for Periods Ending 1971*. Climatological Memorandum, No. 50A. Meteorological Office, Bracknell.

Houghton, J. G. and Ó Cinnéide, M. S., 1977. Distribution and synoptic origin of selected heavy precipitation storms over Ireland. *Irish Geography*, 10, 1–17.

Houghton, J. T., Meira Filho, L. G., Callander, B. A., Harris, N., Kattenberg, A. and Maskell, K. (eds.), 1996. *Climate Change 1995 – The Science of Climatic Change. The Second Assessment Report of the Intergovernmental Panel on Climate Change: Contribution of Working Group I*. Cambridge University Press, 572 pp.

Hurst, G. W. and Smith, L. P., 1967. Grass growing days. J. A. Taylor (ed.), *Weather and Agriculture*. Pergaman, 147–155.

Jackson, M. C., 1977a. Evaluating the probability of heavy rain. *Meteorological Magazine*, 106, 185–192.

Jackson, M. C., 1977b. The occurrence of falling snow over the United Kingdom. *Meteorological Magazine*, 106, 26–38.

Jones, M. B. and Brereton, A. J., 1986. The Energy Balance. T. Keane (ed.), *Climate, Weather and Irish Agriculture*. AGMET, Meteorological Service, Dublin, 85–100.

Jones, P. D., 1979. *Climate Monitor*, 8, 4, Climatic Research Unit, University of East Anglia.

Jordan, C., 1987. The precipitation chemistry at rural sites in Northern Ireland. *Record of Agricultural Research*, (Department of Agriculture for Northern Ireland), 35, 53–66.

Jordan, C., 1988. Acid rain in Northern Ireland. W.I. Montgomery, J. H. McAdam and B. J. Smith (eds.), *The High Country: Land Use and Land Use Change in Northern Irish Uplands*. Institute of Biology (NI Branch) and Geographical Society of Ireland Symposium, Belfast, 47–55.

Jordan, C., 1994. GIS as a tool in aquatic resource management. D. Bond, J Reid, M. Stevens and L. Worrall (eds.), *GIS, Spatial analysis and public policy*. University of Ulster, Coleraine, 95–117.

Jordan, C., 1996. Mapping of rainfall chemistry in Ireland, 1972–1994, *Proceedings of Royal Irish Academy, Biology and Environment* – in press.

Keane, T., 1986. Meteorological Parameters in Ireland. T. Keane (ed.), *Climate, Weather and Agriculture*. AGMET, Meteorological Service, Dublin, 24–60.

Keane, T., 1988. Features of the Irish climate of importance to agriculture: comparison with neighbouring Europe. T. Keane (ed.) *Proceedings of Conference on Weather and Climate*. AGMET, Meteorological Service, Dublin.

Kirkpatrick, A. H., 1988. *A vegetation survey of heath and moorland in Northern Ireland and Co Donegal*. Unpublished DPhil thesis, University of Ulster.

Kirkpatrick, A. H. and Rushton, B. S., 1990. The oceanicity/continentality of the climate of the north of Ireland. *Weather*, 45, 322–326.

Lamb, H. H., 1950. Types and spells of weather around the year in the British Isles: annual trends, seasonal structure of the year, singularities. *Quarterly Journal of the Royal Meteorological Society*, 76, 393–429.

Lamb, H. H., 1972a. *Climate: Present, Past and Future, Vol. 1: I Fundamentals and II Climate Now*. Methuen, London.

Lamb, H. H. 1972b. *British Isles Weather Types and a Register of the Daily Sequence of Circulation Patterns 1861–1971*. Meteorological Office, Geophysical Memorandum No. 116. H.M.S.O., London.

Lamb, H. H., 1977. *Climate: Present, Past and Future, Vol. 2: III Climatic History and IV The Future*. Methuen, London.

Lamb, H. H., 1995. *Climate, History and the Modern World (2nd ed.)*. Routledge, London.

Logue, J. J., 1978. *The Annual Cycle of Rainfall in Ireland*. Technical Note No. 43. Meteorological Service, Dublin.

Logue, J. J., 1995. *Extreme Rainfalls in Ireland*. Technical Note No. 40. Meteorological Service, Dublin.

Mayes, J., 1996. Spatial and temporal fluctuations of monthly rainfall in the British Isles and variations in the mid-latitude westerly circulation. *International Journal of Climatology*, 16, 585-596.

McEntee, M. A., 1976. Preliminary assessment of climate conditions on Irish hills I and II. *Irish Journal of Agricultural Research*, 15, 223-246.

Meteorological Office, 1975. *Flood Studies Report, Vol II, Meteorological Studies*, N.E.R.C.

Miles, M. K., 1977. Atmospheric circulation during the severe drought of 1975–76. *Meteorological Magazine*, 106, 154–164.

Murray, R., 1977. The 1975–76 drought over the United Kingdom – hydrometeorological aspects. *Meteorological Magazine*, 106, 129-145.

Palutikof, J. P., Wigley, T. M. L. and Lough, J. M., 1984. *Seasonal scenarios for Europe and North America in a high-CO_2 warmer world*. U.S. Department of Energy, Carbon Dioxide Research Division, Technical Report TR 012

Parker, D. E., Horton, E. B., Cullum, D. P. N. and Folland, C. K., 1996. Global and regional climate in 1995. *Weather*, 51, 202–210.

Penman, H. L., 1962. Woburn irrigation, 1951–1959; I Purpose, design and weather; II Results for grass; III Results for rotation crops. *Journal of Agricultural Science*, 58, 343-379.

Perry, A. H., 1972. Spatial and temporal characteristics of Irish precipitation. *Irish Geography*, 6, 428–442.

Perry, A. H., 1976. Synoptic Climatology. T. J. Chandler and S. Gregory, (eds.), *The Climate of the British Isles*. Longman, London, 8-38.

Perry, A. H. and Walker, J. M., 1977. *The Ocean-Atmosphere System*. Longman, London.

Prior, D. B. and Betts, N. L., 1974. Flooding in Belfast. *Irish Geography*, 7, 1–18.

Ratcliffe, R. A. S., 1977. A synoptic climatologist's viewpoint of the 1975–76 drought. *Meteorological Magazine*, 106, 145–154.

Rohan, P. K., 1986. *The Climate of Ireland*. 2nd edition. The Stationery Office, Dublin.

Rotmans, J., Hulme, M. and Downing, T. E., 1994. Climate change scenarios for Europe – an application of the ESCAPE model. *Global Environmental Change*, 4, 97–124.

Rowe, M. W., 1990. The storm of 25th January 1990. *Journal of Meteorology*, 15, 197–200.

Rowntree, P. R., 1990. Estimates of future climatic change over Britain, Part 2: results. *Weather*, 45, 79-89.

Rowntree, P. R., Murphy, J. M. and Mitchell, J. F. B., 1993. Climate change and future rainfall predictions. *Journal of the Institute of Water Environ. Management*, 7, 464–470.

Royal Society, 1978. Scientific aspects of the 1975–76 Drought in England and Wales. Summary of discussion meeting held at Royal Society, London, 28 October 1977, on behalf of British National Committee on Hydrological Sciences. *Proceedings of the Royal Society, London*, A 363, 3–137.

Shellard, H.C., 1962. Extreme wind speeds over the United Kingdom for periods ending 1959. *Meteorological Magazine*, 91, 39–47.

Shellard, H. C., 1965. *Extreme Wind Speeds over the United Kingdom for Periods ending 1963*. Climatological Memorandum, No. 50. Meteorological Office, Bracknell.

Smith, L. P., 1976. *The Agricultural Climate of England and Wales*. Technical Bulletin 35, Ministry of Agriculture, Fisheries and Food. H.M.S.O. London.

Stephens, N., 1963. Climate. L. Symons (ed.), *Land Use in Northern Ireland*. University of London Press, 75–92.

Stevenson, J. C., Wark, L. G. and Kearney, M. S., 1986. Vertical accretion in marshes with varying rates of sea level rise. *Estuarine Variability*, 241–259.

Stewart, D. A. and Gibson, C. E., 1987. A model for the heat budget of Lough Neagh. *Record of Agricultural Research*, 35, 67–75.

Sumner, G., 1988. *Precipitation: Process and Analysis*. John Wiley, Bath.

Sweeney, J. C., 1985. The changing synoptic origins of Irish precipitation. *Transactions of the Institute of British Geographers*, N.S., 10, 467–480.

Sweeney, J. C., 1989. The Climatic Influence of the Irish Sea. J. C. Sweeney (*ed.*). *The Irish Sea: A Resource at Risk*. Geographical Society of Ireland Special Publications No. 3, Maynooth, 47–57.

Warrick, R. and Farmer, G., 1990. The greenhouse effect, climatic change and rising sea level: implications for development. *Transactions of the Institute of British Geographers N.S.*, 15, 5–20.

Wilcock, D. N., 1982. Rivers. J. G. Cruickshank and D. N. Wilcock (*eds.*), *Northern Ireland: Environment and Natural Resources*. Queen's University of Belfast and New University of Ulster, Belfast, 43–71.

Woods, D., 1995. *Fluctuations in the growing season in Northern Ireland, 1961–1990*. Unpublished BSc dissertation, Department of Geography, School of Geosciences, Queen's University of Belfast.

CHAPTER 5 - RIVERS, DRAINAGE BASINS AND SOILS

Bailey, A.D. 1979, Drainage of clay soils in England and Wales, (in) J. Wesseling (ed), *Proceedings of the International Drainage Workshop*, 1978, Institute of Land Reclamation and Improvement, Wageningen, The Netherlands, 220–242.

Bailey, A.D and Bree, T, 1981, The effect of improved land drainage on river flood flows, (in) *The Flood Studies Report, Five Years On*, Institution of Civil Engineers, 131–142.

Betts, N, 1992, The North Antrim Flood of October 1990, *Irish Geography*, 25 (2), 138–148.

British Geological Survey and Environment Service, Department of the Environment for Northern Ireland, 1994, *Hydrogeological Map of Northern Ireland*.

Brookes, A, 1988. *Channelized rivers: perspectives for environmental management*, Wiley, 326pp.

Burt, T.P. 1995, The role of wetlands in runoff generation from headwater catchments, (in) J.Hughes and L.Heathwaite, (eds), *Hydrology and Hydrochemistry of British Wetlands*, Wiley, 21–38.

Cochrane, S.R, and Wright, G.D, 1983, The estimation of mean annual floods for Northern Ireland rivers, *Public Health Engineer*, April, 50–53.

Department of Agriculture for Northern Ireland, 1991, *Drainage Works carefully planned and executed minimize environmental damage and encourage regeneration*, 12pp.

Department of the Environment for Northern Ireland and the Department of Agriculture for Northern Ireland, 1993, *Review of the Water Act (Northern Ireland) 1972 – a Consultation Paper*, HMSO, 19pp.

Dooge, J, 1987, Manning and Mulvany; River improvement in 19th century Ireland, (in) G. Garbrecht (ed), *Hydraulics and Hydraulics Research - A Historical Review*, Balkema, Rotterdam, 173-183.

Energy Technology Support Unit, 1993, *Prospects for Renewable Energy in Northern Ireland*, Department of Economic Development, 66pp.

Environmental Protection Division, 1987, *River quality in Northern Ireland, 1985*, Department of the the Environment for Northern Ireland, HMSO, 16pp.

Environment Service, 1993, *Methodology for identifying sensitive areas (Urban Waste Water Treatment Directive) and designating vulnerable zones (Nitrates Directive) in Northern Ireland*, Consultative Document, Department of the Environment for Northern Ireland, 59pp.

Essery, C.I and Wilcock, D.N, 1990(a), The impact of channelization on the hydrology of the upper River Main, County Antrim, Northern Ireland – a long-term case study, *Regulated Rivers, Research and Management*, 5, 17–34.

Essery, C.I, and Wilcock, D.N., 1990(b), Quantifying the hydrological impacts of a major arterial drainage scheme on a 200 km2 river basin, (in) J.C.Hooghart, C.W.S.Posthumus, and P.M.M.Warmerdam, *Hydrological research basins and the environment*, The Netherlands Organization for Applied Scientific Research, 131–140.

Foy, R.H, Smith, R.V, Jordan, C, and Lennox, S.D., 1995, Upward trend in soluble phosphorus loadings to Lough Neagh despite phosphorus reduction at sewage treatment works, *Water Research*, 29,4, 1051–1063.

Galvin, L, 1979, *Land drainage and reclamation*, Transactions of the Institution of Engineers of Ireland, 103, 53–63.

Gardiner, J.L. (ed), 1991, *River Projects and Conservation - a manual for holistic approach*, Wiley, 236pp.

Government of Northern Ireland Parliamentary Paper, 1944, *Report of the Agricultural Inquiry Committee to investigate the future of agriculture in Northern Ireland*, HMSO, Cmd 249, 312 pp.

Green, F.H.W., 1974, Changes in artificial drainage, fertilizers and climate in Scotland, *Journal of Environmental Management*, 2, 107–121.

Green, F.H.W, 1979, *Field drainage in Europe: a quantitative survey*, Institute of Hydrology, Report No.57, Wallingford, Oxford.

Gribbon, H.D, 1969, *The history of water power in Ulster*, David and Charles, 299pp.

Hammond, R.F, Van der Krogt, G, and Osinga, T, 1990, Vegetation and water-tables on two raised bog remnants in County Kildare, (in) G.J.Doyle (ed), *Ecology and Conservation of Irish Peatlands, Royal Irish Academy*, 121–134.

Hanna, J.E and Wilcock D.N, 1984, The prediction of mean annual flood in Northern Ireland, *Proceedings of the Intitution of Civil Engineers*, Part 2, 77, 429–444.

Hewlett, J.D and Hibbert, A.R, 1967, Factors affecting the response of small watersheds to precipitation in humid regions. In: W.E.Sopper and H.W.Lull (eds), *Forest Hydrology*, Pergamon, Oxford, 275–290.

Higgins, A.J. ,1988, *An investigation of saturated hydraulic conductivity and hydraulic characteristics of the soils in the upper River Main basin*, Northern Ireland, unpublished M.Phil. thesis, New University of Ulster, 124pp.

Higginson N.J., and Johnston, H.T., 1989, Riffle-pool formation in Northern Ireland Rivers, *Proceedings of the International Conference for centennial of Mannings formula and Kuchling's Rational Formula*, American Society of Civil Engineers, 10pp.

Ineson, J and Downing, R, 1965, Some hydrogeological factors in permeable catchment studies, *Journal of the Institution of Water Engineers*, 19 (1), 59–86.

Institute of Hydrology. 1980. *Low flow studies report*. United Kingdom Natural Environment Research Council, 3 volumes.

Jordan, C and Smith, R.V., 1987, The effect of phosphorus reduction on P– availability to algae, (in) *Annual report on Research, Development and Technical Work*, Department of Agriculture for Northern Ireland, 41–49.

Lyn, M.A, 1980, The arterial drainage programme, its background, economic and environmental effects, (in) *Impacts of Drainage in Ireland*, National Board for Science and Technology, Dublin, 1–12.

Mulqueen, J, 1978, Hydrology (in) *Proceedings of a Land Reclamation Seminar*, Ballinamore Research Station, The Agricultural Institute, Dublin, 17–28.

Mulqueen, J and Gleeson, T.N, 1981, Some relationships of drainage problems in Ireland to solid and glacial geology, geomorphology and soil types, (in) M.J.Gardiner (ed), *Land Drainage*, Balkema, Rotterdam, 11–32.

Natural Environmental Research Council, 1975, *Flood studies report*. London, 5 volumes.

Richardson, D.H.S., 1987, (ed). *Biological Indicators of Pollution*, Royal Irish Academy, 242pp.

Robinson, M, 1990, *Impact of improved land drainage on river flows*, Report Number 113, Institute of Hydrology, 226pp.

Robinson, M. Eeles, C.W.O, and Ward, R.C, 1990, The research basin and stationarity, (in) J.C.Hooghart, C.W.S.Posthumus, and P.M.M.Warmerdam, *Hydrological research basins and the environment*, The Netherlands Organization for Applied Scientific Research, 113–121.

Royal Commission on Environmental Pollution, 1992, *Freshwater Quality*, HMSO, London, 291pp.

Royal Commission on Environmental Pollution, 1996, *Sustainable Use of Soil*, HMSO, London, 260pp.

Royal Society for the Protection of Birds, National Rivers Authority, and The Wildlife Trusts, 1994, *The New Rivers and Wildlife Handbook*, 426pp.

Smith, S.J, Wolfe-Murphy, S.A, Enlander, I, and Gibson, C.E, 1991, *The Lakes of Northern Ireland: an annotated Survey*, HMSO, Belfast, 32pp.

Ward, R.C. 1984, On the response to precipitation of headwater streams in humid areas, *Journal of Hydrology*, 74, 171–189.

Water Quality Unit, 1993, *Biennial report, 1991/92*, Environmental protection Division, Environment Service, Department of the Environment for Northern Ireland, 76pp.

Wilcock, D.N. 1977, The effects of channel clearance and peat drainage on the water balance of the Glenullin basin, County Londonderry, *Proceedings of the Royal Irish Academy*, 77, B, 15, 253-267.

Wilcock, D.N., 1979, Post-war land drainage, fertilizer use and environmental impact in Northern Ireland, *Journal of Environmental Management*, 8, 137–149

Wilcock, D.N and Essery, C.I, 1984, Infiltration measurements in a small lowland catchment, *Journal of Hydrology*, 74, 191–204.

Wilcock, D.N. and Essery C.I., 1991, Environmental impacts of channelization on the River Main, County Antrim, Northern Ireland, *Journal of Environmental Management*, 32, 127–143.

Wilcock, D.N. and Hanna, J.E, 1987, Derivation of flow duration curves in Northern Ireland, *Proceedings of the Institute of Civil Engineers*, Part 2, 83, 381–396.

Wilcock, D.N and Wilcock F.A., 1995. Modelling the hydrological impacts of channelization on streamflow characteristics in a Northern Ireland catchment. *Proceedings of the International Association of Hydrological Sciences*, Publication No. 231, 41– 48.

Williams, G and Browne, D, 1987, The drainage of Northern Ireland, *Ecos*, 8(3) 8–15

Wilson, P., Griffiths, D, and Carter, C., 1996, Characteristics, impacts and causes of the Carntogher bog-flow, Sperrin Mountains, Northern Ireland, *Scottish Geographical Magazine*, 112, 39–46.

Wood, L.B, and Sheldon, D, 1980, *Water Quality Control* (in) A.M.Gower (ed) Water Quality in catchments, Wiley, 193–228.

Wood, R.B and Gibson, C.E., 1973, Eutrophication and Lough Neagh, *Water Research*, 7, 173–187.

Zollweg, J.A, Gburek, W.J, Pionke, H.B and Sharpley, A.W, 1996, GIS-based delineation of source areas of phosphorus withhin agricultural watersheds of the northeastern USA, (in) *Modelling and Management of Sustainable Basin-scale water resource systems*, International Association of Hydrological Sciences, 231, 31–39.

CHAPTER 9B – ACIDIFICATION OF SOILS

Bull, K.R. 1995. Critical loads – possibilities and constraints. *Water, Air and Soil Pollution*, 85, 201-212.

Charlson, R.J. and Rodhe, H. 1982. Factors controlling the acidity of natural rainwater. *Nature*, 295, 683–685.

CLAG. 1994. *Critical loads of acidity in the United Kingdom.* Critical Loads Advisory Group – Summary Report. Department of the Environment.

Cresser, M.S., Smith, C. and Sanger, L. 1993. Critical loads for peat soils. In : *Critical loads : concepts and applications*, Hornung, M. and Skeffington, R.A. (Eds.), ITE Symposium No 28, HMSO, London, pp 34–39.

Department of the Environment 1990. *This common inheritance.* Britain's environmental strategy (Cm 1200), London, HMSO.

Fowler, D., Cape, J.N., Leith, I.D., Choularton, T.W., Gay, M.J. and Jones, A. 1988. The influence of altitude on rainfall composition. *Atmospheric Environment*, 22, 1355-62.

Grennfelt, P. and Thornelof, E. (Eds.) 1992. *Critical loads for nitrogen.* Report from a UNECE workshop held at Lokeborg, Sweden, 6-10 April 1992 (1992 : 41). The Nordic Council of Ministers, Copenhagen.

Hornung, M., Bull, K.R., Cresser, M., Hall, J., Langan, S.J., Loveland, P. and Smith, C. 1995a. An empirical map of critical loads of acidity for soils in Great Britain. *Environmental Pollution*, 90, 301-310.

Hornung, M., Sutton, M.A. and Wilson, R.B. (Eds.) 1995b. Mapping and modelling of critical loads for nitrogen : a workshop report. *Proceedings of the Grange-over-Sands Workshop 24-26 October 1994*, Institute of Terrestrial Ecology, Edinburgh.

Hornung, M. 1993. Critical load concepts. In : *The Chemistry and Deposition of Nitrogen species in the Troposphere*, A.T.Cocks (Ed.), Special Publication No 115, Royal Society of Chemistry, Cambridge, pp 78-94.

Jordan, C. 1987. The precipitation chemistry at rural sites in Northern Ireland. *Record of Agricultural Research* (Department of Agriculture for Northern Ireland), 35, 53-66.

Jordan, C and Enlander, I. 1990. The variation in the acidity of ground and surface waters in Northern Ireland. *International Revue gesamten Hydrobiology*, 75, 379-401.

Jordan, C. 1994. GIS as a tool in aquatic resource management. In : *GIS, Spatial analysis and public policy*, D.Bond, J.Reid, M.Stevens and L.Worrall (Eds.). University of Ulster, Coleraine, pp 95-117.

Jordan, C. 1996. Mapping of rainfall chemistry in Ireland, 1972-1994. In preparation.

Krug, E.C. and Frink, C.R. 1983. Acid rain on acid soil : a new perspective. *Science*, 221, 520-525.

Mason, B.J. (Ed.) 1990. *The surface waters acidification programme.* Cambridge University Press, Cambridge, pp 522.

Nilsson, J. and Grennfelt, P. (Eds.) 1988. Critical loads for sulphur and nitrogen. *Report of a workshop held in Skokloster, Sweden 19-24 March 1988.* (1988 : 15). The Nordic Council of Ministers, Copenhagen.

Patrick, S., Monteith, D.T. and Jenkins, A. (Eds.). 1995. *UK Acid Waters Monitoring Network: the first five years.* Analysis and interpretation of results April 1988 - March 1993. Published for the Department of the Environment and Department of the Environment Northern Ireland by ENSIS Publishing, London.

Reuss, J.O. and Johnson, D.W. 1986. Acid deposition and the acidification of soils and waters. *Ecological studies* 59, Springer-Verlag, pp 119.

RGAR 1990. *Acid deposition in the United Kingdom 1986-88.* The Review Group on Acid Rain. Department of the Environment Report, London, pp 124.

Sverdrup, H. and Warfvinge, P. 1988. Weathering of primary minerals in the natural soil environment in relation to a chemical weathering model. *Water, Air and Soil Pollution*, 38, 387-408.

Webster, R., Campbell, G.W. and Irwin, J.G. 1991. Spatial analysis and mapping the annual mean concentrations of acidity and major ions in precipitation over the United Kingdom in 1986, *Environmental Monitoring and Assessment*, 16, 1-17.

UNECE. 1994. *Protocol to the 1979 Convention on Long Range Transboundary Air Pollution*, United Nations, Geneva.

SELECTIVE INDEX

acid deposition	171-2	CEC	19
acid rain	171-2	cereals	122
acid soils	171	clay minerals	19
acidity	7	clay soils	89,96,182
acrotelm	88	climatological stations	63
adsorption	181	Clogher Valley	152-3
Agivey	141	cloud	67
agricultural land classification	187-189	Coal Measures	60
air temperature	67-8	columnar basalt	61
alluvium	1	Common Agriculture Policy C.A.P	125-6
ammonium	172	copper	191-2
annual rainfall	1,73,187	Corine classification	99-117
Antrim Plateau	4,61	Corine land cover	166-7
aquifers	190-1	corrosion risk	190
arable land	99,119	Cretaceous Chalk	14
Ards	4	critical load	173-4
Armagh	4,14	crop yields	120
Armoy Moraine	61,136	Cuilcagh	109
aspect	66		
atmospheric circulation	63	daily rainfall	75
		dairy cows	120,122
Ballymoney	136	dairying	122
Bann River	108	Dalradian rocks	58
barley	119	deep ploughing	96
basalt	5,19	deglaciation	154,157
basalt lava	61	Diatomite	149,183
basalt soils	19,140-143,149,181-3,191	diversity	116
base flow index	178	Down	4
Belfast Lough	157	drainage	7,9
Binevenagh	133,139,140	drainage basin	85-6
blanket bog	105	drift geology	61
blanket peat	4,5,9,151	drumlin	2,3,61-2,141,144,149,155, 157,159,163
bog bursts/flows	4,89		
boulder clay	3,61		
Breen Wood	136	emission	175
Brookeborough	161	environmental quality	98
brown earths	1,2,11,183-5,190	erosion	107
brown podzolics	157,164,183,190	ESAs	112-4,129
brown rankers	12	esker	62
brown soils	14	European Union	119
buffering capacity	174	evaporation	85,182
bulk density	182-183	evapotranspiration	1,3,8,77-86
		exceedance	175
Calp rock	150-1,160-1		
capillarity	181	Fairy Water	108-9
carbon density	165,167,169	farm business	123
carbon pools	116, 165,169	farm holdings	121-2
carbon stores	110	farm income	122
Carboniferous rocks	58-60,150-1,153,160,190	farming industry	119
Carboniferous Sandstone	147,155	farms	121-2
carry-over	3,148-161	fen peat	155
Castlederg	144-5	field capacity	182
catchment	85-86	field capacity days	187
catchment hydrology	88	field drainage	96,98
catotelm	88	Fine Earth	3

211

fish kill	89
flax	119
flood peaks	93
floods	85-90
flow duration	92
food prices	125
forcing functions	85
forestry	106,110,112,119
Foyle	4
frost	70
gales	66
Geochemical Atlas	19,191
geological structure	57
geology	5,15-17,57-62
Giants Causeway	4
glaciation	159
Glenelly	114
gley classification	9,12
gleying	3,7,8,9,11,178
gleys	1,11,12,183
global warming	63,110
granite	6,147
granite soils	157,163,183
graphical databases	134
grazing season	73
grey-brown podzolics	14
ground water	86
groundwater vulnerability	89,190
growing season	2,70-1
heath	114
high productivity pastures	101,103
horses	120
horticulture	110,155,159
HOST	177-181
humic rankers	13
hydraulic conductivity	88
hydrogeology	179
hydrological sequence	5
impermeable soils	96
infiltration	87
infiltration capacity	87
Intake	149
integrated air capacity	178
intervention	125
labour force	122-4
Lack	153
Lagan	4,7
Lagan Valley	157
lagg	108
lakes	94-5
land classification	134
land cover	99-117
land drainage	98
land management	187
land taxation	187
land use	119
Land Use Capability	187
Late-glacial	3,159
lazy beds	7
leachate	3,7
leaching	3,7
leaching potential	190

lead	191
Less Favoured Areas	116,126-8
lignite	61
limestone	6
limestone gravel	161
livestock	121
Lough Beg	149
Lough Erne	4
Lough Neagh	1,4,5,7,95,108,148, 154,157
Lough Neagh Clay	14,61,148,154,157
Lower Bann	141
Lower Lough Erne	150
lowland peat	106
Lyles Hill	149
machine cutting	108
manganese mottling	9
man-made soils	12
Marl	159
meltwater	57
mica-schist	6,138,139,145,147
modal soil profiles	20-56,16
moisture retention	182
mole drains	7,96
Mooorland Scheme	114
moorland	114
moraine	61,62
mottling	9
mountains	4
Mournes	4,6,8
Munsell Colour	17
Myroe Levels	138
nickel	192
nitrate	172
non-glaciation	149
oats	119
Old Red Sandstone	48,58,147,152-3
oldest rocks	58
Omagh	108
Omagh Fault	146-7
organic alluvium	167
organic carbon	165,182
osmosis	181
overland flow	85,87
parent material	2,3,5,6,14
part-time farming	129
pasture	101
peat	4-7,12,14,136,138,142, 145,151,163,166178-181
peat bogs	105
peat catchment	88
peat conservation	108
peat depth	166
peat land	105
peatland area	110
pelosol	11
permeability	88
phosphorus	9,19,120,191
pigs	120
plant nutrients	7
Pleistocene	2,3
podzol	14

212

podzolisation	8,9	soil porosity	181
podzols	1,2,9,11,163,167	soil profile	3,11,15
polder	8,98,138	soil series	1,16,18,134
Pomeroy	147	soil survey	2,165
pore space	8	soil temperature	70
Post-glacial	3	soil water	86
potassium	19	soil wetness	189
potatoes	119	soil-forming factors	1,2
potential evapotranspiration	77	soil-forming process	8
poultry	120	solar radiation	66
precipitation	73-76	Sperrins	4,5,6,14,114, 138-9,145-6
		Stewartstown	149
Quaternary	2,3,61,62	stone drains	7
		Strangford Lough	8
rainfall	73-6,87	Strangford Lough	159
rainfall stations	65	stream flow	90
raised bog	5,108	sulphate	172
rankers	1,2,6,11-13,163	sunshine	66
reclamation	8	superficial deposits	619
red sand	149		
Red Triassic Sandstone	159	Tardree	149
relief	4	temperature	67
rhyolite lava	61,149	temperature extremes	69
Ring Dyke	7	textual database	134
River Bann	95	throughflow	86
river channel	85	till	3,61
river hydrology	90	topography	4
rivers	85-98	transpiration	86
Rivers Agency	90	Trias Sandstone	147,148,155,157
runoff cycle	86	Triassic Marl	159
		Triassic rocks	60
sandstone	6	Tyrone	4
sediment loads	89		
shale till	157	underdrainage	7
shear strength	182	unsaturated zone	85
sleech	62		
Slieve Beagh	109	water balance	76-9
Slieve Croob	157	water deficit	76
snowfall	7\6	water quality	93-4
Soil Attribute Database	16,17	watertable	9
soil carbon	165,169	weather types	64
soil classification	11	weathering	3,19,88
soil corrosivity	190	wilting point	182
soil databases	134	wind direction	65
soil drought	189	wind speed	65
soil erosion	2	woodland	112-114
soil formation	11	WRAP	178
soil geochemistry	191		
soil maps	133-164,187	yields	120
soil moisture	72,76,181-2		
soil moisture deficit	77	zonal soils	2
soil polygon	134	zonality	2